T0201065

MANAGING THE DOCUMENTATION MAZE

MANAGING THE DOCUMENTATION MAZE

Answers to Questions You Didn't Even Know to Ask

JANET GOUGH

DAVID NETTLETON

A JOHN WILEY & SONS, INC., PUBLICATION

Published by John Wiley & Sons, Inc., Hoboken, New Jersey
Published simultaneously in Canada

For general information on our other products and services or for technical support, please contact
our Customer Care Department within the United States at (800) 762-2974, outside the United States
at (317) 572-3993 or fax (317) 572-4002.

Wiley also publishers its books in a variety of electronic formats. Some content that appears in print
may not be available in electronic formats. For more information about Wiley products, visit our web
site at www.wiley.com.

Library of Congress Cataloging-in-Publication Data:

Gough, Janet.
   Managing the documentation maze : answers to questions you didn't even know to ask / Janet Gough,
David Nettleton.
       p. cm.
   Includes index.
   ISBN 978-0-470-46708-4 (cloth)
   1. Medical records–Automation.   2. Medical records–Data processing.   I. Nettleton, David, 1963–
II. Title.
   R864.G68 2010
   610.285–dc22

                                   2009038777

Printed in the United States of America

10 9 8 7 6 5 4 3 2 1

CONTENTS

INTRODUCTION

If you didn't write it down, it didn't happen—that's the position of the FDA and other regulators. So what does that mean for companies? They need to keep records of everything they do. Good documentation is proof of product integrity; it's the confirmation of compliance; and it's just plain common sense. So how do they manage it all?

This book provides no-nonsense, practical answers to questions about documentation and records management. It addresses the regulations and how they apply, and it discusses sound documentation systems, electronic and hybrid. An entire chapter is devoted to making the transition to an electronic system, including how to validate and document the process. One chapter is dedicated to compliance, but many chapters will help you to understand regulatory issues and efficient compliance strategies.

Additional chapters address standard operating procedures, nonclinical records in laboratories and manufacturing, and clinical and regulatory documentation. A full chapter is dedicated to writing, reviewing, and approving documents so that they are consistent and readable.

Finally, this book offers advice on how to maintain a system once it is in place, and it tells how to keep it ready for scrutiny during any inspection. A final chapter provides additional resources for document management professionals.

ABOUT THE AUTHORS

Janet Gough is a consultant to the pharmaceutical, medical device, and biotech industries with expertise in systems development, documentation, and training. She has been a director of technical communications for a biotech company where she developed and oversaw the function. Now she assesses company-specific needs and designs and implements Part 11 compliant document management systems. She assists companies in determining needs and preparing documents in accord with QSR, GMP, GLP, GCP, ICH, and other binding regulations. She is currently a faculty member of an industry training organization where she teaches writing and document management courses.

janet.gough@gmail.com
www.gxpdocumentation.com

David Nettleton, Computer System Validation's principal, is an industry leader and author for 21 CFR Part 11, Annex 11, HIPAA, software validation, and computer system validation. He assists in the development, purchase, installation, operation, and maintenance of computerized systems used in FDA compliant applications. He has completed more than 185 mission critical laboratory, clinical, and manufacturing software implementation projects. He also conducts training in systems and validation.

dnettleton@computersystemvalidation.com
www.computersystemvalidation.com

CHAPTER 1

UNDERSTANDING THE REGULATIONS

INTRODUCTION

Therapeutic products in development or in the marketplace require extensive and complete documentation. The position of the regulators in the United States and abroad is this: "If you didn't write it down, it didn't happen." What this caveat means is that companies must produce documentation through every phase of product development, manufacture, and distribution. But producing the documents is hardly enough if a company cannot lay its hands on the documents it needs when it needs them. Good documentation thus requires good controls. The way a company determines what controls it must put in place is first by understanding what documents it must have as proof of sound testing and control of its products to demonstrate that they are both safe and effective. Government agencies dictate what companies must do. Companies, in turn, institute good practices that show adherence to the agency requirements. This chapter addresses the regulatory environment and answers the following questions:

1. What agency in the United States oversees therapeutic products, and what authority does it have?

2. What do I need to know about the regulations?

3. How do regulations come to be?

4. What is the history of the regulations for therapeutic products in the United States?

5. Where can I find the actual regulations?

6. What is the purpose of the regulations, and must companies comply with all of them?

7. How can we know which regulations apply to us and what documentation we need?

8. Are clinical trials always in three phases?

9. Doesn't adherence to the regulations slow companies down?

10. Do the regulations tell you how to achieve compliance?

Managing the Documentation Maze, By Janet Gough and David Nettleton
Copyright © 2010 John Wiley & Sons, Inc.

11. What is the FDA position on regulations and guidances that are not US-based?

12. What is Sarbanes–Oxley, and does it require document controls?

13. What is Part 11, and how does it drive industry practices?

14. Does industry have a say in what goes into a law?

15. How does the FDA keep laws current?

16. What is an electronic record?

17. What is the purpose of 21 CFR Part 11?

18. Does 21 CFR Part 11 apply only to those systems that employ electronic signatures?

19. What is an electronic signature?

20. Is an electronic signature binding the way a handwritten signature is?

21. Which FDA programs does 21 CFR Part 11 apply to?

22. What is a predicate rule?

23. Is my consultant correct in saying that only files generated in a software program and signed electronically are Part 11 files?

24. Are all Word files subject to Part 11?

25. Do small companies also have to comply with 21 CFR Part 11?

26. Are there any other regulations that drive electronic record keeping besides Part 11?

27. Why should companies impacted by 21 CFR Part 11 look at HIPAA regulations?

28. Are there any other regulations that will affect patient records?

29. Why is Part 11 necessary, since the preexisting regulations call for record controls?

30. Does Part 11 override the other regulations for records management?

31. What's the difference between a final rule and a guidance document?

32. Does industry have to comply with guidance documents?

33. What are "best practices" and "industry standards"?

34. How do industry standards develop?

35. What is the scope of industry standards?

36. Are the HIPAA regulations predicate rules for medical records maintained electronically according to 45 CFR Part 164?

37. Does 21 CFR Part 11 apply only to electronic records that are inspected by the FDA?

38. Should an international company have worldwide procedures in place that address 21 CFR Part 11?

39. Why is the FDA revising Part 11?

40. What exactly does "risk" mean?

41. How will companies know which systems pose risk?

42. What will the changes to Part 11 entail?

43. When the FDA withdrew guidance documents in August of 2003, how did that affect Part 11?

44. When will the FDA require electronic submissions?

45. Do other countries accept validation standards based on US regulations?

46. Are the US industry standards for computerized systems comparable to industry standards elsewhere?

47. Is the scope of 820 larger than devices?

48. The regulators are good at telling industry what to do, but do they follow their own dictates?

49. How do companies comply individually?

50. If the regulations call for certain compliance practices, does that mean all companies pretty much do the same thing?

51. How does the FDA issue new regulations?

52. How can companies keep abreast of changes in the regulations and their interpretation?

53. What are warning letters, and how do companies get them?

54. What does it mean to redact sensitive information in warning letters?

55. What happens if a company doesn't fix its problems?

56. What is a consent decree?

57. What is an injunction?

58. What are common citations for companies?

59. What is the direction the industry is taking?

60. Are there plans to harmonize electronic records and signature requirements between the US and the European Union regulatory agencies?

61. Besides submissions and the documentation investigators request during inspections, what other types of documentation do regulators look at?

1.   What Agency in the United States Oversees Therapeutic Products, and What Authority Does It Have?

The FDA is an agency within the United States Department of Health and Human Services. It consists of a variety of offices or centers, and the rules can vary from center to center, so companies need to understand which arm of the FDA governs their activities. Here are some of the centers and offices:

- Office of the Commissioner (OC) (1)
- Center for Biologics Evaluation and Research (CBER) (2)
- Center for Devices and Radiological Health (CDRH) (3)
- Center for Drug Evaluation and Research (CDER) (4)
- Center for Food Safety and Applied Nutrition (CFSAN) (5)
- Center for Veterinary Medicine (CVM) (6)

- National Center for Toxicological Research (NCTR) (7)
- Office of Regulatory Affairs (ORA) (8)

The agency has the authority to issue and enforce laws that affect therapeutic products that cross state lines (interstate commerce).

2. What Do I Need to Know About the Regulations?

Therapeutic product development, manufacturing, and distribution to the marketplace are highly regulated. Title 21 of The Code of Federal Regulations (CFR) addresses the requirements for therapeutic products from discovery through life in the marketplace. Parts within Title 21 address specific requirements for various product types. For instance, Parts 210 and 211 address solid dose pharmaceuticals, Part 820 addresses devices, and Part 606 addresses biologics. All the regulations require controlled documentation. In developing and making therapeutic products, other regulations come into play. If a company has a work environment that could potentially be injurious to its employees, it must adhere to the appropriate parts of Title 29 of the CRF, the Occupation, Safety and Health Administration (OSHA) requirements for worker safety (9). Similarly, if a clinical trial site that maintains patient records keeps them electronically, it must be in accordance with parts of Title 45 of the CRF that address Health Insurance and Portability.

3. How Do Regulations Come to Be?

The FDA proposes a final rule (law), usually as the result of industry discussion or agency observations. The proposed rule is published in the *Federal Register* (*FR*) (10) and industry responds with dialogue about the proposed rule. When there is resolution about a proposed rule, the FDA publishes the final rule and gives a date when the final rule will be effective. This allows industry time to implement and modify activities as necessary to comply.

4. What Is the History of the Regulations for Therapeutic Products in the United States?

In the early history of the United States, drugs could be bought and sold like any other commodity. Once problems with drugs surfaced, the government began to legislate therapeutic product development and marketing. Specific laws marking milestones over a period of 100-plus years have brought us to the regulated world of pharmaceuticals, devices, biologics, and biotechnology we know today. (See Box 1.1.)

5. Where Can I Find the Actual Regulations?

The FDA website (www.fda.gov) is a good place. You can search and bookmark the entire Code of Federal Regulations using a standard web browser. Regulatory agencies for other countries also have websites.

BOX *1.1*

REGULATORY EVOLUTION IN THE UNITED STATES

The first US federal regulation dates back to 1848 when American soldiers in Mexico died after ingesting adulterated quinine to treat malaria. As a result of these deaths, the government passed the **Drug Importation Act**, which required customs inspections on drugs coming from overseas. In 1862, President Abraham Lincoln appointed a chemist to serve in the new Department of Agriculture. This was the start of the Bureau of Chemistry, the precursor to the Food and Drug Administration (FDA).

In 1902, **The Biologic Control Act** became law after 13 children died from a contaminated antitoxin for diphtheria. This act gave the government regulatory power over antitoxin and vaccine development. Shortly after, in 1906, the government passed the **Food and Drugs Act** to authorize the government to monitor food purity and safety of medicines.

In 1911, the **Sherley Amendment** was enacted. This amendment prohibited false and fraudulent label claims of therapeutic effectiveness. This law was unsatisfactory, however, since a promoter had to be proven to be deliberately fraudulent. In addition, the law covered labeling, but not advertising.

The control of narcotics under the Food and Drugs Act was also unsatisfactory, since it required manufacturers to state only the quantity of any alcohol, opium, morphine, or cannabis in the product. After babies died or suffered addiction from teething remedies containing opium, the government passed the **Harrison Narcotic Act** in 1914. This law required physicians and pharmacists to record the dispensing of narcotics.

In 1927, the government formed the **Food, Drug, and Insecticide Administration.** This agency was reorganized in 1930 as the **Food and Drug Administration**.

Several other events were significant in developing binding regulations designed to protect humans and animals. The 1932, Tuskegee Study of Untreated Syphilis in the Negro Male, conducted under the auspices of the US Public Health Service, deprived infected men of effective treatment so as not to interrupt the project.

Then in 1937, 107 people died after taking "elixir of sulfanilamide," which turned out to be an antifreeze solution. The FDA removed the product from the market, not because it caused fatalities but because it was mislabeled. In 1938, the government passed the **Food, Drug, and Cosmetics Act.** This Act expanded the role of the FDA to control of cosmetics and devices. It also mandated that safe tolerances be established for unavoidable poisonous substances such as pesticides; authorized standards of identify, quality, and fill of containers for foods; authorized factory inspections; added the injunctions as an act of the FDA in addition to penalties of seizure and prosecution; required drugs intended for humans to bear labels warning against habit forming; and defined drug, device, cosmetic, label, and labeling terminology that still applies today. Shortly after, in 1941, the government added the **Insulin Amendment** to the Food Drug and Cosmetic Act. This amendment added standards to ensure purity, quality, strength, and identify of insulin-containing products for diabetes treatment, and it required batch certification. The **Public Health Service Act** of 1944 further tightened controls by calling for regulation of biological products and control of communicable diseases. And then in 1945, the **Penicillin Amendment** was added; this amendment required FDA testing and certification of the safety and efficacy of all penicillin products since production technologies were uncertain. This legislation led to subsequent amendments in 1963 to cover any other antibiotics or derivatives.

(Continued)

It was after World War II, however, that testing in humans received acute focus. During the war, experiments were done in large scale on unconsenting humans. The Nuremberg War Crime Trials brought these atrocities to light, and the result was the **Nuremberg Code**, which cited 10 standards for ethical human research, not just in the United States.

In 1952, the FDA added the **Durham–Humphrey Amendment** to the Code of Federal Regulations. This amendment clarified the obligations of pharmacists in dispensing drugs by defining the types of drugs that cannot be used safely without medical supervision. It restructured the sale of such drugs to prescription by a licensed physician, and it defined which drugs required prescription drug label and which could be over-the-counter (OTC). The label "Caution: Federal Law Prohibits Dispensing Without a Prescription" was required on all prescription drugs. The law also prohibited unauthorized refills.

The FDA began to pay closer attention to drug manufacturing activities. In 1953, FDA passed the **Factory Inspection Amendment**, which updated and clarified the FDA's authority to inspect. This amendment established the 483 form, which is issued at the close of inspections. It also removed the requirement for FDA to announce inspections. Then in 1958, the FDA added the **Food Additives Amendment**, which required makers of new food additives to establish safety standards before exposure to the public. The Delany Proviso prohibits approval of food additives shown to induce cancer in humans or animals. The law authorized the FDA to evaluate the safety of all new ingredients, including those in dietary supplements. The FDA then published a list entitled "Substances Generally Recognized as Safe"(GRAS) in the *Federal Register*. This was followed by the **Color Additive Amendments** of 1960, which authorized the FDA to establish the conditions of use for color additives in foods, drugs, and cosmetics and required manufacturers to test products for safety.

A wake-up call for better monitoring of development activities in the clinic came in 1962, when thousands of babies were born with defects, the result of their mothers taking thalidomide while pregnant. The drug had never been approved for marketing in the United States, but was undergoing research in American women. Of these women, nine gave birth to defective infants. This event induced the FDA to require notification of investigational use of drugs, which up until this time had not been required. The result was the **Kefauver–Harris Amendment to the Food, Drug, and Cosmetic Act.** This act also required manufacturers to institute Good Manufacturing Practices (GMPs); made FDA approval of the NDA a prerequisite for marketing; placed prescription drug advertising under the FDA's supervision, while allowing the FTC to continue supervision of OTC advertising; required registration and periodic inspections (at least once every two years) of manufacturing facilities; required manufacturers and distributors of new drugs to submit adverse event reports; required assurance of informed consent of research subjects; made qualification of drug investigations subject to review; and required manufacturers to include full information on adverse events and contraindications to provide a balanced picture for health-care professional. The last requirement led to the creation of package inserts.

At about the same time, President John F. Kennedy announced the **Consumer Bill of Rights** in a message to Congress. This Bill of Rights said that the people have the right to safety, the right to be informed, the right to choose, and the right to be heard. In the same period, in 1964, the World Medical Association issued the **Declaration of Helsinki,** and physicians were tasked with embracing this statement: "The health of my patients will be my first consideration." The declaration has been amended four times, and the CFR has incorporated the basic elements.

In 1966, the **DESI Review and Fair Packaging and Labeling Act** called for evaluating the effectiveness of 4000 drugs approved on the basis of safety alone between 1938 and 1962. This Act became known as the Drug Efficacy Study Implementation (DESI) Review. Also in 1966, the **Fair Packaging and Labeling Act** was passed; this act required consumer products in interstate commerce to have honest and informative labeling.

In 1970 the **Poison Prevention Packaging Act** was passed. This legislation required special packaging of controlled substances and prescription drugs for enhanced safety, especially for children. Certain products for children required "child-resistant packaging." The Consumer Product Safety Commission assumed responsibility for enforcing this statute. Also in 1970, the Environmental Protection Agency was established and assumed control of pesticide tolerances.

The **Drug List Act** of 1972 provided the FDA with a current inventory of all marketed drugs, and it required manufacturers to submit a semi-annual list of all drugs introduced or discontinued since the last submission. Four years later the requirement to list new medical devices was added. Also in 1972, the government passed the requirement for **Over-the-Counter Drug Review.** Formal OTC drug reviews were required to ensure safety, effectiveness, and correct labeling of drugs sold without prescription.

In 1972, the National Institutes of Health transferred the regulation of biologics to the FDA. This was followed by the **National Research Act,** which created the National Commission for the Protection of Human Subjects of Biomedical and Behavioral Research. Additional legislation has continued to promote ethical treatment of health-care recipients.

In 1974, the **National Research Act** was signed into law, creating the National Commission for the Protection of Human Subjects of Biomedical and Behavioral Research. This committee had the purpose of identifying the basic ethical principles on which clinical research should be founded.

In 1976, landmark legislation was passed to ensure the safety and effectiveness of medical devices. The **Medical Device Amendments** established a risk-based classification system for devices: class I, II, or III, with class I being the least risky and class III having the most risk. Pre-amendment devices were grandfathered. New devices had to show substantial equivalence to a pre-amendment device to establish class I or II status. Class III devices would require a Pre-Market Approval (PMA) from this point on.

The **Vitamins and Mineral Amendments** of 1976 thwarted FDA efforts to establish standards for limiting the potency of vitamins and minerals in food supplements or regarding them as drugs. These amendments were precursors to legislation 18 years later that permitted the unrestricted use of dietary supplements.

In 1978, the FDA published the current Good Manufacturing Practices, 21 CFR Parts 210 and 211.

In 1979, the National Commission for the Protection of Human Subjects of Biomedical and Behavioral Research issued the **Belmont Report**. This report set forth basic ethical principles and guidelines for the protection of human research subjects: respect for persons; beneficence (an obligation to do no harm); and justice (fair and equal distribution of clinical research burdens and benefits). The FDA and the Department of Health and Human Services (HHS) subsequently incorporated the principles in the Belmont Report into laws regarding clinical research. These laws relate to the protection of human subjects, the responsibilities of Institutional Review Boards (IRBs), requirements for an NDA, responsibilities of investigators, control of drugs, record keeping, and record retention.

In the next years, these were passed into law: **The Tamper-Resistant Packaging Act** (1982), a result of cyanide-laced Tylenol reaching the market; and the **Orphan Drug Act** (1983), which provided incentives for drug makers to develop drugs for rare diseases or

(Continued)

conditions. The **Drug Price Competition and Patent Term Restoration Act** (1984), which is also known as the Hatch Waxman Act, permitted the FDA to approve generic versions of brand drugs without repeating the extensive research. This act established the Abbreviated New Drug Application (ANDA). It also permitted brand drug makers to apply for up to five years of additional patent protection.

In 1988, the FDA became an agency of the Department of Health and Human Services. Since that time, the International Conference on Harmonisation (ICH) has been formed. A significant ICH goal is to maintain safeguards on quality, safety, efficacy, and regulatory obligation for the protection of the public. The **Clinical Laboratories Improvement Amendments** (CLIA) of 1988 established standards to improve the quality of clinical laboratory testing in US laboratories that conduct testing in humans for health assessment for the diagnosis, prevention, or treatment of disease. Then in 1990, representatives from Europe, Japan, and the United States met at the International Conference on Hamonization of Technical Requirements for Registration of Pharmaceuticals for Human Use (ICH). The objective was to effect better use of human, animal, and material resources, to eliminate duplication of testing, and to remove delays in development.

The **Safe Medical Devices Act** (SMDA) of 1990 expanded and strengthened some provisions of the 1976 Medical Device Amendments. This act codified the 510(k) process and refined the definition of "substantial equivalence"; required user facilities to report injuries; required post-marketing surveillance on implants; established procedures for tracking; and authorized the FDA to order recalls and impose fines. Also in 1990, the **Nutrition and Labeling Education Act** (NLEA) became law. This act required consistent nutritional labeling.

The **Generic Drug Enforcement Act** of 1992 was the result of the generic drug scandal, when generic manufacturers were caught bribing personnel at the FDA and falsifying data. Offenders were subject to drug approval denial for 18 months, suspension of drug distribution, civil penalties, and debarment.

The **Prescription Drug User Fee Act** (PDUFA) was passed in 1992 to accelerate FDA review of applications. Drug and biologics manufacturers would henceforth pay fees for review; these fees would support the hiring of more reviewers at the agency.

In 1992, the **Medical Device Amendments** clarified four provisions of the medical device regulation: tracking; postmarket surveillance, medical device reporting; and the repair, replacement, or refund stipulation.

Shortly after, in 1994, the **Dietary Supplement Health and Education Act** (DSHEA) became law. Under this legislation, dietary supplements were no longer subject to premarket safety evaluations. However, it authorized the FDA to promulgate GMPs and outlined permissible usage claims and nutritional support statements.

In 1996, medical devices became subject to the Quality System Regulation (QSR). In 1996, as well, The Department of Health and Human Services enacted the **Health Insurance Portability and Accountability Act** (HIPAA) into law. The next year, the **Food and Drug Administration Modernization Act** reauthorized the Prescription Drug User Fee Act of 1992 for five more years and instituted reforms in agency practices. These actions provided the forward momentum for broad changes in the health-care industry, but the specifics of the regulation were still being written. Shortly thereafter, also in 1997, **Electronic Records; Electronic Signatures** was enacted.

In 2000, the FDA and the National Institutes of Health (NIH), in response to the death of an 18-year-old receiving gene therapy, renamed and transferred the Office for Human Research protections (OHRP) (formerly the Office for Protection from Research Risks [OPRR]) from the NIH to the Office of the Assistant Secretary of the Department of Health and Human Services (HHS). This move placed more emphasis on the protection of human subjects.

In 2002, the **Medical Device User Feed and Modernization Act** (MDUFMA) was enacted. This legislation parallels the PDUFA and applies to PreMarket Approvals (PMAs) and Biologic Licensing Agreements (BLAs), certain supplements, and 510 (k)s.

In September 2007, the president signed the **Food and Drug Administration Amendments Act** (FDAAA). This act became effective on March 25, 2008. The FDAAA is the most comprehensive revision to the FD&CA, particularly in pharmacovigilance. This act requires a risk evaluation and mitigation strategy (REMS) to ensure that the benefits of a medicine outweigh its risks, adjustments to safety labeling, postmarketing studies, and the payment of monetary penalties for violations of REMS.

The Government does not issue laws without forethought. The Office of the Federal Register issues the *Federal Register* (*FR*), a weekly disclosure publication that informs citizens of their rights and obligations by providing access to the official text of approved regulations and descriptions of federal organizations, programs, and activities. It also publishes texts of proposed regulations and changes to existing regulations. This gives industry the opportunity to react and share dialogue with the government agency that has ownership of the proposal. Reviewers can comment on content and wording, the date the regulation goes into effect, and the penalties for noncompliance. Comments are reviewed in a government forum, and the final regulation becomes the "final rule."

Once enacted, laws are published in the Code of Federal Regulations, issued annually on April 1. Laws are enforceable by the respective divisions within the Department of Health and Human Services. It's important to note, however, that once a final rule appears in the *FR*, companies are responsible for instituting compliance. Thus, keeping abreast of the regulations requires constant vigilance.

The *CFR* contains regulations of specific government departments and agencies. The *CFR* has 50 "Titles," each assigned to a different unit of government. Title 21, Food and Drugs, contains regulations mandated by the FDA. Title 45, Public Welfare, falls under the auspices of the National Institutes of Health (NIH). Each title of the *CFR* is then divided into chapters, and each chapter is divided into parts and subparts.

Remember, too, that as new regulations are enacted, they do not supersede existing regulations unless the government has rescinded them. New regulations in essence become adjuncts to the ones already in place. Companies must adhere to predicate rules and remain vigilant about industry best practices for compliance.

6. What Is the Purpose of the Regulations, and Must Companies Comply with All of Them?

The regulations are in place to ensure that therapeutic products are both safe and effective for their intended use. Companies must adhere to those regulations that apply to their products and business model.

7. How Can We Know which Regulations Apply to Us and What Documentation We Need?

Each company must fully understand its business model. The regulations that drive laboratory activities are different from those that govern manufacturing and distribution. It's really a matter of understanding where the company is on the continuum of product development, manufacturing, and distribution. Research (concept development) is not covered by regulations, but once a company moves into development, following the regulations is mandatory. To bring a product to market requires extensive nonclinical testing and confirmation of safety, followed by testing in humans, typically in three trial phases. Companies need to document all their product-related activities as they apply to the business model wherever they are on the continuum. See Figure 1.1.

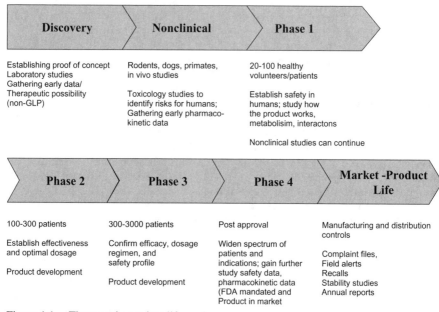

Figure 1.1.   Therapeutic product life cycle.

8.   Are Clinical Trials Always in Three Phases?

That has been the traditional model. But companies are now designing adaptive trials that are more efficient. In addition, some companies conduct Phase 0 trials in a limited number of volunteers to gather preliminary data on pharmacokmetics and pharmacodynamics. And the regulators very often require that companies conduct Phase 4 studies that track product in the marketplace. Companies also use Phase 4 studies to address product comparisons for effectivity versus price.

9.   Doesn't Adherence to the Regulations Slow Companies Down?

That is often the argument for not putting controls in place. Yet the sooner a company institutes controls and has proof of controls in its practices and documentation, the easier it is for the company to move forward. Good controls speak to more than the regulations. Good controls support good business practices. A lack of controls early on can make for time-consuming and costly corrections going forward.

10. Do the Regulations Tell You How to Achieve Compliance?

The regulations are prescriptive, not descriptive. That is, they tell you what you must do, but they don't tell you how to do it. Each company must figure out for itself "how it happens here."

11. What Is the FDA Position on Regulations and Guidances that Are Not US-Based?

The United States strongly recommends adherence to the International Conference on Harmonisation Guidelines (11). These guidelines endorse common standards for reporting on therapeutic products. The United States, the European Community, Japan, and Australia are countries that have embraced these standards. The advantage of adhering to ICH guidelines is that they satisfy the requirements for more than a single country, so reporting is uniform, and companies significantly reduce the need to prepare separate documentation for each regulator.

12. What Is Sarbanes–Oxley, and Does It Require Document Controls?

Sarbanes–Oxley is legislation passed in 2002 by the Securities and Exchange Commission (12). It requires disclosure of financial interests of executives of publicly held companies to prevent conflicts of interest. Many companies now manage legal records in their document management systems.

13. What Is Part 11, and How Does It Drive Industry Practices?

21 CFR Part 11 Electronic Records; Electronic Signatures is a final rule for electronic record keeping. It became effective in 1997. It is a vaguely worded law, currently undergoing revision. It addresses how to maintain electronic records and employ electronic signatures. See the Appendix, page 261, for the text of the law.

14. Does Industry Have a Say in What Goes into a Law?

Yes. Before a law becomes effective, the FDA proposes the law in the *Federal Register* (*FR*), which is published daily. Industry has an opportunity to respond, and the comments are subsequently published in the FR. The final law is usually the result of dialog between industry and the agency.

15. How Does the FDA Keep Laws Current?

The FDA can amend existing laws. In September of 2007, for instance, the President of the United States signed the Food and Drug Administration Amendments Act of 2007 (13). This law became effective on March 25, 2008. This act represents a comprehensive revision to the Federal Food, Drug, and Cosmetics Act. It focuses heavily on pharmacovigilence. With this act, the FDA has the authority to (a) require a risk evaluation and mitigation strategy (REMS) if it thinks it will help ensure that the benefits of a new medicine outweigh its risks, (b) order safety labeling adjustments, (c) require postmarketing studies, and (d) impose civil monetary penalties for violations of new REMS provisions, postmarketing study/clinical trial requirements, or labeling violations.

16. What Is an Electronic Record?

Electronic records cover a wide scope. The Federal Register, in the mid-1990s, defined an electronic record as "any combination of text, graphics, data, audio, pictorial, or other information representation in digital form that is created, modified, maintained, archived, retrieved, or distributed by a computer system." Basically, any electronic data in any medium is an electronic record. An electronic record is any type of data retained in any format on any nonvolatile medium (semipermanent, such as a hard drive or removable media). Today's modern technology allows for electronic records to be included in PCs, laptop computers, memory sticks, personal digital assistants, and a wide range of portable devices.

17. What Is the Purpose of 21 CFR Part 11?

21 CFR Part 11 allows electronic records to be used in place of or in addition to paper records. It also allows an electronic signature, which is most often a user name and password, to be used in place of a handwritten signature. An electronic signature may also be a biometric signature, such as a fingerprint scan. Computer systems that are used to manage electronic records must be validated regardless of whether approval is made by handwritten or electronic signatures.

18. Does 21 CFR Part 11 Apply Only to Those Systems that Employ Electronic Signatures?

The regulation applies to all systems that manage records electronically. Electronic signatures are optional, but most companies choose to use them because they allow for the replacement of paper records.

19. What Is an Electronic Signature?

It is an electronic equivalent of a handwritten signature, usually a combination of a user name and password, but it can also be a biometric marker. Electronic signature is often used generically to mean an electronic form of approval. This now includes digital signatures and certificates, handwriting capture instruments

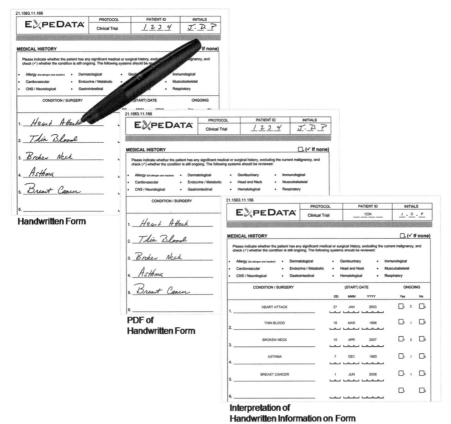

Figure 1.2. Expidata digital pen.

such as pens that remember what is written, or devices that capture handwriting. (See Figure 1.2.)

20. Is an Electronic Signature Binding the Way a Handwritten Signature Is?

Yes. Part 11 establishes an e-signature as the legally binding equivalent of a handwritten signature. Furthermore, in June of 2000, Congress passed the E-Sign Act which gave electronic signatures in other industries the same legal standing as pen and paper.

21. To Which FDA Programs Does 21 CFR Part 11 Apply?

It applies to all FDA programs, whether they address pharmaceutical, biologic, device, or combination products. Even if Part 11 applies, the applicable predicate rules do as well.

22. What Is a Predicate Rule?

A predicate rule is any Title 21 Code of Federal Regulation to which 21 CFR Part 11 applies. Predicate rules identify requirements for records and signatures. Since these predicate rules were written before computer systems were in widespread use, the predicate rules are generally thought to refer to paper records and handwritten signatures. Part 11 allows handwritten records and signatures to be replaced with electronic records and electronic signatures.

23. Is My Consultant Correct in Saying that Only Files Generated in a Software Program and Signed Electronically Are Part 11 Files?

Absolutely not. The regulation addresses minimum standards for electronic records and it applies to "records that are required to be maintained under predicate rules and that are maintained in electronic format in place of paper format" and to "records that are required to be maintained under predicate rules, that are maintained in electronic format in addition to paper format, and that are relied on to perform regulated activities" (14). That means your Word files are electronic records because you do not recreate an entire document every time you revise it; you use the electronic file. It also seems that many people think an electronic signature somehow makes an electronic record subject to Part 11 while a file used on a company's intranet is not subject to Part 11. Part 11 has no requirement for e-signatures, and it is actually optional in the regulation. Most electronic records subject to Part 11 never get an e-signature. The argument is totally misguided.

24. Are All Word Files Subject to Part 11?

Many, but not all, Word files are subject to Part 11. Part 11 applies to any documents that are required by any of the regulations. GXP is an abbreviation for all good practices for the industry, such as Good Manufacturing Practices (GMPs), Good Clinical Practices (GCPs), and Good Laboratory Practices (GLPs). Thus a Word file for a document such as an SOP is a GXP electronic record, and so are PDF files created from them. Electronic files need to be retained in a validated software Electronic Data Management System (EDMS) either purchased or built in house, or kept in a qualified network drive that has secure, limited access folders to ensure security of the files.

25. Do Small Companies Also Have to Comply with 21 CFR Part 11?

Yes. Here's what the FDA has said: "Because widespread use of electronic technology is relatively recent, the significance of official, legally binding electronic records may not be fully appreciated by everyone (15)." Part 11 has a positive impact on nearly all organizations subject to the rule, including small business. Right now, 93% of device firms are small businesses with less than 500 employees, and so are about 500 pharmaceutical companies.

26. Are There Any Other Regulations that Drive Electronic Record Keeping Besides Part 11?

21 CFR Part 11 Electronic Records; Electronic Signatures became law in 1997, and industry standards have evolved since then. In 2003, the Department of Public Welfare passed 45 CFR Parts 160, 162, and 164, as part of the Health Insurance, Portability, and Accountability Act (HIPAA) regulations, into law (16). The requirements of these regulations parallel those of 21 CFR Part 11. HIPAA requires the same types of electronic controls as Part 11, so the HIPAA-driven industry can use the already established standards developed by the Part 11-driven industry. In short, industry standards affect all electronic records regardless of industry. Since Part 11 and HIPAA are related for electronic records, industry standards in one will affect the other. (See Table 1.1.) See the Appendix, page 331, for the full text of the law.

27. Why Should Companies Impacted by 21 CFR Part 11 Look at HIPAA Regulations?

HIPAA will affect many of the Part 11-driven industries. Companies conducting clinical trials, for instance, need to adhere to the HIPAA regulations for electronic record keeping, since clinical trials rely on patient records.

28. Are There Any Other Regulations that Will Affect Patient Records?

Yes. The Department of Health and Human Services is facilitating the development of a nationwide "interoperable" (exchangeable) health data system that will allow sharing of patients' health information electronically (17). The expectation is that it will be law by 2014, so clearly we are moving toward electronic data overall.

29. Why Is Part 11 Necessary, Since the Preexisting Regulations Call for Record Controls?

An argument that industry does not require Part 11 is that the predicate rules call for management of documents, and it is implied that the same principles apply to electronic records as to paper records. However, with industry mandated to submit documentation to the FDA electronically, Part 11 provides a roadmap for achieving consistency within the industry. Also, since the predicate rules do not mention electronic records, Part 11 provides details related to electronic records that cannot be found elsewhere.

30. Does Part 11 Override the Other Regulations for Records Management?

No. All the regulations apply. The predicate rules, those in place outside Part 11, do not go away with Part 11. In fact, the argument that Part 11 is unnecessary stems from the realization that the preexisting laws already cover the requirement for documentation, whether it's paper or electronic.

TABLE 1.1. Comparison Matrix of Part 11 and the HIPAA Regulations

What Part 11 Says	What It Means	The HIPAA Parallel
Subpart A—General Provisions § 11.1 Scope. (a) The regulations in this part sets forth the criteria under which the agency considers electronic records, electronic signatures, and handwritten signatures executed to electronic records to be trustworthy, reliable, and generally equivalent to paper records and handwritten signatures executed on paper.	Before this regulation, paper records were the compliance focus. With Part 11, electronic records became equivalent to paper records. While paper records are inherently static, electronic records change to keep them current. Electronic records are thus the focus of compliance. By making electronic signatures equivalent to handwritten signatures, the FDA allows electronic signatures to replace handwritten signatures. Companies can retain paper systems, use hybrid paper and electronic systems, or use fully electronic systems.	**164.306 Security standards: General rules**
(b) This part applies to records in electronic form that are created, modified, maintained, archived, retrieved, or transmitted under any records requirements set forth in agency regulations. This part also applies to electronic records submitted to the agency under requirements of the Federal Food, Drug, and Cosmetic Act and the Public Health Service Act, even if such records are not specifically identified in agency regulations. However, this part does not apply to paper records that are, or have been, transmitted by electronic means.	The FDA sets the scope of electronic records to include all agency regulations. All previous and future regulations do not have to be rewritten to include a statement of acceptance for electronic records. This section specifically excludes facsimiles from the definition of an electronic record. Paper that is transmitted by fax remains a paper record, and the faxed copy is equivalent to a paper copy.	**160.103 Definitions for disclosure and electronic media and protected healthcare information**

TABLE 1.1 *(Continued)*

What Part 11 Says	What It Means	The HIPAA Parallel
§ **11.2** (e) Computer systems (including hardware and software), controls, and attendant documentation maintained under this part shall be readily available for, and subject to, FDA inspection.	Industry standards recognize a computer system to be hardware, operating system, software utilities, application software, user instructions, training materials, and validation documentation. The FDA is stating that it will inspect the computer system and the supporting procedural infrastructure related to it.	**164.304 Definitions (for computerized systems)**
Subpart B—Electronic Records § 11.10 Controls for closed systems. Persons who use closed systems to create, modify, maintain, or transmit electronic records shall employ procedures and controls designed to ensure the authenticity, integrity, and, when appropriate, the confidentiality of electronic records and to ensure that the signer cannot readily repudiate the signed record as not genuine. Such procedures and controls shall include the following:	A procedural infrastructure is necessary: Facilities Security, Network Security, Computer System Back-up, Data Archiving, Computer System Maintenance Event Recording, Electronic Signatures. System-specific procedures are also required.	**164.306 Security standards: General rules** **164.308 Administrative safeguards** **164.310 Physical safeguards** **164.312 Technical safeguards** **164.314 Organizational requirements** **164.316 Policies and procedures and documentation requirements**
(a) Validation of systems to ensure accuracy, reliability, consistent intended performance, and the ability to discern invalid or altered records.	Computer System Validation and Computer System Change Control are applicable SOPs.	**164.310**
(c) Protection of records to enable their accurate and ready retrieval throughout the records retention period.	Data Archiving and Electronic Record Retention are applicable SOPs.	**164.316**

(Continued)

TABLE 1.1 *(Continued)*

What Part 11 Says	What It Means	The HIPAA Parallel
(d) Limiting system access to authorized individuals.	The procedure for granting system-specific access is similar to that included in a Network Security SOP. The procedure may employ handwritten signatures or electronic signatures to authorize issuance of security privileges that allow user access to a system. In some systems the request and issuance of user access is contained within the computer system itself.	**164.306** **164.308** **164.310** **164.312** **164.314** **164.316**
(e) Use of secure, computer-generated, time-stamped audit trails to independently record the date and time of operator entries and actions that create, modify, or delete electronic records. Record changes shall not obscure previously recorded information. Such audit trail documentation shall be retained for a period at least as long as that required for the subject electronic records and shall be available for agency review and copying.	This requires an automated audit trail. Record changes must not obscure previously recorded information; previous data values must reside in the audit trail. The regulation does not define the audit trail as an electronic record but establishes a requirement to retain audit trail data for the same duration as the electronic records.	**164.312**
(g) Use of authority checks to ensure that only authorized individuals can use the system, electronically sign a record, access the operation or computer system input or output device, alter a record, or perform the operation at hand.	This expands the requirements of section (d) above. Controlled access rights to individual system functions must be granted to authorized users. This is often accomplished by creating roles that have predefined security access and then assigning the role to individual users.	**164.306** **164.308** **164.310** **164.312** **164.314** **164.316**

TABLE 1.1 (*Continued*)

What Part 11 Says	What It Means	The HIPAA Parallel
(h) Use of device (e.g., terminal) checks to determine, as appropriate, the validity of the source of data input or operational instruction.	In the typical computer environment there are virtual connections between the input devices (client workstations) and the computer that retains the electronic record (server). Each time data transmits between the input device, such as a computer workstation, barcode reader, or instrument, the receiving software application must confirm that that it is appropriate for that device to be transmitting data at that time. Typically, this is done by capturing the network address of the input device at the time of log-on and then verifying that data received comes with the same network address. If a user logs on at one computer workstation and then starts transmitting data from another workstation, the receiving software application cannot confirm that the same person is responsible for the data and the data should be rejected.	**164.306** **164.308** **164.310** **164.312** **164.314** **164.316**
(i) Determination that persons who develop, maintain, or use electronic record/ electronic signature systems have the education, training, and experience to perform their assigned tasks.	This requirement restates the predicate rule for training. It specifically addresses the skill of the computer user to ensure that reliable electronic records are generated and maintained.	**164.308**
(j) The establishment of, and adherence to, written policies that hold individuals accountable and responsible for actions initiated under their electronic signatures, in order to deter record and signature falsification.	The Electronic Signature SOP is applicable. While FDA requires only one authorized notification of electronic signatures use, all users must understand and certify that their electronic signature is equivalent to their handwritten signature.	**164.308**

(*Continued*)

TABLE 1.1 *(Continued)*

What Part 11 Says	What It Means	The HIPAA Parallel
(k) Use of appropriate controls over systems documentation including: (1) Adequate controls over the distribution of, access to, and use of documentation for system operation and maintenance.	Usually a master document delineates document control and distribution. Therefore, system-specific SOPs that describe the operation and maintenance of the system inherently have the appropriate controls.	**164.316**
(2) Revision and change control procedures to maintain an audit trail that documents time-sequenced development and modification of systems documentation.	A standard part of SOPs is the version change history. Use of existing, well-established SOP documentation practices ensures compliance with this regulation.	**164.316**
§ **11.30 Controls for open systems.** Persons who use open systems to create, modify, maintain, or transmit electronic records shall employ procedures and controls designed to ensure the authenticity, integrity, and, as appropriate, the confidentiality of electronic records from the point of their creation to the point of their receipt. Such procedures and controls shall include those identified in § 11.10, as appropriate, and additional measures such as document encryption and use of appropriate digital signature standards to ensure, as necessary under the circumstances, record authenticity, integrity, and confidentiality.	So far the regulation has addressed closed systems. When users don't have direct control over the source of data, such as when the Internet is used, the FDA makes additional requirements. The objective is to make the open system as secure as a closed system. Industry standards employ end-to-end encryption in addition to all other security features. Digital signatures may provide a pathway for the next evolution of computer system security, but this technology is still immature.	**164.306** **164.308** **164.310** **164.312** **164.314** **164.316**

TABLE 1.1 (*Continued*)

What Part 11 Says	What It Means	The HIPAA Parallel
§ 11.300 Controls for identification codes/ passwords. Persons who use electronic signatures based upon use of identification codes in combination with passwords shall employ controls to (a) ensure their security and integrity. Such controls shall include: (a) Maintaining the uniqueness of each combined identification code and password, such that no two individuals have the same combination of identification code and password.	The electronic signature must be unique and assigned to only one user. The combination of the user name and password must be unique. There is no requirement to have passwords be unique.	**164.306** **164.308** **164.310** **164.312** **164.314** **164.316**
(b) Ensuring that identification code and password issuances are periodically checked, recalled, or revised (e.g., to cover such events as password aging).	Electronic signatures, like logon criteria, must be changed at regular intervals. Password aging is required for logon criteria and therefore meets the requirement for electronic signatures.	**164.306** **164.308** **164.310** **164.312** **164.314** **164.316**
(c) Following loss management procedures to electronically deauthorize lost, stolen, missing, or otherwise potentially compromised tokens, cards, and other devices that bear or generate identification code or password information and (b) issue temporary or permanent replacements using suitable, rigorous controls.	Regardless of the hardware and software components for making an electronic signature, procedures must be in place to issue and maintain them. For user name and password components, this doesn't have additional requirements.	**164.306** **164.308** **164.310** **164.312** **164.314** **164.316**

(*Continued*)

TABLE 1.1 (*Continued*)

What Part 11 Says	What It Means	The HIPAA Parallel
(d) Use of transaction safeguards to prevent unauthorized use of passwords and/or identification codes, and to detect and report in an immediate and urgent manner any attempts at their unauthorized use to the system security unit, and, as appropriate, to organizational management.	The system must allow only authorized persons to use electronic signatures at the appropriate time and in association with objects within the scope of each user's allowed responsibilities. The system must detect any security breaches. In most systems, failure to provide a log-on password after a certain number of predefined tries causes a lockout. A system administrator must unlock the account.	**164.306** **164.308** **164.310** **164.312** **164.314** **164.316**
(e) Initial and periodic testing of devices, such as tokens or cards, that bear or generate identification code or password information to ensure that they function properly and have not been altered in an unauthorized manner.	When devices are used as electronic signature components, procedures must account for all devices and ensure their proper operation on a continual basis.	**164.306** **164.308** **164.310** **164.312** **164.314** **164.316**

31. What's the Difference Between a Final Rule and a Guidance Document?

A final rule is a law; it is akin to a directorate in Europe. You must comply with the laws that apply to your operations. The FDA and other regulatory agencies often issue guidance documents to help industry comply; guidance documents are not legally binding, but they reflect the agency's current thinking on a subject.

32. Does Industry Have to Comply with Guidance Documents?

No. Guidance is simply advice. That said, most companies adhere to best practices, and these usually reflect the intent of the guidance documents. In the beginning of every guidance, there is a disclaimer that says the guidance is not binding.

33. What Are "Best Practices" and "Industry Standards"?

Best practices and industry standards are synonymous. These terms refer to how industry as a whole interprets and complies with the laws.

34. How Do Industry Standards Develop?

It takes about five years to establish industry standards after a rule becomes final. Dialogue between industry leaders and the regulators help shape standards; professional discourse among industry members provides a "give and take" and sharing of "tried and true" practices for industry to reach a modicum of standardization in accord with an applicable regulation.

35. What Is the Scope of Industry Standards?

Industry standards develop across all branches of the federal government and are primarily related to the FDA, the National Security Agency (NSA) (18), and the Health Insurance and Portability Act (HIPAA). Industry standards for record keeping evolve as computer hardware and software evolve in order to address exploited and anticipated vulnerabilities.

36. Are the HIPAA Regulations Predicate Rules for Medical Records Maintained Electronically According to 45 CFR Part 164?

The HIPAA regulations are predicate rules to Part 164, just as the FDA has predicate rules and 21 CFR Part 11.

37. Does 21 CFR Part 11 Apply Only to Electronic Records that Are Inspected by the FDA?

No. The regulation also applies to electronic records not submitted to the FDA that are relative to the company's design, development, manufacture, packaging, distribution, and tracking of its products.

38. Should an International Company Have Worldwide Procedures in Place that Address 21 CFR Part 11?

FDA doesn't expect a company to have a set of procedures that are applicable to every site the company may have. Each facility should have procedures specific to the operations that take place there. Bear in mind that, for FDA-regulated products marketed in the United States, the company must have SOPs that cover the predicate rules as well Part 11.

39. Why Is FDA Revising Part 11?

The current Part 11 does not call for grandfathering of systems in place prior to the issuance for this law. That means that companies have to validate any systems housing electronic records. Industry was slow to comply, since validation of working systems could be extremely costly. Revision of Part 11 will make it more understandable and efficient for companies to comply with the intent of the law. The revision to Part 11 will address risk.

40.   What Exactly Does "Risk" Mean?

Risk means that the users of the software and their management understand what hazards can potentially occur, and what the potential effect could be.

41.   How Will Companies Know Which Systems Pose Risk?

It is up to companies to determine which systems pose risk. For instance, software that reports on adverse events during a trial is high risk because it is directly related to safety. The methodology for identifying risk in systems is often called "Gap Analysis and Remediation Planning." A gap analysis compares the industry standards with the actual functionality of the system. While a general level of risk can be determined to prioritize computer systems, a detailed measure of risk is performed during the validation of the computer system when the Hazard Analysis document is created.

42.   What Will the Changes to Part 11 Entail?

The current Part 11 law is vague and uses language that has changed since 1997. The rewrite for Part 11 will most likely update the language and make more clear how electronic records and predicate rules work together. The concepts of Part 11 will not change. Computer systems will need to be validated to ensure they are secure and are capable of maintaining data integrity.

43.   When the FDA Withdrew Guidance Documents in August of 2003, How Did that Affect Part 11?

Much of industry took the withdrawal of guidance documents as an indicator that Part 11 was also going away. This was a misconception; the law remains in place.

44.   When Will FDA Require Electronic Submissions?

The FDA established December 31, 2007 as the last day to file paper submissions. While the agency has waived requirements on a per-case basis since then, it has announced that after June 2009 it will no longer issue waivers.

45.   Do Other Countries Accept Validation Standards Based on US Regulations?

Yes, the current industry standard for computer system validation employs the risk-based approach. This approach contains all of the essential documentation components required for process-level validation of software supplied by software vendors, which is also known as commercial or configurable off-the-shelf (COTS) validation.

46. Are the US Industry Standards for Computerized Systems Comparable to Industry Standards Elsewhere?

Yes, industry standards for computer systems are uniform throughout the world. Most computer systems are accessible via the Internet from almost anywhere, so standards must be kept uniform to ensure security and reliability.

47. Is the Scope of 820 Larger than Devices?

In the FDA Guidance Document Part 11 Electronic Records; Electronic Signatures part C, there is a reference to 21 CRF 820.70. The scope of 21 CRF 820 is devices and is the most recent of the GMP regulations. As such, it is the benchmark for the other regulations. The expectation is that 820 will be the standard for any revisions to the other GMP regulations.

48. The Regulators Are Good at Telling Industry What to Do, But Do They Follow Their Own Dictates?

Yes. The FDA validates agency systems that read and maintain regulatory submissions from industry.

49. How Do Companies Comply Individually?

Companies adhere to the regulations that apply to them, and they document what they do in procedural documents such as Standard Operating Procedures and Work Instructions. They train the workforce in those procedures, and they keep records of all activities as they occur. They maintain quality through self-audits, monitoring, and corrective action. Everything is documented, and documentation provides the proof of quality and compliance.

50. If the Regulations Call for Certain Compliance Practices, Does that Mean All Companies Pretty Much Do the Same Thing?

The concepts of what has to be done to be compliant are about the same from company to company, but the methods for implementation vary widely.

51. How Does the FDA Issue New Regulations?

The FDA publishes proposed rules in the *Federal Register*, a daily publication available online. Industry can then comment on the regulations. Generally, there is a common consensus, and then a proposed rule becomes final rule, or law. The FDA publishes rules that establish or modify the way it regulates drugs, biologics, radiation-emitting electronic products, and medical devices. "These rules are not created arbitrarily or in a vacuum. They are formed with the public's health in mind," according FDA's own website (19).

52.   How Can Companies Keep Abreast of Changes in the Regulations and Their Interpretation?

Changes in regulations are far less common than changes in industry standards, and the regulations are always available in the Code of Federal Regulations. However, the FDA rulings between an existing CFR and the next annual copy indicate how the FDA is interpreting industry practices. Companies must be vigilant in watching trends in regulations and interpretations. Good sources are newsletters, such as the Drug GMP Report and the Pink Sheet. The FDA website also publishes warning letters that can serve as a source for gauging the perspectives of the agency. Daily newsletters, available online, also provide information. And companies attend conferences and seminars to stay current. (See Chapter 13.)

53.   What Are Warning Letters, and How Do Companies Get Them?

The agency inspects companies who are developing and manufacturing therapeutic products. They inspect not just the physical facility, but the records and information the organization must maintain. Drug and device manufacturers are inspected about every two years. Suppliers can expect inspections less frequently. If the investigators find noncompliance, they will issue a Form 483 that cites violations at the end of the inspection. If the citations are serious, the agency will send a warning letter to the company, and the company must respond with a plan for corrective action. The FDA posts warning letters on its website, and they become a matter of public record (19).

54.   What Does It Mean to Redact Sensitive Information in Warning Letters?

Redaction is annotation to conceal certain parts of sensitive documents. In FDA-posted warning letters, redaction is simply a black strike through of information that may be proprietary to the company who has received the warning letter.

55.   What Happens If a Company Doesn't Fix Its Problems?

If a company fails to fix its problems, it can expect another warning letter at the next inspection. Repeated failures to implement corrective action can lead to a consent decree, and beyond that they can lead to an injunction and even prosecution.

56.   What Is a Consent Decree?

A consent decree means that the agency is looking over your shoulder very carefully and that it is often onsite to monitor how you fix things. A consent decree is usually very costly in terms of fines to the agency.

57.   What Is an Injunction?

Injunction means the company can no longer conduct business. If violations are serious—such as for willfully fraudulent activities—there may also be legal action.

58. What Are Common Citations for Companies?

Companies are often cited for lack of adequate documentation and document controls. Statements such as "failure to adequately document …" or "failure to develop adequate written procedures …" appear in warning letters posted on the FDA website (19).

59. What Is the Direction the Industry Is Taking?

Globalization is definitely here. That means that US companies are now conducting clinical research outside the country. Many clinical trials are now in progress in Brazil, Russia, India, and China (BRIC). In the United States, the FDA has issued a Critical Path Initiative with the goal of modernizing the critical path of medical product development to move products through the development process and to patients more quickly. This initiative has six priorities: biomarker development; streamlining clinical trials; bioinformatics; efficiency in manufacturing; development of antibiotics and countermeasures to combat emerging infections and bioterrorism; and developing therapies for children and adolescents (20).

60. Are There Plans to Harmonize Electronic Records and Signature Requirements Between the US and the European Union Regulatory Agencies?

Available EU documents that delineate electronic record keeping embody the same principles and controls that 21 CFR Part 11 spells out. Annex 11 for Computerised Systems provides a Guide to Good Manufacturing Practice for Medicinal Products.

61. Besides Submissions and the Documentation Investigators Request During Inspections, What Other Types of Documentation Do Regulators Look At?

Regulators watch what industry is doing; they read press releases, web information, and publications; attend industry conferences to sit through presentations and view posters and exhibits; and monitor web blogs and peer communications forums.

REFERENCES

1. Office of the Commissioner (OC), www.fda.gov/oc/
2. Center for Biologics Evaluation and Research (CBER), www.fda.gov/Cber/
3. Center for Devices and Radiological Health (CDRH), www.fda.gov/cdrh/
4. Center for Drug Evaluation and Research (CDER), www.fda.gov/CDER/
5. Center for Food Safety and Applied Nutrition (CFSAN), www.cfsan.fda.gov/
6. Center for Veterinary Medicine (CVM), www.fda.gov/cvm
7. National Center for Toxicological Research (NCTR), www.fda.gov/NCTR/
8. Office of Regulatory Affairs (ORA), www.fda.gov/ORA/
9. Occupation, Safety and Health Administration (OSHA), www.OSHA.gov/
10. *Federal Register*, www.accessdata.fda.gov/scripts/oc/ohrms/index.cfm
11. International Conference on Harmonisation, www.ich.org/

12. Sarbanes–Oxley Act, www.sec.gov

13. Food and Drug Administration Amendments Act, 2007, www.fda.gov/regulatoryinformation/ legislation

14. Guidance for Industry: Part 11, Electronic Records; Electronic Signatures, gov/cder

15. Supporting Statement for Electronic Records; Electronic Signatures 21 CRF Part 11, docket 05N-0045.

16. Health Insurance, Portability and Accountability Act, www.dhhs./gov/privacy/index.html, Department of Health and Human Services, www.ihe.net/Technical_Framework/upload/IHE-PHDSC_Public_ Health_White_Paper_2008-07-29.pdf

17. National Security Agency (NSA), www.nsa.gov/

18. www.nsa.gov

19. www.fda.gov

20. Critical Path Initiative, www.fda.gov/oc/initiatives/criticalpath/initiative.html

CHAPTER 2

PEOPLE, PROCESS, AND DOCUMENTATION

INTRODUCTION

Robust documentation systems have commonality, regardless of their structures. They result from a healthy understanding of the data and documents the company needs to control, a workable and user-friendly process, and a knowledgeable workforce that functions as a team in sustaining and growing the system. Unfortunately for some companies, document management has been perceived as an administrative function or even a filing system. Such a viewpoint has proven to be problematic, since documentation is the life blood of companies developing or producing therapeutic products. Without documentation, a company has no proof of its activities or the integrity of its products. Thus it is critical that companies allot the appropriate resources to producing and managing documentation. Such resources include people knowledgeable in document controls as well as managerial and budgetary support. With these resources, the company can build and maintain a system that functions optimally within the culture of the company. This chapter answers the following questions:

1. What exactly does the term "documentation" mean?
2. What is the purpose of documentation?
3. Why is a document system necessary?
4. How do you define what a document system actually is?
5. How does the term "informatics" differ from document management?
6. To what unit of a company does document management belong?
7. What do systems require in terms of people?
8. How do you determine who is best qualified to build and manage a system?
9. How can we build a team to implement better document management?
10. How important are the "human factors" in developing a system?
11. What is the scope of document management?
12. What are the key roles for developing an effective document management system?

---

13. Shouldn't IT be the final say in how a system is built and operates?
14. How do you define "commodity"?
15. Can we assign administrative assistants to manage the documents?
16. Are there consistent industry standards for building a document management system?
17. How can we determine what kind of a system we need?
18. What is a record/document life cycle?
19. Is there an official guidance document that identifies what documents require controls?
20. What types of documents do companies generate?
21. What kinds of reports do companies generate?
22. What are source data?
23. What is the purpose of source data?
24. How do source data support other documentation?
25. Do source data require controls?
26. What category of documentation do laboratory notebooks fall into?
27. What are "backbone" or "skeleton" documents?
28. What's the difference between a procedure and a manual?
29. Should drug manufacturers have quality manuals?
30. If a company creates a "work aid" to augment an SOP, does it require controls?
31. Can a company create a document that is not legally binding?
32. Do companies need to keep documentation on activities that are not directly related to a product?
33. Should contracts be part of a standard document management system?
34. Is e-mail documentation?
35. How should e-mail with attached files be handled?
36. Are forms documentation needing controls?
37. What is the difference between a template and a form?
38. What's the purpose of a template?
39. Do we have to control our templates?
40. Do we have to create our own templates?
41. What does "boilerplate" text mean?
42. Why is boilerplate text useful, and must you control it?
43. Our company stresses the importance of publication, but why is it important?
44. How can we best disseminate information outside the company?
45. How is publishing different from advertising?

46. Who is the audience for disclosure?

47. Why are regulators interested in what a company publishes?

48. Is an abstract submitted to a professional meeting considered documentation?

49. Are journal articles and letters to the editor considered documentation?

50. What's the purpose of a poster?

51. Do slide shows for presentations count as documentation?

52. What is a white paper?

53. What are submission documents?

54. Does training require documentation?

55. How can we best build a system that works for us?

56. What is the best way to categorize documents?

57. How do we start document controls?

58. Are flow charts useful, and do they fall into the category of documentation?

59. Does it matter what type of flow diagram we create?

60. How can we make sure that people use our systems correctly?

61. Does everybody have to be trained in a document system before they can use it?

62. How much training is sufficient?

63. Where can we find the requirements for training?

1. What Exactly Does the Term "Documentation" Mean?

Here's a definition from 21 CFR Part 820, the newest of the good manufacturing practices regulations (1996): "Documentation means any records—written, electronic or automated—that provide information describing, defining, specifying, reporting, certifying, or auditing a manufacturer's FDA regulatory activities, requirements, verifications, or validations (1).

2. What Is the Purpose of Documentation?

Documentation tells what you did, what you will do, and what you do regularly. In sum, it tells the "story" of the company and its products.

3. Why Is a Document System Necessary?

Without a document system, there would be chaos. The sheer magnitude of documentation that this industry generates necessitates controls. You must be able to generate documents/records efficiently and be able to lay your hands on your documentation when you need it. Failure to exercise control can result in liability. A look at the FDA website's warning letters can tell you that faulty or missing documentation and controls are frequently cited violations.

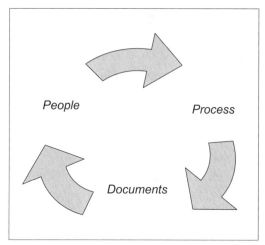

Figure 2.1. Document triumvirate.

4. How Do You Define What a Document System Actually Is?

A document system provides a way for people to identify the need for documents and records and then generate, review, revise, approve, and archive them in a consistent and controlled manner. Good systems come into being because those people putting them into place understand the triumvirate of intersections: people, technology, and process. (See Figure 2.1.)

5. How Does the Term "Informatics" Differ from Document Management?

Informatics is the science of information, information processing, and information systems. In a sense, informatics is what we do when we build, manage, and use document management systems. Informatics includes interaction of people with systems that store, process, and communicate information. Informatics includes computations and database processing that is not document-oriented.

6. To What Unit of a Company Does Document Management Belong?

There's no one answer to this question. A lot depends on the resources of an organization, but clearly, no matter where the function resides, it is a part of quality overall. Many companies like to keep document management separate from QA proper so that QA can audit objectively.

7. What Do Systems Require in Terms of People?

Successful systems require people who understand the system because they have been trained on it, and they function as part of a team in making the system work optimally. There should also be strong management and budgetary allowance for the entire system: the system itself, system configuration, system function and maintenance, and system growth and evolution. In other words, support needs to be in place for the system throughout its entire life cycle, inception through retirement.

8.   How Do You Determine Who Is Best Qualified to Build and Manage a System?

Document systems work best when organizations understand their personnel resources and use them effectively to establish their systems. People need to work together over the life cycle of a system to establish the system and keep it current and running optimally. Companies can use both their internal resources and contracted personnel. It's important, however, that contract personnel do not "drive" the development of the system, but rather support it.

9.   How Can We Build a Team to Implement Better Document Management?

In a healthy corporate culture, there is a focus on shared purpose and continual quality improvement, and both management and the workforce are committed to positive growth. There is an air of leadership, not domination.

10.   How Important Are the "Human Factors" in Developing a System?

Document systems work best when they develop in organizations that are open to collaboration. In other words, are people encouraged to work together on joint efforts to accomplish a specific task or project? If they are involved, they are usually less likely to obstruct progress. Often the greatest obstacle to progress is the human one because people are reluctant to accept change. Yet implementing a new system means change is coming. How the organization handles such an event depends on its mental health.

11.   What Is the Scope of Document Management?

Systems should encompass all GXP documentation and records. That means there may be many components to managing documents. Consider clinical trial data compilation, which takes place during a study. Data may be gathered and maintained by a Contract Research Organization (CRO), but the sponsor is still responsible for it. Thus, a clear roadmap of where data reside and how they are brought to the company for analysis and reporting is essential. Similarly, records that a Contract Manufacturing Organization (CMO) produces in support of manufacturing activities need controls, as do batch records for products a company makes at its own facilities. In short, the range of records and documents a company needs to control correlates to the business model—when the firm is in developing or manufacturing a product. (See Chapter 1, Figure 1.1, for the document continuum.)

12. What Are the Key Roles for Developing an Effective Document Management System?

Project management is critical. A managerial level person can assemble a team to determine user needs and then develop a plan. The team needs to understand the goals of the system and determine how the system will work. Communication with and input from all areas of the company that will use the system help ensure that the system is user-friendly and sufficient for the company's needs. Information Technology (IT) is also critical in implementing the system, whether it is a commercial software system or an in-house system. And of course, there must be trainers to ensure that the users understand the system once it is in place.

13. Shouldn't IT Be the Final Say in How a System Is Built and Operates?

IT is essential, to be sure. But IT should not be the driver. IT is a service, a commodity. The users of the system, those with responsibility for the documentation processes, should have the most to say about how a system is built and operates.

14. How Do You Define "Commodity"?

A commodity is a good or service. IT in this sense is a commodity because it provides service for a system; IT should not be either the designer or the "owner" of a system.

15. Can We Assign Administrative Assistants to Manage the Documents?

Some companies believe that document management is simply a filing system and that the function is secretarial in nature. This is a faulty perception. If you do assign the task to administrative assistants, they must understand the system and how it operates and then adhere to the standard procedure for managing it. It's best to assign a designation to the roles, such as Document Manager or Document Coordinator. Furthermore, the person(s) entrusted with managing the system must have the authority to enforce its standards. While large companies have people assigned to document management, many do not understand the complexities of document management and fail to empower people for the task. This is particularly true in small, young companies where people have to "wear many hats." Currently, there is no established role for document management in this industry, but there is every expectation that it will become a core discipline, much as project management did over a decade ago.

16. Are There Consistent Industry Standards for Building a Document Management System?

Not at press time, although many thought leaders in industry have proposed common standards. The Pharmaceutical Research and Manufacturing Association (PHrma) has not endorsed a common standard, nor have organizations such as the Drug Information Association (DIA) or Parenteral Drug Association (PDA). Companies still have to assess their document needs and available resources and determine a system that will work for them. All computer systems used in regulated environments should adhere to the industry standards related to 21 CFR Part 1, Electronic Records; Electronic Signatures.

17. How Can We Determine What Kind of a System We Need?

There is no "one size fits all." Each company has to determine its own needs based on the business model and the documentation it needs to control. Each document or record has a life cycle.

18. What Is a Record/Document Life Cycle?

Life cycle refers to the steps the process of generating, reviewing, revising, approving, and archiving a document or record. (See Figure 2.2.)

19. Is There an Official Guidance Document that Identifies What Documents Require Controls?

No. Each company needs to determine which records it will control. The best approach is to think "product-related activities." What source data are relative to the

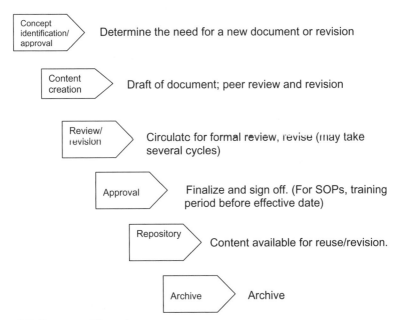

Figure 2.2. Document life cycle.

product? What documents do you generate as a result of the data you gather? What processes are related to your products? What peripheral activities do you carry out to make sure your processes are effective? All documentation of this nature needs control.

20.   What Types of Documents Do Companies Generate?

Documents must accurately reflect the activities of the company, so each company must determine the documentation needs based on the business model. Most documentation falls into categories: source data, procedural "backbone" documents and forms, investigations, protocols and reports, and records that support the infrastructure.

21.   What Kinds of Reports Do Companies Generate?

Companies in any stage of development typically produce protocols and reports, whether for Good Laboratory Practices (GLPs) or Good Clinical Practices (GCPs). If a company seeks product approval, submission documents are requirements. Once in the marketplace, there are annual reports that go to the regulators. Other reports reflect a whole range of activities such as vendor evaluation and product qualification; equipment, facility, and process validation; out of specification and deviation reporting; and more, depending on the business model.

22.   What Are Source Data?

Source data, or raw data, are original records that companies collect to support their products. The Good Laboratory Practices (GLP) guidelines identify source data as follows: Raw (source) data "means any laboratory worksheets, records, memoranda, notes, or exact copies thereof, that are the result of original observations and activities of a non-clinical laboratory study and are necessary for the reconstruction and evaluation of the report of that study" (2). Laboratory notebook entries are examples of source data. So are QC test results for incoming, in process, or finished products. Validation test results are also source data. In short, source data are original, first-time data entries. Source data may include paper and electronic records. Typically, they feed into other documents.

23.   What Is the Purpose of Source Data?

Source data are the foundation for much of what occurs in the activities of a company. If not retained, there is no way to confirm that data have been incorporated and assessed correctly in reports in other documents. Source data can be critical in investigations.

24.   How Do Source Data Support Other Documentation?

Source data feed into other documents. For instance, if a company plans to do a preclinical GLP study, the first step is to prepare a protocol, a "how to" for the study to explain what the study is going to do and why. Once a study is underway, study members adhere to the protocol, perform testing, and collect data (source data). When the study is done, study members prepare a report and tell what they did, and what the outcome was, based on the data from the study.

25. Do Source Data Require Controls?

Yes. Source data are the basis for development and manufacturing activities. In fact, you should be able to reconstruct clinical trial outcomes from the source data collected during the trials. Similarly, product in the marketplace should be traceable back to manufacturing batch records.

26. What Category of Documentation Do Laboratory Notebooks Fall into?

Notebooks contain source data. Any notebooks that record information from the earliest development activities require management. While controls are not mandatory for notebooks generated during pure research activities, companies often put controls in place because notebooks can be the source of intellectual properties and first proof of concept.

27. What Are "Backbone" or "Skeleton" Documents?

Backbone, or skeleton, documents are the procedures that companies must put in place. They include policies, SOPs, work instructions, quality manuals, methods, plans, forms, and reference documents—in short, any procedural documents that tell "how it happens here" on a regular basis.

28. What's the Difference Between a Procedure and a Manual?

A procedure tells how a system or process works. It may also provide instructions with defined start and stop points for carrying out a process. A manual is a big picture document that gives a "bird's-eye view" of any entire function, such as quality. An example of a manual is an operator's handbook for using a medical device.

29. Should Drug Manufacturers Have Quality Manuals?

In the United States, it is only a requirement for devices. 21 CFR 820, the newest of the GMP regulations, calls for quality manuals. Quality manuals are also required in Europe. Many drug manufacturers are adopting quality manuals, however, because they make good business sense in that they give a comprehensive quality overview. Furthermore, it seems likely that a revision to Part 211 will be based on Part 820, thereby requiring quality manuals.

30. If a Company Creates a "Work Aid" to Augment an SOP, Does It Require Controls?

Yes. Any document that is a "how to" requires control. It doesn't matter what you call these documents; they are legally binding, and the regulations mandate controls. They may be called work instructions, methods, workstation instructions, or any other name.

31. Can a Company Create a Document that Is Not Legally Binding?

Yes. Companies often create reference documents such as style guides, but when they do, they must be careful to identify them as "guidances" or "references" and not procedural documents.

32. Do Companies Need to Keep Documentation on Activities that Are Not Directly Related to a Product?

If the activity supports the infrastructure that must be in place to produce a product or service, documentation of that activity must be controlled. Examples are valida-tion and qualification records. Validation and qualification are integral to making sure processes, equipment, and software function as they should. The end result of validation and qualification is a packet of documentation, which must be in place and must be controlled.

33. Should Contracts Be Part of a Standard Document Management System?

They can be, but there is no requirement. Many contracts are retained in a company's legal department in a system separate from the standard document system. SOPs that drive contractual work should spell out where they reside.

34. Is E-Mail Documentation?

Not all e-mail satisfies the definition of documentation. All e-mail that is relative to product development or the product life cycle should be considered to be documenta-tion. For instance, if a company is conducting a clinical trial, correspondence between the sponsor and a principal investigator may hold information that can determine the direction of a trial. Companies also use e-mail to transport files, and such e-mail may be treated differently than e-mail that contains direct information. Companies must have procedures in place that address the maintenance of e-mail documentation.

35. How Should E-Mail with Attached Files Be Handled?

E-mail that is transmitted over the Internet is insecure. Users must be careful not to send confidential information in e-mail attachments. If e-mail is used to transmit confidential information, the attached file must be encrypted. The user should inform the person receiving the e-mail of the password via telephone, not via e-mail.

36. Are Forms Documentation That Needs Controls?

Yes. Forms are critical for capturing source data. Forms can be controlled the same way you control SOPs and other versioned documents.

37. What Is the Difference Between a Template and a Form?

The terms template and form are often thought of as the meaning same thing. A template defines a set of content labels that are uniform, but the content itself is very flexible. A form defines specific fields of information to be captured and typically has restrictions on the flexibility of the content.

38. What's the Purpose of a Template?

Templates ensure consistency. Good templates help in both the writing process and the outcome. Effective templates can be instructional, with each section telling what to write where.

39. Do We Have to Control Our Templates?

Yes. Templates and forms are similar and require the same types of controls. Templates must be current. It's important to make sure that people are not generating documents based on uncontrolled copies of templates, such as from copies of templates residing on individual PCs.

40. Do We Have to Create Our Own Templates?

Many companies do and in so doing create their own "signature look." However, many commercial companies provide templates for the document types that the industry generates.

41. What Does "Boilerplate" Text Mean?

Boilerplate text is information that is built into templates so that messages are consistent from document to document. For instance, a clinical trial report template can have boilerplate text for product and facilities descriptions, product indication, and disease states.

42. Why Is Boilerplate Text Useful and Must You Control It?

Boilerplate text helps ensure that the company talks about its product(s) in the same way from document to document, regardless of the author. When boilerplate text is included in a template, it is controlled by the template version.

43.   Our Company Stresses the Importance of Publication, But Why Is It Important?

Companies publish to share information with the industry and the public. Publication is also known as disclosure. Disclosure covers abstracts, journal articles, letters to editors of industry publications and newspapers, press releases, websites, and slide presentations and posters for professional meetings. Companies must be careful not to inadvertently make a claim they cannot support, or release proprietary information that they should not.

44.   How Can We Best Disseminate Information Outside the Company?

There are many vehicles for disseminating information outside the company. In addition to the vehicles mentioned above, companies can also provide information brochures for trade shows, newsletters, product dossiers, scripts for professional forums, and the like. Many companies establish publication plans to strategically place information for optimal exposure.

45.   How Is Publishing Different from Advertising?

Publishing is informational, not promotional, and that is how it differs from advertising. Companies cannot promote products that have not received marketing approval from the regulators, but publishing offers a way of (a) reporting on the status of products in development and (b) telling what the company knows to be true about a particular therapeutic product at the current time. Companies cannot make claims for off-label use of an approved product, but they can disseminate information to professionals who need it. Companies also publish to maintain a presence in the scientific arena. The FDA is strict in enforcing what companies can and cannot say about their products in publications and has published a guidance for industry: The Good Reprint Practices (3).

46.   Who Is the Audience for Disclosure?

The audience for disclosure is outside the company and is diverse: industry professionals including competitors; the public; potential partners, clients, and owners; analysts and investors; and regulators.

47.   Why Are Regulators Interested in What a Company Publishes?

The regulators watch what companies say about their products in development as well as those in the marketplace. It is important that companies not make unsubstantiated claims about their products.

48.   Is an Abstract Submitted to a Professional Meeting Considered Documentation?

Yes. Careful companies keep close controls on abstracts for industry meetings. Abstracts should be subject to review and approval prior to release to ensure that the author does not say anything that the company's science or technology has not established or that may be patent pending.

49.   Are Journal Articles and Letters to the Editor Considered Documentation?

Yes. These, too, should be subject to document controls. Industry professionals and the regulators read journals to keep abreast of what companies are doing. Again, letters and articles must not state a position that is not official for the company or that is not substantiated by the company's science or technology.

50.   What's the Purpose of a Poster?

Posters are visuals that appear in poster halls at conferences. They are usually 4×6 feet in size and are dramatic. They include the abstract submitted and accepted for the meeting, and the poster itself expands upon the abstract, often with diagrams, drawings, and photos.

51.   Do Slide Shows for Presentations Count as Documentation?

Careful companies make sure that nothing leaving the company is uncontrolled. That applies to talks as well as written media. Establishing a slide library for the company's products and positions is prudent, since such a resource ensures consistency in the messages that reach the industry and public.

52.   What Is a White Paper?

There is no consensus as to what a while paper actually is. The government issues white papers via statements and guidances. Companies may post white papers on their web pages. In essence, a white paper describes some issue in detail. It is often not related to an established process or document type and as a result is often not included in change control. This presents a problem in that white papers are often lost or misused. It is best to control white papers and document how they evolve into standard documents if they do.

53.   What Are Submission Documents?

Submission, or regulatory, documents go to an audience outside the company—the regulators. These documents sum up development activities for products in the clinic. These documents include Investigational New Drug Applications (INDAs), Investigational Device Exemptions (IDEs), Clinical Trial Authorizations (CTAs), and dossiers for market approval. *Note*: Chapter 9 addresses submission documents in more depth.

54. Does Training Require Documentation?

Yes. The regulations require training and proof of training. Records are tangible proof that you have conducted training. These records must be controlled and accessible.

55. How Can We Best Build a System that Works for Us?

A good starting point is to create a comprehensive document plan that identifies everything the company must control. Then determine who "owns" each piece. The important thing is that people designing and managing systems understand the comprehensive overview of what is subject to controls and how those controls work.

56. What Is the Best Way to Categorize Documents?

The best way to make sense of the document maze is to categorize documents by type. That way you can avoid creating "silos" of information for each area of the company. The silo model by its nature creates a great deal of replication of documentation. If documents are classified by category, they can be accessible and usable throughout the firm, as they apply. For instance, SOPs as a general category helps prevent overlaps and contradictions because all the SOPs reside in the same place. It's scope that determines where they apply.

57. How Do We Start Document Controls?

The best thing to do is to identify the types of documents and records you produce. Then chart the flow for the different document types. Word and PowerPoint both have simple tools that can help determine your document cycle.

58. Are Flow Charts Useful, and Do They Fall into the Category of Documentation?

Flow charts present processes visually, and as such they are very helpful. They are indeed documentation and must accurately reflect the process they address. Like SOPs, they require controls for generating, reviewing, revising, approving, updating, and archiving.

59. Does It Matter What Type of Flow Diagram We Create?

No. There are many good models. The "Swim Lane" model is effective, and you can create such a diagram in landscape or portrait. Many software programs are available for creating flow charts, and Word has the capacity to construct visuals as well.

60. How Can We Make Sure That People Use Our Systems Correctly?

Training is a requirement for all activities related to therapeutic product develop-ment, manufacturing, and distribution. All training occurs according to standard operating procedures, which reflect all the processes the company follows. Training must occur before anyone carries out a process.

61. Does Everybody Have to Be Trained in a Document System Before They Can Use It?

Users of a system must be trained before they use the system. That does not mean that all users must be trained before the system is operational, but only that no one uses a system without undergoing training first.

62. How Much Training Is Sufficient?

Initial training in a process is a requirement. But the regulations also call for periodic update training. And if an employee fails to follow a procedure correctly, then retraining is necessary. If a procedure changes, then training on the changes to the procedure is necessary. (See also Chapter 6 and Box 2.1.)

BOX *2.1*

SOME REGULATIONS THAT ADDRESS TRAINING

21 CFR PART 211—CURRENT GOOD MANUFACTURING PRACTICE FOR FINISHED PHARMACEUTICALS

Subpart B—Organization and Personnel

§ 211.25 *Personnel Qualifications*

Each person engaged in the manufacture, processing, packing, or holding of a drug product shall have education, *training*, and experience, or any combination thereof, to enable that person to perform the assigned functions. *Training* shall be in the particular operations that the employee performs and in current good manufacturing practice (including the current good manufacturing practice regulations in this chapter and written procedures required by these regulations) as they relate to the employee's functions. *Training* in current good manufacturing practice shall be conducted by qualified individuals on a continuing basis and with sufficient frequency to assure that employees remain familiar with CGMP requirements applicable to them.

Each person responsible for supervising the manufacture, processing, packing, or holding of a drug product shall have the education, *training,* and experience, or any combination thereof, to perform assigned functions in such a manner as to provide assurance

(Continued)

that the drug product has the safety, identity, strength, quality, and purity that it purports or is represented to possess.

§ 211.34 *Consultants*

Consultants advising on the manufacture, processing, packing, or holding of drug products shall have sufficient education, *training,* and experience, or any combination thereof, to advise on the subject for which they are retained. Records shall be maintained stating the name, address, and qualifications of any consultants and the type of service they provide.

21 CFR PART 606—CURRENT GOOD MANUFACTURING PRACTICE FOR BLOOD AND BLOOD COMPONENTS

Subpart B—Organization and Personnel

§606.20 *Personnel*

(b) The personnel responsible for the collection, processing, compatibility testing, storage or distribution of blood or blood components shall be adequate in number, educational background, *training,* and experience, including professional *training* as necessary, or combination thereof, to assure competent performance of their assigned functions, and to ensure that the final product has the safety, purity, potency, identity and effectiveness it purports or is represented to possess. All personnel shall have capabilities commensurate with their assigned functions, a thorough understanding of the procedures or control operations they perform, the necessary *training* or experience, and adequate information concerning the application of pertinent provisions of this part to their respective functions.

21 CFR PART 820—QUALITY SYSTEM REGULATIONS

Subpart B—Quality System Requirements

§ 820.20 *Management Responsibility*

(2) *Resources.* Each manufacturer shall provide adequate resources, including the assignment of *trained* personnel, for management, performance of work, and assessment activities, including internal quality audits, to meet the requirements of this part.

§ 820.25 *Personnel*

(b) *Training.* Each manufacturer shall establish procedures for identifying *training* needs and ensure that all personnel are *trained* to adequately perform their assigned responsibilities. *Training* shall be documented. As part of their *training,* personnel shall be made aware of the device defects which may occur from the improper performance of their specific jobs.

Subpart G—Production and Process Controls

§ 820.70 *Production and Process Controls*

(d) *Personnel.* Each manufacturer shall establish and maintain requirements for the health, cleanliness, personal practices, and clothing of personnel if contact between such personnel and product or environment could reasonably be expected to have an adverse effect on product quality. The manufacturer shall ensure that maintenance and other personnel who are required to work temporarily under special environmental conditions are appropriately *trained* or supervised by a *trained* individual.

ICH Q7A Good Manufacturing Practice Guide for Active Pharmaceutical Ingredients

3 *Personnel*

3.1 Personnel Qualifications

3.10 There should be an adequate number of personnel qualified by appropriate education, *training* and/or experience to perform and supervise the manufacture of intermediates and APIs.

3.12 *Training* should be regularly conducted by qualified individuals and should cover, at a minimum, the particular operations that the employee performs and GMP as it relates to the employee's functions. Records of *training* should be maintained. *Training* should be periodically assessed.

63.   Where Can We Find the Requirements for Training?

The predicate rules all address the requirements for training and records of training. The predicate rules require that people undergo training relative to their responsibilities.

REFERENCES

1. 21 CRF Part 820, Quality Systems Regulation, Definitions.
2. 21 CRF Part 58, Good Laboratory Practices, Definitions.
3. Good Reprint Practices for the Distribution of Medical Journal Articles and Medical or Scientific Reference Publications on Unapproved New Uses of Approved Drugs and Approved or Cleared Medical Devices US., www.fda.gov/oc.

CHAPTER 3

PRINCIPLES OF DOCUMENT MANAGEMENT

INTRODUCTION

Companies are bound by the regulations that call for generation of records and documentation of all product-related activities. As a result, managing the ongoing accrual of information has become a formidable task for the industry. What is certain is that the demand for records is not lessening. Indeed, it is becoming greater as companies merge, split off, partner, and outsource. The ability to produce records electronically as well as in paper form brings challenges in imposing order and avoiding chaos. How then can organizations make sense of the documentation requirements? A good understanding of what document management really means is the best starting point. This chapter answers the following questions:

1. What is data mining?
2. What is product life-cycle management, and what does it mean in terms of documentation?
3. What is a software document management system?
4. What kinds of systems are there for electronic record keeping?
5. Can we outsource document management?
6. If we use SaaS, how can we be sure the system is compliant and secure?
7. If we use contract research or manufacturing, can we rely on the contractors' documentation practices?
8. Are there specific regulations that state there must be a central records facility with a designated individual to manage it?
9. Is there an efficient way for us to control a set of external standards such as the Canadian Medical Devices Regulations, Medical Devices Directive from the EU, and guidance documents?
10. Can a company be paper-based?
11. What is a manual system?
12. What exactly is a hybrid system?

13. In a hybrid system, which copy is the official, paper or scanned?
14. If using a hybrid system, how does the handwritten signature link to the electronic record?
15. If we use electronic signatures, is compliance with Part 11 mandatory?
16. Are electronic signatures always necessary for the creation of electronic records?
17. Do e-records and e-signatures require the same controls?
18. What is a legacy system?
19. What are the characteristics of a good system?
20. How can we make paper documents secure?
21. Are we required to buy expensive software to manage our documents?
22. Is it possible to build software in-house that employs electronic signatures?
23. Why do companies get electronic systems?
24. As a new company that must put controls in place, can we go electronic immediately?
25. What is meant by interoperability?
26. What are privileges?
27. What is taxonomy?
28. Why is taxonomy important?
29. What is metadata?
30. What are attributes?
31. How long should it take a user to get into a system and enter data?
32. How do we demonstrate data integrity?
33. What are the considerations for using Excel for data processing?
34. Can we use wireless LAN for data acquisition and transmission?
35. Must document numbers identify the document category—say, manufacturing or shipping?
36. Can we reserve numbers for planned documents, say if we are doing a series of development reports for a specific project, and want the numbers to be sequential?
37. What's the purpose of document numbering?
38. If a document is retired, can a company reuse the number from that document?
39. Is there any standard terminology to identify document status?
40. What is document change control?
41. Do we have to keep copies of the drafts that went through a review cycle?
42. Besides Standard Operating Procedures (SOPs), what documents should be version-controlled?

43. Do SOPs and Work Instructions (WIs) require the same levels of control?

44. Do manuals require the same review cycles as procedures (SOPs)?

45. Must we manually sign our SOPs?

46. What documents actually do require signatures?

47. What color ink should we sign original documents with?

48. What kind of ink should we use?

49. Should we build a folder system for our hybrid system based on the Common Technical Document?

50. What's the best way for build a folder structure for a hybrid system?

51. How can we store images?

52. If we scan documents how can we be sure the results are acceptable?

53. Will the FDA accept copies of scanned documents?

54. How can we scan large format documents?

55. What is OMR?

56. What is OCR?

57. What is barcoding?

58. What's the difference between a QC check and a QA check?

59. If a company changes its templates, should it reformat the reports done in previous formats?

60. Since Part 11 requires date and time stamps, should we put date and time stamps on PDF documents posted on our company server?

61. How should we keep our paper documents?

62. Do the regulations say anything about keeping documents in fire-safe cabinets?

63. Do we have to keep our records onsite?

64. Must we retain original documents, or can we keep copies?

65. Are document storage facilities a safe place to put documents?

66. What happens if a data storage facility "loses" some of our documents? What should we do?

67. Can we rely on our contractors to adequately maintain the documentation they produce for us?

68. How can we bring documents in-house from our contractors?

69. If a contractor goes out of business, what happens to our documents?

70. How long do we have to keep our documents and records?

71. How can we safely delete e-mail?

72. What is a typical schedule for deleting e-mail?

73. If we have an electronic system and we bring legacy documents into it, can we destroy the paper copies?

74. Does a hybrid system require validation?

75. How can we make sure our hybrid system is secure?

76. If we decide to go fully electronic, should we still keep a hybrid system?

77. Our staff travels a lot. How can they access our system remotely?

78. What controls does remote access require?

79. How can we control documents that must leave the company, say to go to a clinical trial site?

80. Once we implement a system, can we expect it to function indefinitely?

81. As we generate documents, how can we make sure the right people review and approve documents?

1.   What Is Data Mining?

Data mining is a systematic examination using statistical or mathematical tools to glean information about a product or event. Data mining assesses patterns, time trends, and likelihood.

2.   What Is Product Life-Cycle Management, and What Does It Mean in Terms of Documentation?

Product life-cycle management is showing control over the entire life cycle of a product from first proof of concept, through development, approval to market, manufacturing, distribution, and to the end of its availability. The documentation generated during the life cycle of the product shows its history.

3.   What Is a Software Document Management System?

In its simplest form, a software document management system may be a set of folders on a server with user privileges set to allow for draft, review, approval, and distribution of documents. A document management system usually contains computer and software applications that provide a wide range of functionality: routing of documents in a workflow, training before the effective date (for procedures), automated version control, distribution, and retirement of superseded versions, archiving, record retention, electronic signatures, and searching of metadata and full text content.

4.   What Kinds of Systems Are There for Electronic Record Keeping?

Companies use Electronic Document Management Systems (EDMS) for managing documents. Laboratory Information Management Systems (LIMS) are tools for managing laboratory data. Electronic Data Capture (EDC) systems are often used to capture clinical trial data. Serious Adverse Event Reporting (SAE) systems track adverse events.

5. Can We Outsource Document Management?

There are third parties who provide Software as a Service (SaaS), and companies can subscribe to these services. In SaaS, software is hosted on a central server, and a company accesses the software using a web browser. The SaaS provider may be the same company as the software vendor, or the SaaS provider may be a different company. A third party manages the software installation and maintenance. It is important to note that many SaaS providers claim that subscribers need to do little more than some minor configuration and user training, but the fact is that the users must still perform computer system validation of the configurable off-the-shelf (COTS) system.

6. If We Use SaaS, How Can We Be Sure the System Is Compliant and Secure?

Many companies look to SaaS solutions because they want to reduce the need for IT services to install and maintain the system. Additionally, many companies look to SaaS solutions to avoid having to perform validation. The fact is that in order to be compliant, there are few differences between an in-house system and a SaaS solution. In both cases, evaluating the software vendor is necessary. With a SaaS solution, there should also be an evaluation of the hosting facility. In both cases, documentation of installation must be generated. And in both cases, the users are responsible for performing a complete set of validation activities to ensure correct operation and security, understanding of risk, and development of new processes and SOPs. The 10-step risk-based approach is applicable to all COTS systems hosted either in-house or via SaaS (1). [See Chapter 5 for more information about the 10-step risk-based approach to validation.]

7. If We Use Contract Research or Manufacturing, Can We Rely on the Contractors' Documentation Practices?

Yes, but with a caveat. Companies must determine if their contractors' systems meet the requirements for good documentation practices and determine documentation deliverables contractually. They must assess contractors before entering into arrangements with them, and then they must exercise quality assurance with regular audits. Beware, however, of contractor claims of compliance. There is nothing to prevent a claim, and many software companies are unaware of regulations and industry standards. The regulated company always has full liability when using contract organizations.

8. Are There Specific Regulations that State There Must Be a Central Records Facility with a Designated Individual to Manage It?

No, but the regulations do say that records must be secure and accessible, and industry puts controls in place to ensure that they are. Security means that original documents cannot be altered, defaced, or copied injudiciously.

9. Is There an Efficient Way for Us to Control a Set of External Standards Such as the Canadian Medical Devices Regulations, Medical Devices Directive from the EU, and Guidance Documents?

Many companies keep a category of documents called "Reference" and control them as such. This category houses documents such as those you've identified. But it's important that these documents remain current—if there's a revision to a regulation or guidance, the company needs to update accordingly. Doing so doesn't require version controls as your SOPs and work instructions do, but you should have a procedure that tells who is responsible for keeping these documents current and maintaining them.

10. Can a Company Be Paper-Based?

Technically, it could if it printed out a document and manually signed it without saving the electronic file and then retyped each update or revision as necessary. This is the "typewriter rule" that is often referred to when discussing Part 11. Word processing affords the ability to create electronic files, and it would be foolish not to retain electronic copies of documents. In a nutshell, like using typewriters, being paper-based is outdated and inefficient.

11. What Is a Manual System?

A manual system used to be a paper-based system, where everything worked off paper and copies of paper. Now all documents are produced in word processors, and these files require controls. Using both paper as the official documents and electronic files of them means that you have a hybrid system. A typical example of a hybrid system is a one where a user generates an electronic record and then prints and signs it according the requirements in the predicate rules.

12. What Exactly Is a Hybrid System?

Most document systems in place today are hybrid systems. Documents are generated in word processors that create files, which are electronic records. These files are printed and signed. It is common to post documents to a server or web server in PDF form, and these files require controls. Additionally, the signed paper original documents may be scanned to create files for distribution, or the word processing files may be used for distribution. When electronic files and paper documents are used together, you have a hybrid system.

13. In a Hybrid System, Which Copy Is the Official, Paper or Scanned?

Some companies have a paper-based system, but have scanned copies of documents. These are the ones they work with, since they keep their paper copies off site.

Companies often call these documents "unofficial copies." "Unofficial" is probably not the best choice of words here, and "working copy" would be more appropriate. What is most important is this: You must make sure these copies are exact replicas, and you must have an SOP that tells how you make sure they are. Again, a system that uses both paper and electronic versions of documents is a hybrid system.

14. If Using a Hybrid System, How does the Handwritten Signature Link to the Electronic Record?

Part 11 does not require that electronic records have electronic signatures, and the file may be printed and signed manually. The e-record must include the time and date of printing and have a means to link the e-record to the manually signed paper version. Signing a printout of the electronic record does not exempt the electronic record from Part 11 compliance.

15. If We Use Electronic Signatures, Is Compliance with Part 11 Mandatory?

Yes. Implicit in the use of electronic signatures is that you have an electronic record-keeping system. As such, the system must be in compliance with all the provisions of Part 11.

16. Are Electronic Signatures Always Necessary for the Creation of Electronic Records?

Part 11 does not determine what records require signatures. The predicate rules tell what records companies must sign. Not every e-record requires a signature. The best place to find what needs a signature and who should provide it is in regulations that govern your product type.

17. Do E-Records and E-Signatures Require the Same Controls?

Yes. The controls are the same, since an e-signature is a form of e-record. Logging into a system and signing a document both require a password known only to the user.

18. What Is a Legacy System?

A legacy system is any system that does not meet the requirements of 21 CFR Part 11. The primary requirements of Part 11 are (a) software functionality such as security and audit trails and (b) documentation of computer system validation. The definition of a legacy system no longer relates to when the system was implemented in relation to the promulgation of 21 CFR Part 11 in 1997.

19. What Are the Characteristics of a Good System?

Good systems, whether electronic or hybrid, are robust. They have the following characteristics:

- Flexibility, able to adapt to company changes
- Scalability, able to grow with the company
- Searchability, able to locate a document and its history quickly
- User friendliness, easy to access and use
- User accountability
- Ability to track changes
- Security
- Controlled versioning and distribution

20. How Can We Make Paper Documents Secure?

Paper documents must be stored in an environmentally protected area (fire, water, humidity) with access limited to authorized persons. Paper documents by their very nature are less secure than electronic records. Paper documents can be scanned to make graphic files, which are a legal equivalent to the paper documents. The files can then be stored onsite or offsite. All stored files should have access controls equivalent to those for the original paper files.

21. Are We Required to Buy Expensive Software to Manage Our Documents?

No. Many companies do, but it's possible to manage documents quite well without commercial software. Companies can build their own systems internally based on operating system folder permissions. What they cannot typically do with such in-house systems is perform an electronic signature. Some companies have also developed their own software for managing documents; such systems, while hybrid, require full validation, as do systems using commercial software.

22. Is It Possible to Build Software In-House that Employs Electronic Signatures?

It's possible, but it's not common. There are many excellent software packages available, and the prices have come down dramatically. The advantage of commercial software is that it ensures help-desk capability that companies may not be able to provide over time for in-house built software.

23. Why Do Companies Get Electronic Systems?

Electronic systems provide a secure repository for storing documents and a tool for version control. These are the two primary reasons companies bring systems in-house. These systems also provide the ability use electronic signatures and perform extensive search capabilities.

24.   As a New Company that Must Put Controls in Place, Can We Go Electronic Immediately?

Yes, but with the following caveat: You must have a system in place to approve the documents that the process of implementing an e-system requires. That means you must have established SOP templates and validation templates. You must also have SOPs on SOP management and training as well as how the system works. In other words, you can't use the new system to manage the documents used to validate and initiate the new system. Chapter 7 explains these procedures further.

25.   What Is Meant by Interoperability?

The US National Committee offers three levels of interoperability: basic, which lets a message go from one computer to another, but the second computer cannot interpret the data; functional, which ensures messages between computers can be interpreted at the level of data fields; and semantic, which provides common interpretability so that information within the data fields can be used intelligently. In general, interoperability allows different levels to work together on a limited common level (2).

26.   What Are Privileges?

Privileges, or permissions, are assigned to users according to the need for users to be able to access certain documents and records. For instance, there is no need for an administrative assistant to be able to view all toxicology documents, so privileges to see those documents may be limited to toxicology scientists. Similarly, not everyone should have the ability to post or remove versions of documents; these privileges are generally assigned to people who manage systems.

27.   What Is Taxonomy?

Taxonomy identifies a system for the document cycle and a folder hierarchy for reposing documents. Taxonomy literally means classification and division into ordered groups or categories. Thus, when a company structures a document management system, it applies principles of taxonomy. Many people think taxonomy is a collection of file folders where companies park their documents. But in total, it is a system that allows both storage of documents and categorization of documents.

28.   Why Is Taxonomy Important?

Good taxonomy is built on the relationships between document types, so that documents are searchable by category and available for reuse. Without consistent taxonomy, companies run the risk of multiple groups building silos for storing documents and creating duplication of effort and confusion as to which document is the "official" one. Furthermore, when a company adheres to a document retention plan, it may apply the rules for document destruction to one document, but fail to realize another copy still exists under another name.

29.   What Is Metadata?

Metadata is simply data about data. If you think about it, records have information about products or activities, and they also have controls such as core numbers, version numbers, and effective dates. These components are metadata. Metadata may be key words, statistics about documents or document access, related documents, or other types of information.

30.   What Are Attributes?

Attributes, or properties, are a set of information associated with an object for the purpose of characterizing its content and enabling system functions. For example, the attributes for an SOP file may include the file name, document identifier, access permissions, version, and approval levels.

31.   How Long Should It Take a User to Get into a System and Enter Data?

The main reason for using computer systems is to increase productivity, and that means being able to access information quickly. A user should be able to launch the application, obtain secure access, locate the desired record, and begin entering data within a minute.

32.   How Do We Demonstrate Data Integrity?

Data integrity is part of any process, so when you validate the process, you demonstrate data integrity.

33.   What Are the Considerations for Using Excel for Data Processing?

You can use Excel to store data and for specific applications. Excel applications for GXP records require validation.

34.   Can We Use Wireless LAN for Data Acquisition and Transmission?

A wireless LAN is no different than a wired LAN except that it must be made secure with encryption.

35. Must Document Numbers Identify the Document Category — Say, Manufacturing or Shipping?

No. There is no such requirement. In the past, companies would designate certain documents for packaging or production, for instance, and give them identifying numbers. This system worked when documents were controlled manually, and there were documents designated for individual areas in the company. With electronic access and electronic searching, the need for area identifiers is no longer necessary. Some companies have a flat system that works simply by document type. In such a system, all SOPs bear an alphanumeric number that identifies the document category, the document number, and the version as in SOP 001-1.

In such a system, numbers are assigned sequentially. When a document goes through a revision process, it retains its core number, and the version moves up: SOP 001-2. Other companies number documents by levels, so SOPs and manuals may be level 100, protocols and reports level 200, submissions level 300, and disclosure level 400. In such a system, an SOP might bear a number such as SOP 106-2, and a submission document might have a number such as SUB 303-1. It doesn't matter how a company sets up its numbering system, only that it is consistent and workable. A suggestion is to keep it as simple as possible.

36. Can We Reserve Numbers for Planned Documents, Say If We Are Doing a Series of Development Reports for a Specific Project and Want the Numbers to Be Sequential?

It's best not to. Numbering provides a placeholder for a document and is an attribute of it. If numbers are issued and a document is never generated, there is a hole in the numbering system that must then be explained: What happened to this document?

37. What's the Purpose of Document Numbering?

Document numbers are metadata about documents; they provide a search vehicle and a control mechanism. Document numbers must always be unique. They are the key to all the documents that have been developed over time.

38. If a Document Is Retired, Can a Company Reuse the Number from that Document?

No. A document number should never be reissued. A number is part of an existing document, whether it is active or otherwise. Even if a number has been assigned to a planned document, and that document is never issued, the number serves as a place holder for it and provides evidence that you had planned such a document at a point in time.

39. Is There Any Standard Terminology to Identify Document Status?

Each company needs to determine its own controlled vocabulary and remain consistent in its use. In the document life cycle (see Chapter 2), each phase should have determiners that indicate status of a document or record. For instance, an SOP system may identify the status of documents as follows:

> Planned (document need identified, and a number and title issued)
>
> Draft (document in review)
>
> Approved (document completed review cycle, and signed off)
>
> Effective (document in effect, after training)
>
> Active (document in the working system)
>
> Superseded (previous version of an existing document)
>
> Retired (document no longer in use in any version)

40. What Is Document Change Control?

Document change control is a process in which changes to documents are managed to ensure proper review, approval, and training occur before the document is made effective. Previous versions or revisions of documents are controlled and retained so that only the current effective documents are available for use by the workforce.

41. Do We Have to Keep Copies of the Drafts that Went Through a Review Cycle?

No. The regulations require a review cycle, and companies typically explain their process for generating, reviewing, and approving documents in a Standard Operating Procedure (SOP). That you have a system and adhere to it is the requirement. (See Box 3.1.)

42. Besides Standard Operating Procedures (SOPs), What Documents Should Be Version-Controlled?

All documents in place that are subject to revision need version controls. Procedural documents, sometimes known as "backbone documents," include work instructions, work aids, and laboratory methods; quality manuals; plans such as chemical hygiene, product development, device validation, project management, and more. Forms, templates, and reference documents such as style guides and organization charts are included too. In short, any documents that tell "how it happens here" and that are subject to revision need version controls. Version control is the same as change control.

BOX *3.1*

REQUIREMENTS FOR DOCUMENT CHANGE CONTROL

PART 820—QUALITY SYSTEM REGULATION

Subpart D—Document Controls Sec. 820.40 Document Controls

Each manufacturer shall establish and maintain procedures to control all documents that are required by this part. The procedures shall provide for the following: (a) *Document approval and distribution.* Each manufacturer shall designate an individual(s) to review for adequacy and approve prior to issuance all documents established to meet the requirements of this part. The approval, including the date and signature of the individual(s) approving the document, shall be documented. Documents established to meet the requirements of this part shall be available at all locations for which they are designated, used, or otherwise necessary, and all obsolete documents shall be promptly removed from all points of use or otherwise prevented from unintended use.

(b) *Document changes.* Changes to documents shall be reviewed and approved by an individual(s) in the same function or organization that performed the original review and approval, unless specifically designated otherwise. Approved changes shall be communicated to the appropriate personnel in a timely manner. Each manufacturer shall maintain records of changes to documents. Change records shall include a description of the change, identification of the affected documents, the signature of the approving individual(s), the approval date, and when the change becomes effective.

43. Do SOPs and Work Instructions (WIs) Require the Same Levels of Control?

Any document that instructs people as to how a process is performed, regardless of what a company calls it, needs to be a change-controlled document. Change-controlled documents typically include a unique document number, title, page numbers, effective date, version history, and at least two approval signatures. Examples of change-controlled instruction documents include SOPs, work instructions, workstation instructions, policies, and procedures.

44. Do Manuals Require the Same Review Cycles as Procedures (SOPs)?

Industry has adopted, for the most part, a two-year review cycle for SOPs. Manuals typically undergo review when something changes, but many companies have adopted the practice of reviewing quality manuals every year as a safeguard to remaining current.

45. Must We Manually Sign Our SOPs?

That is a common practice in hybrid systems, but there is no mandate that SOPs actually bear signatures. What companies must do, however, is have proof of the review and approval process. In e-systems, the review and approval records bear electronic signatures that do not show on the document itself. In all cases, the key is to be able to provide approval information along with the document itself even if they are separate entities.

46. What Documents Actually Do Require Signatures?

There is no regulatory requirement that documents bear signatures, but traditionally companies have signed submission documents manually. In e-systems, the signature does not appear on the document itself. By industry conventions, usually the first page or the last page of a document is designated for signatures or there is an indication that signatures are available elsewhere.

47. What Color Ink Should We Sign Original Documents with?

There is no official requirement, and the FDA will accept all ink colors that are permanent. It used to be that black was the standard because blue did not photocopy well. Now blue is the standard because photocopying quality is good, and blue permits identification of the original from the copies.

48. What Kind of Ink Should We Use?

Use indelible ink that cannot be erased or smeared when rubbed or when wet. Gel pens are a poor choice because signatures or comments can be altered or obliterated.

49. Should We Build a Folder System for Our Hybrid System Based on the Common Technical Document?

While the Common Technical Document (CTD) is the standard for organizing information in submissions, it's not the best model for a document management system. It is not user-friendly because not everyone in a company understands the structure, nor should they have to. Furthermore, such a model does not address where to put documentation that does not go into a submission.

50. What's the Best Way to Build a Folder Structure for a Hybrid System?

There are many infrastructures that work. The best ones result when there is a good understanding of the business model and the requirements for the documentation that model generates, including the documentation that contractors may produce. Builders of such systems need to figure out who needs to have access to certain folders and who doesn't. A system requires administrators with the authority to post documents to folders and to archive them as necessary.

BOX *3.2*

RECOMMENDED RESOLUTIONS FOR SCANNING

- Text 300 dpi
- Photos 600 dpi resolution
- Gels and karyotypes: 600 dpi; 8-bit grayscale depth
- Plotter graphics: 300 dpi
- HPLC: 300 dpi

---

51.   How Can We Store Images?

Black-and-white images are most commonly stored as standard TIFF files or as PDF files. The general idea is to store images in a format that retains their readability while using the smallest file size (lowest scan resolution acceptable).

52.   If We Scan Documents, How Can We Be Sure that the Results Are Acceptable?

Scanned documents must undergo quality controls. That means a document that becomes an e-document must have 100% quality control (QC) check. 100% QC doesn't mean that a person has to read every word to verify that a scanned document is an exact replica, but that the first and last word on every page match those on the original, and all pages must be verified. The scan must be clearly legible with good resolution. (See Box 3.2.)

53.   Will FDA Accept Copies of Scanned Documents?

Yes, provided they are an exact replica of the original, and the company can prove verification.

54.   How Can We Scan Large-Format Documents?

There are scanners specifically designed for large-format documents. These types of scanners may be available at your local printing company. An alternative is to reduce the large format in size and then scan it with a regular scanner. If you reduce a document in size, you will need to do a QC check on the reduction image as well as on the final scanned image.

55.   What Is OMR?

OMR is optical mark recognition. It refers to the ability of a system to recognize marks commonly used on forms, such as filled in circles and check marks.

56. What Is OCR?

OCR is optical character recognition. It refers to the ability of a system to "read" a scanned document (graphic) and determine the text. OCR is error prone, and all OCR documents need to be 100% verified in order to determine that every single word is correct. Special care must be taken for symbols, abbreviations, units of measure, and other nonword contents.

57. What Is Barcoding?

Barcoding increases the ability to store and archive documents by affixing a small pattern of lines that can be read by a laser or an optical scanner. The barcode corresponds to a record in a database.

58. What's the Difference Between a QC Check and a QA Check?

QC checks data with respect to specifications; QA checks overall quality and compliance.

59. If a Company Changes Its Templates, Should It Reformat the Reports Done in Previous Formats?

It's not necessary. Typically, a document is based on the template that is current when the document is started. In some cases, it may be wise to update the document based on an updated template. If, for instance, a document is part of a larger project and documentation for that project is to be disseminated, then reformatting for consistency may be wise.

60. Since Part 11 Requires Date and Time Stamps, Should We Put Date and Time Stamps on PDF Documents Posted on Our Company Server?

It makes sense. If you put a time and date stamp on read-only PDF documents that people can print out, you don't have to worry about uncontrolled documents if you have a statement/watermark that says "valid only on print date" or "valid only for 24 hours after printing," or something to that effect.

61. How Should We Keep Our Paper Documents?

Paper is harder to control than electronic records. Paper documents need tight controls so that they cannot be accessed and lost or destroyed. A common model is for companies to keep paper documents and records in controlled, limited access areas, with a designated document manager. If a copy of a document or record is required, it is produced and identified as a copy. Then the document requestor signs for the copy of the document, and returns it for destruction when it is no longer needed. There is always a record of such an activity, typically in a log. However, most companies find it best to scan a paper document to make an electronic copy that can be distributed via access permissions. This creates a hybrid system that requires controls on the scanning and distribution process.

62. Do the Regulations Say Anything About Keeping Documents in Fire-Safe Cabinets?

No. The regulations simply say that they must be safe. In the past, industry has protected paper records from fire and water damage using special file cabinets and document rooms. Several companies have already experienced the grief that comes from flooded document repositories where original records were obliterated. Today, most paper documents are scanned to make electronic copies that are stored offsite, so the need for paper file safeguards is much less important.

63. Do We Have to Keep Our Records Onsite?

No. But records need to be accessible in a "timely manner." Industry has interpreted "timely manner" to mean within 24 hours.

64. Must We Retain Original Documents, or Can We Keep Copies?

The regulations do not require retention of original records; retention of true copies such as scans, microfilm, microfiche, or other accurate reproductions of original records is acceptable. Records must be available at the location where they were generated, but not necessarily stored there in hard copy.

65. Are Document Storage Facilities a Safe Place to Put Documents?

The sheer volume of documentation, particularly for devices which can have shelf lives of decades, means that firms will engage contract storage entities. There have been instances of data loss in these facilities, however. Today, most paper documents are scanned to make electronic copies that are stored offsite, so the need for paper file safeguards is much less important.

66. What Happens If a Data Storage Facility "Loses" Some of Our Documents? What Should We Do?

In an ideal world, contract organizations, such as those that retain records for companies, would not lose documents. Yet they do; they have had fires and floods that have damaged documents. The best a company can do when this happens is try to reconstruct the data that were lost and then document the event. This is a primary reason that companies scan documents, since the scan can be used to create a new paper copy.

67. Can We Rely on Our Contractors to Adequately Maintain the Documentation They Produce for Us?

Some contractors retain the original copies of documents they produce for their clients, and they send them a verified copy. All arrangements for retention of original documents should be according to contractual agreements. The documents belong to you, even if the contractor retains them for purposes of convenience. Make sure that your agreements spell out the particulars, including your ability to access these documents. And, of course, maintain an audit plan that evaluates records. Many companies today allow the contracted company to retain the paper copes but retain electronic copies of the documents themselves.

68. How Can We Bring Documents In-House from Our Contractors?

If you have had the foresight to give your document numbers to your contractors, you can import the documents directly into your system with a QC check. If the contractor's documents bear numbers that are inconsistent with your system, you may elect to affix the documents to a reference sheet that identifies the document and its contents and bears a number that works in your system. Such a method allows you to search the system and locate the contractor-generated documents easily.

69. If a Contractor Goes Out of Business, What Happens to Our Documents?

State laws typically require any contractor going out of business to deliver any documentation in its possession to the owner. While this may be the law, it often is not the practice. It is best for owners to periodically get original documentation from contractor. It is best that all documentation be accessible electronically in real time.

70. How Long Do We Have to Keep Our Documents and Records?

Most companies enlist the input of legal to determine policy for retaining records, even though the CFR specifically addresses requirements for records retention:

21 CFR Part 65.115 Institutional Review Board Records

21 CRF 58.195 Nonclinical Laboratory Research Records

21 CFR 211.198 Drug Product Complaint Files

21 CFR 312.57 and 312.62 Clinical Investigation Records

21 CFR 320.36 Records of Bioequivalence

21 CRF 211.180 Manufacturing Records for Drugs

21 CRF 600.12 Manufacturing Records for Biologics

21 CRF 820.180 Manufacturing Records for Devices

71.   How Can We Safely Delete E-Mail?

Electronic mail is often stored on a central e-mail server with copies stored on local PCs or on file servers that are divided into user accounts. Companies should have procedures in place that identify how e-mail is stored and deleted. Often backup tapes and other media used for data restoration contain e-mail as well. Companies should consider all copies of e-mail when establishing procedures.

72.   What Is a Typical Schedule for Deleting E-Mail?

Many factors affect data retention for e-mail. If a company uses e-mail for transporting files, or any pertinent information in e-mail is transferred to another file, or a copy of the e-mail is retained outside of the e-mail system, then data retention may be short, perhaps only 90 days. When e-mail contains data as a supportive record that is retained only in the e-mail system, then the entire system is subject to longer retention times. Those retention times may be as long as the other types of documents the company uses, perhaps as long as 15 years or more. It is vitally important to have SOPs that instruct users about what should and should not be included in e-mail and how data contained in e-mail are to be retained. It is possible to set different retention times for categories of e-mail to limit the volumes of data to be retained, backed up, and archived.

73.   If We Have an Electronic System and We Bring Legacy Documents into It, Can We Destroy the Paper Copies?

Once electronic copies of paper documents are made and verified, those electronic copies are backed up and stored offsite, and the system has been fully validated according to 21 CFR Part 11, the paper documents may be destroyed. Most companies elect to retain the paper copies for a short period of time after those events as a safeguard. This period is usually one to two years.

74.   Does a Hybrid System Require Validation?

A hybrid system uses electronic records and paper records. Usually the electronic records are printed in order to obtain approval signatures. This is the least efficient type of system because both electronic and paper retention is required in the long term. Hybrid systems often have problems related to the electronic and paper records not matching and often result in regulatory issues. If a hybrid system uses company-generated software, then it requires validation, just as a purchased software program does. If it is a folder structure generated on the company's server not using a software application, it requires qualification, just as equipment does.

75.   How Can We Make Sure Our Hybrid System Is Secure?

Since a hybrid system contains electronic records, security must follow the 21 CFR Part 11 industry standards for security, data transfer, audit trails, and validation.

However, if it is a folder structure generated on the company's server not using a software application, many industry standards related to software applications do not apply. A hybrid system based only on a folder structure requires secure controlled access and qualification.

76. If We Decide to Go Fully Electronic, Should We Still Keep a Hybrid System?

Once a company obtains an electronic system to replace a paper system, there is often insecurity about giving up paper. Many companies don't take full advantage of their electronic systems and decide to initially use the system in a hybrid manner. While doing so may satisfy emotional security, the process actually creates real insecurity related to the very nature of a hybrid system. Companies should invest in a comprehensive validation process that provides everyone with confidence in the fully electronic system and eliminate hybrid use.

77. Our Staff Travels a Lot. How Can They Access Our System Remotely?

Companies often allow remote access to their internal network by creating a Virtual Private Network that includes strong encryption. This means that the transport of information over the Internet is secure from any location. However, there are still significant security issues related to the remote computer being stolen or having unauthorized access. User and IT staff must work together to create strong passwords to prevent unauthorized use.

78. What Controls Does Remote Access Require?

Remote access requires encryption of all data transmitted, passwords for access, restrictions on simultaneous access, access logging, and all other 21 CFR Part 11 industry standards for security. Many companies elect to increase security by using devices such as ID generators or dongles so that unauthorized users have to gain both a physical device and a private password to gain access to the system.

79. How Can We Control Documents that Must Leave the Company, Say to Go to a Clinical Trial Site?

Documents that contain confidential information must be encrypted prior to electronic transport. There are many techniques, and it is best to coordinate the method with both parties. Document files may be encrypted and sent via e-mail with a password issued verbally over the telephone. Virtual Private Network connection may be made between offices. Secure Internet transfers such as https (secure web access) and ftps (secure file transfer protocol) may be employed to allow remote access to company electronic documents.

TABLE 3.1.   Minimum Required Signatures[a]

Code	Signatory[b]	Minimum
CLI (clinical)	Author Regulatory Vice President Drug Development Operations Medical Director Quality Assurance	3
REG (Regulatory)	Author Regulatory/Quality Assurance Head of relevant area	3
DEV (Development)	Author Study Director VP Development Director of Chemistry (CMC) Regulatory/Quality Assurance	3
RES (Research)	Author Study Director VP Research Regulatory/Quality Assurance	3
FRM (Form)	Author Head(s) of impacted area(s)	2
REC (Record)	Author Heads of impacted areas	2
SOP (Standard Operating Procedure)	Author Heads of impacted areas Regulatory/Quality Assurance	3
INS (Instruction)	Author Head of generating department Head of impacted department(s)	3
MET (Laboratory Method)	Author Head of generating department Supervisor Head of impacted departments	3
PRD (Product)	Author Quality Assurance Director of Chemistry	3
REF[c] (Reference)	Author Heads of impacted areas	2
MAN (Policies, plans, manuals)	Author Head of generating department Head of impacted department(s) Regulatory/Quality Assurance	3

[a]These are minimum acceptable signatures for document approval. Any number of additional approvers may be included within each category according to document demands.

[b]Any of these signatories may designate a qualified alternate.

[c]Generation of REF approval/history/cover sheets excluded; documents generated by contractors are brought into the system by the document coordinator.

80. Once We Implement a System, Can We Expect It to Function Indefinitely?

Systems require maintenance, and they must adapt and grow as the company changes. Change control and revalidation are always ongoing.

81. As We Generate Documents, How Can We Make Sure the Right People Review and Approve Documents?

In a manual system, it's prudent to (a) have a minimum required signature list that identifies document types and (b) review and approve signatories for each type of document. For instance, if an SOP mentions a QA check in a product procedure, and QA doesn't see the SOP in the review cycle, there's trouble brewing. A minimum required signature list should indicate that the "heads of impacted departments" review SOPs. Table 3.1 is an example of a document signatory list. In electronic systems, signatories can be configured in the system itself.

REFERENCES

1. Nettleton, David, and Gough, Janet. *Risk Based Software Validation: 10 Easy Steps*, 2nd edition, 2006. Baltimore, MD: Parenteral Drug Association; and River Grove, IL: Davis Horwood International.
2. http://health-careit.advanceweb.com, accessed March 28, 2009.

CHAPTER 4

DECIDING TO GO ELECTRONIC AND FINDING A VENDOR

INTRODUCTION

Making the transition to electronic document management is inevitable for this industry. With the FDA mandating electronic submissions, and the European Union slated to do the same, e-systems will be the norm rather than the exception. Yet change can be fraught with apprehension. Humans have used paper for 3000 years, and, quite frankly, they are comfortable with it. So, going electronic can mean venturing into new territory. Yet the transition to e-system can be palatable and even rewarding, provided that the company understands how e-systems function and the advantages they offer. This chapter discusses what options are available and how to find the best vendor for the company's culture. It answers the following questions:

1. If a company wants to install an electronic record-keeping system, can it contact FDA for advice and recommendations for software?

2. Does FDA provide 21 CRF Part 11 certification of software vendors and computer systems?

3. Why aren't vendors of COTS software regulated by the FDA?

4. Since we've had paper for hundreds of years, why should we go electronic?

5. What kinds of software constitute document management applications?

6. How do these software systems for document management work, and what are the variations?

7. Are all software programs for this industry compatible only with personal computers (PCs)?

8. Are electronic notebooks subject to document management?

9. Is a handheld patient device that records patient data considered software?

10. What are web-based software applications, and should we consider using them?

11. How do web-based software applications work?

Managing the Documentation Maze, By Janet Gough and David Nettleton
Copyright © 2010 John Wiley & Sons, Inc.

12. If a company wants to go completely electronic, must it keep signed paper copies of documents approved outside the system?

13. What makes software systems different from paper systems?

14. Is it possible to transfer a file from one electronic system to another and still retain all the metadata and history?

15. Where do you begin the process of selecting document management software?

16. How do you determine what you need the software to do?

17. Who should be on the user requirements team?

18. What skill sets does the user requirements team have to have?

19. Can't we just let IT make the decision about what's best for the company?

20. If we anticipate the company will grow, what should we take into consideration when purchasing an e-system?

21. What does the vendor selection process entail?

22. What software features we should look for?

23. What types of search capabilities should a document management system have?

24. How do the user requirements figure into process?

25. What should we do if we find a vendor that offers a nice feature that we didn't think about?

26. How do we select the best software vendor?

27. Aren't large, well-established software vendors preferable to smaller, newer software vendors?

28. What does it mean when a vendor of an e-system says the product is validated?

29. Should we audit the vendor?

30. What is Secure Software Certification and what does it mean?

31. How does a software developer actually get certification?

32. Why is Secure Software Certification important?

33. How do users benefit from Secure Software Certification?

34. Does Secure Software Certification provide any benefits to a developer?

35. Does Secure Software Certification save the developer money?

36. How can we find out if a software developer has certification?

37. If the software vendor has validated the software, do companies that buy that software have to validate the software?

38. Do we have to buy a seat for every user?

39. What is a source code escrow agreement?

40. Is it possible for software to be 21 CFR Part 11 compliant even if the vendor is not 21 CFR Part 11 compliant?

41. If a company does a thorough job selecting a software vendor and finds that the vendor is fully compliant with the binding regulations and offers a validated product, must the company still perform user validation of the system?

42. What's the difference between developer and user validation?

43. Can the software vendor provide user training or must companies train users?

44. When should a company reaffirm a software vendor's capability?

45. What are typical configurations for software?

46. What is the return on investment (ROI) for computer system validation?

47. Because we couldn't agree on a vendor, our IT department has built software in-house, but what do we do to begin using it?

48. What is computer system user validation?

49. Is the only reason to perform validation because it is required by the regulations?

50. How long should validation take?

51. Should computer system validation be an IT responsibility?

52. When a company has limited resources, can it farm out the validation responsibility to temporary staff or consultants?

53. Is there anything that companies can do to make the validation process less laborious?

54. What is IQ/OQ/PQ?

55. Is qualification the same as validation?

56. Why isn't IQ/OQ/PQ sufficient for software validation?

57. What's the objective of installation qualification in software validation?

58. What needs to be qualified?

59. How is validating software different from validating a piece of equipment with computerized controls?

60. How are validation and verification different?

61. Can the validation use simulated data to test the system?

62. Do scanners require validation?

63. If a company builds a database internally to keep laboratory source data, what does it need to do to make sure it's functional?

64. What is prospective validation?

65. What is retrospective validation?

66. What's the difference between prospective validation and retrospective validation?

67. What is retrospective evaluation?

68. When is validation required?

69. What is user validation?

70. What types of software need validation?

71. What types of software do not need validation?
72. Who is responsible for computer system validation?
73. Where is COTS computer system validation performed?
74. When is COTS computer system validation performed?
75. What types of changes can be made to a computer system without revalidation?
76. Why is computer system validation performed on COTS software?
77. What is a validation document template?
78. How is validation documentation written?
79. How are large software systems validated?
80. What are the validation requirements when COTS software is customized?

1. If a Company Wants to Install an Electronic Record-Keeping System, Can It Contact the FDA for Advice and Recommendations for Software?

The FDA audits many companies that have electronic record keeping, so it has first-hand knowledge of which systems work well and why. However, the FDA never endorses specific products or evaluates software aside from that used in regulated companies who use software. To recommend products or even publish a list of "acceptable" software would be unfair to software developers and is not permitted by law.

2. Does the FDA Provide 21 CRF Part 11 Certification of Software Vendors and Computer Systems?

No. FDA does not regulate software vendors and provides no certifications of any kind. Regulated companies must ensure that software vendors are compliant with Part 11 industry standard product features and have sufficient developer validation documentation. Companies should evaluate software vendors well in advance of making purchasing decisions. The FDA inspects regulated companies to ensure that computer systems are compliant with Part 11 industry standard product features and have sufficient user validation documentation. The regulated company assumes all liability for products it uses. Companies are thus wise to select software vendors that are compliant because doing so greatly lessens the burden on the company to achieve compliance.

3. Why Aren't Vendors of COTS Software Regulated by the FDA?

The FDA can only regulate companies that produce goods that fall under the FDA's legislative jurisdiction. Regulated companies do, however, assume all liability of their vendors, and therefore companies basically regulate what vendors provide; this activity in turn makes them indirectly regulated by the FDA. This is an interesting "trickle down" concept that really works quite well.

4.   Since We've Had Paper for Hundreds of Years, Why Should We Go Electronic?

Electronic records are more secure, more accessible, more efficient, and more reliable. The use of electronic records and signatures is less expensive than paper records and results in increased productivity.

5.   What Kinds of Software Constitute Document Management Applications?

Document management encompasses a wide range of software functionality and includes operating systems and software applications. Document management may be as simple as network folders with restricted security access. Often document repositories are included in applications that perform other specific functions. For example, a Laboratory Information Management System (LIMS) may allow for files to be stored with the laboratory records in order to promote access to relevant supporting data. Electronic Record-Keeping Systems (ERKS) and Electronic Document Management Systems (EDMS) often provide document review and approval workflow, electronic signatures, document archiving, metadata and keyword searching, and even full text searching.

6.   How Do These Software Systems for Document Management Work and What Are the Variations?

They pretty much work the same way from product to product. It used to be that software developers built in features specific to their products, but the features wars are pretty much over. The variables in applications are mostly attributable to the configurable options that these software packages provide. Most document management software works with windows that allow users to enter data. Box 4.1 shows how a new or revised SOP goes through the review process, which covers reviewer and approver sign-off and training. All signatures link to the document electronically.

7.   Are All Software Programs for This Industry Compatible Only with Personal Computers (PCs)?

PCs primarily refer to desktop and laptop computers that use the MS Windows operating systems. PCs also include those that operate on Macintosh, Linux, or other operating systems. Some software programs only come in a version that will run on PCs that use the MS Windows operating system. Much of the computer software that is available commercially operates on web browsers, so they are usually compatible with all PCs using any operating system.

BOX *4.1*

AN SOP REVIEW CYCLE

Step 1

1. The Document Administrator (DA) identifies the document ready for review.

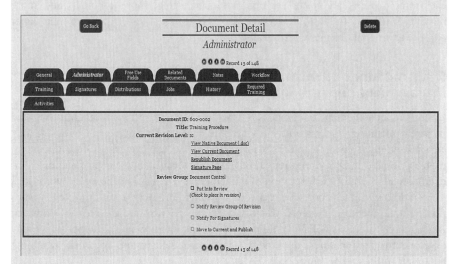

2. The DA enters the reason for change and new revision number along with any prior notes about the document.

3. A list of assigned reviewers displays. The DA can add and delete from the list as necessary. The DA dispatches notification to reviewers.

4. Reviewers receive e-mails and can check the document out of the repository for changes.

5. Reviewers summarize reason(s) for change and check the document. Only one reviewer at a time can make changes. The process repeats as necessary until there is concurrence.

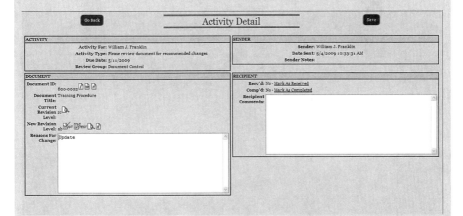

Step 2

6. When reviewers are finished, an e-mail goes to DA for accepting and rejecting changes.

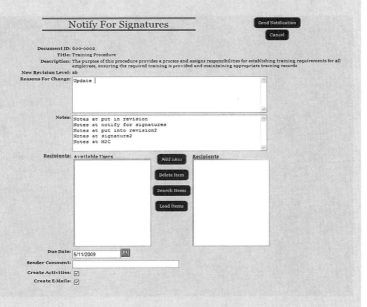

7. The DA converts the document to PDF and notifies approvers to sign.

(Continued)

Step 3

8. Approvers receive e-mails and preview the document in PDF. If it is acceptable, they sign off by entering their password. The electronic signatures link directly to the document.

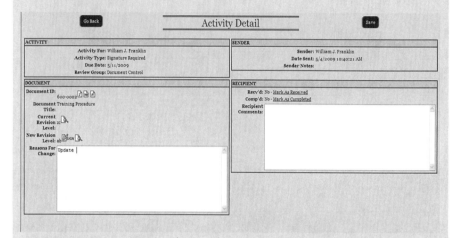

9. Once all approvers have signed off on the document, notification goes to the DA.

10. The DA can now set the effective date and, if required, update the reason for change(s). The system automatically calculates the next revision date in two years.

11. If the document is part of a training matrix, the system notifies individuals to train on the procedure.

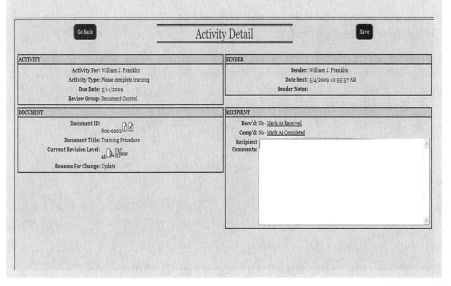

12. Trainees receive e-mail notification of training. Training can be any number of formats. If there is a test, trainees must pass before updating of training records occurs.

13. Electronic signatures of trainees link to the training record.

(*Continued*)

14. When training is complete, the DA posts the document to the hierarchy and retires the previous version, if any. This cycle of review is complete.

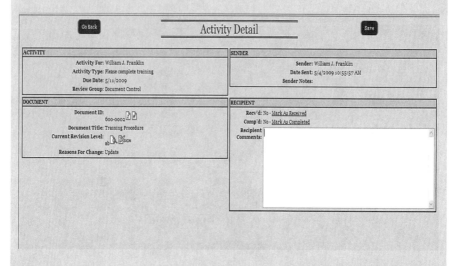

Note: While the system has automatically flagged the document for the next review cycle in two years, anyone can identify a need for a change, and the document can go into review again as necessary.

Source: Screen Shots courtesy of Quality Systems Integrators, Eagle, PA. www.qsi.com.

8. Are Electronic Notebooks Subject to Document Management?

Electronic notebooks are themselves type of document management systems. They are specialized systems dedicated to replacing paper laboratory notebooks but often allow for many document management functions. The primary advantages of electronic notebook applications relate to the central storage, date, and time stamping of entries, as well as search capabilities for intellectual property.

9. Is a Handheld Patient Device that Records Patient Data Considered Software?

Any medical device that contains software is subject to the same software controls as a stand-alone software program.

10. What Are Web-Based Software Applications, and Should We Consider Using Them?

Software applications that use an Internet browser as the user interface are web-based applications. These are highly desirable because there is considerable uniformity, security, and accessibility from just about any location in the world. In addition, the browser interface is often independent of both the operating system and the computer hardware, thereby reducing validation requirements and promoting collaboration worldwide.

11.   How Do Web-Based Software Applications Work?

Web-based applications work in a manner similar to that of mainframe computers that preceded PCs. A central server manages programs, security, and data. The server can be located anywhere in the world, and the standard Internet Protocols allow for access via a common text Uniform Resource Locator (URL). When a user enters the URL in the address bar of the Internet browser, the server is contacted, and the required programs and data are transferred to the Internet browser. Nothing is stored on the local computer, and nothing other than the Internet browser is required. Such a system eliminates complex software installations related to multiple locations, operating systems, and hardware requirements.

12.   If a Company Wants to Go Completely Electronic, Must It Keep Signed Paper Copies of Documents Approved Outside the System?

No. If a company brings its documents into an e-system (typically by scanning paper documents) and that e-system is validated and compliant, with backup and recovery procedures, the company can destroy the paper copies. (See Chapter 5.)

13.   What Makes Software Systems Different from Paper Systems?

Any comments made in an e-system remain with the document for the life of the document. The systems provide an audit trail for each record, and the systems are more secure than paper systems. The main difference is in the way that people perceive working with electronic versus paper records. People tend to prefer paper because it is tangible and familiar. With time, people often change their perception and ultimately come to prefer electronic records.

14.   Is It Possible to Transfer a File from One Electronic System to Another and Still Retain All the Metadata and History?

The transfer of data between systems is often difficult because the systems retain data in different formats. This is improving due to interchange standards such as Extensible Markup Language (XML). Sometimes custom programming is required to transfer data. In all cases, data transfers must be verified by the users of the computer system.

15. Where Do You Begin the Process of Selecting Document Management Software?

All software acquisitions start with the users compiling user requirements. The risk-based approach to computer system validation calls for approval of a User Requirements document in order to uniformly assess software vendors. Once an organization has agreed to a set of requirements, it researches software vendors to determine which have suitable products. Often users learn of available features and want to update their User Requirements. This is a positive document evolution that is encouraged but must be controlled so that the scope of the project is not breeched.

16. How Do You Determine What You Need the Software to Do?

A team with members representing every area of the company that will use the system is the best way to start the process. This team then determines what features the software should provide to satisfy the company's needs. It is best to have the team work together where they can all view the User Requirements document as it develops. This interaction eliminates multiple reviews and differences of opinion that are detrimental to consensus.

17. Who Should Be on the User Requirements Team?

The user requirements team should consist of at least one representative from every group that will use the software. The team should include process experts and management. In addition, Information Technology should be represented because IT is are responsible for technical details that differ from the responsibilities of the users.

18. What Skill Sets Does the User Requirements Team Have to Have?

The user requirements team must be knowledgeable of the current practices and the overall concept for changes in those practices. The team should be aware of all 21 CFR Part 11 computer system industry standard features because these need to be included in the User Requirements. In addition, users must understand how the new computer system will impact and work with existing computer systems.

19. Can't We Just Let IT Make the Decision About What's Best for the Company?

Information Technology (IT) does not own computer systems and processes related to them. They are not qualified to represent the interests of the diverse user community. IT should help with technical details related to hardware, operating systems, software platforms, software utilities, and other system components, but they should not define the functional requirements.

20.   If We Anticipate the Company Will Grow, What Should We Take into Consideration When Purchasing an E-System?

Typically, it is best to buy a system that will meet your needs within a short period of time, often three to five years in the future. System size, complexity, and cost go together, so companies want to buy the right size system when they need it.

21.   What Does the Vendor Selection Process Entail?

Once users have created and approved a set of User Requirements, the process for vendor and product selection begins. First, the user requirements are sent to multiple vendors in the form of a Request for Information (RFI). Vendors respond by indicating how their products meet or do not meet the requirements. The users compile all responses and select only those vendors that have desirable matches. This "short list" of vendors will now have detailed discussions and product demonstrations with the users. The users will update their User Requirements as necessary. The users send out the revised user requirements and make a Request for Proposal (RFP) or Request for Quote (RFQ). The vendors respond with detailed information about the products and services required to meet the requirements including expected costs. This response is legally binding and is the basis for any resulting contracts.

22.   What Software Features We Should Look for?

When the users create requirements, they should include features that will implement the processes they desire, meaning processes that will promote efficiency and productivity. In addition, users must include 21 CFR Part 11 industry standards for security, data transfer, audit trails, and electronic signatures if desired.

23.   What Types of Search Capabilities Should a Document Management System Have?

One of the greatest advantages of an electronic document management system is the ability to locate documents and related information between documents. There are many types of information that relate to a document. The document title, file name, key words, and text contained in the document are all possible search criteria. Searching is much more complex than just finding matches to search text entered by users. The hierarchy of how documents are stored, cross-referenced, and related is an essential part of the system. Users must be aware and take responsibility for establishing rules and practices to promote information retrieval. For example, key words may sound like a great idea; but if everyone uses different words, the ability to retrieve meaningful results reduces significantly.

24.   How Do the User Requirements Figure into Process?

User requirements start and end the validation process. Each user requirement is a sentence or phrase that indicates what the computer system will do. It is not so vague as to be indefinable. The requirements are the starting point for finding a vendor and ultimately validating the system, since they determine what software solution is needed. The requirements are also the ending point because the last validation steps determine if real production use of the system matches the original user requirements.

25.   What Should We Do If We Find a Vendor that Offers a Nice Feature that We Didn't Think About?

The initial set of User Requirements is the starting point for the evolution of the features set. As the users learn more about what is available, they can update their document. When they have completed the first round of research, the users get back together and update the User Requirements to include features that are required and identify features that are not required but desired. Users must be careful not to require more than is necessary and cost effective. It is easy to want everything, so users and management must work together to identify the appropriate set of features. The revised User Requirements are used in the second research phase that is associated with the best vendors selected from the first phase, often referred to as the "short list."

26.   How Do We Select the Best Software Vendor?

Users should select a software vendor based on recommendations, thorough research, product demonstration, reference checks, vendor audits, and contract terms. If an attractive vendor does not meet the selection criteria noted above, users should be willing to select another vendor.

27.   Aren't Large, Well-Established Software Vendors Preferable to Smaller, Newer Software Vendors?

Not necessarily. Often innovation and compliance are easier to achieve at a smaller company. Furthermore, some software vendors with products that have been on the market for a length of time may not be keeping up with industry standards in their products. Thus it's best to investigate a range of vendors to find the most suitable match between company needs and available, compliant software. Many new vendors have an advantage over established vendors in that they start off using the most modern technologies. Of course, these newer vendors also have a disadvantage in that they lack experience and diversity.

28.   What Does it Mean When a Vendor of an E-System Says the Product Is Validated?

There are two validations that should take place for every system. The software vendor performs the first validation. The buyer confirms this validation through an audit or evaluation process. The users of the computer system, based on their specific intended use, perform the second validation. Since no two user groups use software exactly the same, there is no way to buy a system that is fully validated for the users' purpose. In addition, vendors often claim to be validated, but this does not mean that have performed the required steps that constitute validation, and it does not mean that their products contains the required feature sets related to regulatory compliance.

29. Should We Audit the Vendor?

Users should audit (evaluate) a software vendor well in advance of product purchase. Users should use the experience as an indicator of how a long-term relationship will unfold. Some companies provide developer checklists and base the audit on those, without actually going on site. However, if you look for a vendor that has a Secure Software Certificate, you can be assured that the vendor has already been subject to scrutiny, and there is no need to do a physical audit. If the developer is certified, there is no need for the customer to perform an on site audit, which decreases the sales cycle overhead. Beware of vendors that claim to be certified. Certification must be made by a reliable independent company that specializes in software compliance. (See Figure 4.1.)

Figure 4.1.   Secure software certificate.

30.   What Is Secure Software Certification and What Does It Mean?

When companies look to purchase software from a vendor, they need to know the quality of the vendor and the quality of the software products before they make a purchasing decision. For many companies, an onsite audit of the vendor is required by internal policy or regulatory requirements. But without adequate resources to go out and perform the audit, expertise in software development to accurately ascertain the software vendor's quality level, and experience performing software vendor audits, most companies are unable to get an accurate measure of a software vendor. Secure software is based on the development of "hardened" software products (1). Secure software can only originate from vendors that (a) establish secure software development methods and (b) include robust software security features in their products. For software to be truly secure, it must be protected against threats such as these:

- Natural disasters
- Hacker/cracker intrusion
- Terrorism
- Corporate espionage
- Insider threats
- Malicious software (e.g., viruses and spyware)
- Software that is secure will address a wide array of business needs and concerns, including:
  - Theft of intellectual property
  - Physical security
  - Data backup and restoration
  - Risk management
  - Destruction of data
  - Disaster recovery and business continuity
  - Improper disclosure of confidential data
  - Alteration of data (intentional and accidental)
  - Regulatory compliance
  - Legal protection, due diligence
  - Loss of productivity
  - Fraud and misuse

31.   How Does a Software Developer Actually Get Certification?

In 2005 a new approach was developed to bring together software vendors and end users. Software vendors can obtain a Secure Software Certificate only after complying with a rigorous set of software development, software quality, and software features that are required for certification. All software vendors must meet the same high standards to receive certification. An experienced independent auditor audits the software vendor. Once a software vendor is certified, end users can confirm the certification and receive a full copy of the audit report.

32. Why Is Secure Software Certification Important?

Many companies fail to audit the software vendor before they purchase software, when the quality measurements can be used in the vendor selection process. Some companies incorrectly audit the software vendor after the purchase, installation, or even production use of the software. The majority of companies that do perform an audit are ill-equipped to perform a comprehensive evaluation that yields an accurate assessment of vendor and product quality. Certification provides end users with a uniform standard of software quality for comparisons across multiple software vendors. End users have confidence in certified vendors from the start, saving even more time that could be potentially wasted researching and bidding with an unqualified vendor.

33. How Do Users Benefit from Secure Software Certification?

End users save time and effort performing onsite audits of vendors, because a trusted third party has already completed this all-important step of the process. Completion of this step reduces productivity losses, because the time allotted for auditing may now be used elsewhere. End users can obtain the Secure Software Certificate Audit Report for approximately 80% less than the cost of conducting the audit themselves. The audit report meets regulatory requirements and saves both time and money.

34. Does Secure Software Certification Provide Any Benefits to a Developer?

Software vendors benefit from the certification process because it matures their software development and validation processes. This, in turn, leads to fewer software defects, fewer customer complaints, and less time resolving technical support issues. To receive a Secure Software Certificate, software vendors and valued-added resellers (VARs) must pass an audit by an independent and trusted third party. Only those audits that receive superior results receive a Secure Software Certificate. As an added assurance, customized software from VARs can only be certified once the original software/vendor has been certified.

35. Does Secure Software Certification Save the Developer Money?

Software vendors themselves are proponents of the Secure Software Certificate process. Vendors that receive certification are differentiated from other vendors. In addition, software vendors can receive a Secure Software Certificate for less money than hosting audits regularly.

36. How Can We Find Out If a Software Developer Has Certification?

Software vendors that receive a Secure Software Certificate are able to use the logo on their product marketing and web designs, showing customers that they are certified from the start. The Secure Software Certificate website lists all certified vendors, allowing potential customers to verify that fact at the click of a mouse.

37. If the Software Vendor Has Validated the Software, Do Companies that Buy that Software Has to Validate The Software?

Once the users have confirmed that the software vendor has performed software validation, either by performing an audit or reviewing the vendor's Secure Software Certificate, the users must perform computer system validation (CSV). All software requires two validations: The first is by the developer, the second by the users.

38. Do We Have to Buy a Seat for Every User?

There are many licensing schemes in use today. Named users, concurrent users, and processing power are a few examples. Named users means that a license is used for each user able to log in to the software. Often licenses are sold in groups where the first group is one price and a subsequent group is another price. Concurrent users means you can have an unlimited number of users but only a set number may log in at exactly the same time. This is often advantageous over named users because it allows many users to time share an application. Processing power is a licensing scheme where the number of microprocessors or the number of microprocessor cores determine the pricing, which is independent of the number of users. Often the licensing for a software application is different and in addition to licensing schemes for database and other software components. Together, users and IT staff need to work together to establish the best licensing method to follow in both the short and long terms.

39. What Is a Source Code Escrow Agreement?

A source code escrow agreement is a formal contract between a configurable off-the-shelf software (COTS) vendor and the client that allows the client to receive the source code if the software vendor goes out of business. Per the agreement, the software vendor promises to send a trusted third party a copy of the source code and the third party agrees to provide the client with that source code in the event that the software vendor goes out of business. Escrow agreements may be free or cost hundreds or even thousands of dollars per year. Most often, source code agreements are not as useful as they first appear. Here are some examples of the problems related to source code agreements: it is difficult at best to ensure that the software vendor is updating the escrow; it takes many years for a software vendor to be legally declared out of business; and once a client has the software source code, there is no trained staff and no documentation to assist in the continual maintenance of the software.

40. Is it Possible for Software to Be 21 CFR Part 11 Compliant Even If the Vendor Is Not 21 CFR Part 11 Compliant?

For the product to be compliant, both the vendor and product must meet the regulatory requirements. The vendor makes a product, and therefore the quality of the vendor directly affects the quality of the product. Compliance of the product and vendor is only part of the regulatory requirement. Users must also validate the software, and the processes used with the software must meet the current industry standards for 21 CFR Part 11.

41. If a Company Does a Thorough Job Selecting a Software Vendor and Finds that the Vendor Is Fully Compliant with the Binding Regulations and Offers a Validated Product, Must the Company Still Perform User Validation of the System?

Yes. Software must be validated from the user perspective. Even if software has been developed expressly for a unique need and is not commercially available, it should still undergo developer validation to ensure its suitability for its designated function. Developer validation does not eliminate the need to perform user validation, however, because the system must also be validated for its intended use, and only the users know exactly what that is. Vendor claims do not ensure compliance with the regulation because they are not regulated and never undergo audits from regulatory agencies. Furthermore, even when a vendor shows evidence of validation, it is validation from the developer's perspective, not that of the end users'. The regulations require validation from the users' perspective.

42. What's the Difference Between Developer and User Validation?

Developer and user validation have commonalties, but the primary difference is that they are validating for different objectives. The developer wants to confirm that the product provides all the functionalities it has built in. The user wants to confirm that the product does what it is intended to do in the actual company environment. (See Table 4.1.)

TABLE 4.1.   Developer and User Validation

Developer	User
Product requirements	User requirements
Software development and validation project plan	Validation project plan
—	Installation protocol
—	Installation report
Design specifications	Functional specifications
Technical specifications	—
Code review report	—
Build notes	—
—	Hazards analysis
Software quality assurance testing protocol	User testing protocol
Software quality assurance testing report	User testing protocol
Release notes	System release report
—	Validation completion report

43.   Can the Software Vendor Provide User Training or Must Companies Train Users?

Vendor training can be useful, but it is not sufficient. Users must undergo training on the processes specific to the application of the software and their individual tasks. Only users know what that application is. One solution is to have the vendor train a select group of people and then have those people design the training in how the system will be used in house and deliver it to the users.

44.   When Should a Company Reaffirm a Software Vendor's Capability?

Unlike raw material suppliers, software vendors do not require regularly scheduled audits. Software vendors provide updates to software as part of the software maintenance agreement. Whenever high-risk or high-complexity major software releases occur, a company should determine if a software vendor audit will help them to minimize their revalidation effort. A company should audit a software vendor whenever additional software products are purchased from a vendor.

45.   What Are Typical Configurations for Software?

Because software is configurable, companies have the ability to tailor a system to suit their specific needs. Typical configurations are the following:

- E-mail interface
- Data backup options
- User password change frequency

- File types for export
- ID format for users
- Temporary passwords makeup
- Confirm identification
- Day/date identification
- Time and date stamps
- Watermarks
- User groups/roles and access privileges/permissions
- Department identification
- Document numbering
- Document versioning
- Reports from the system (e.g., audit trail, user ID, groups, outstanding change requests)
- Workflow options
- Edit checking logic
- Form layout

46.   What Is the Return on Investment (ROI) for Computer System Validation?

Computer system validation has an excellent return on investment (ROI). The cost of compliance is low compared to potential costs for losses. During validation, there is verification of procedures for disaster recovery and back up. Validation also reduces labor costs by increasing employee efficiency and effectiveness. The validation process educates the system users, which saves money by making the organization more productive. The use of electronic records is less expensive than the use of paper records with respect to creation, organization, searching, and retention. Finally, electronic signatures are much more secure than handwritten signatures. Based on experience, a 30K validation typically has an ROI in 9 to 12 months. As for vendors, their 30K investment is usually returned in one new sale, not to mention new market advantage.

47.   Because We Couldn't Agree on a Vendor, Our IT Department Has Built Software In-House, But What Do We Do to Begin Using It?

In-house built software requires validation, just as commercial software does. Since IT is the software vendor, they must perform software validation from the developer's perspective. The users must also validate the software from their perspective. If IT and the users work together on the development and validation efforts, they may be able to combine their efforts into a single project that addresses both perspectives.

48.   What Is Computer System User Validation?

Computer system validation is validation of an entire system by the actual system users. A computer system includes all software, operating system, utility programs, hardware, procedures and instructions, training, and interfaces that comprise business processes. Validation means that the users confirm that the system does what they expect it to do and that users can detect when the system doesn't do what it is expected to do.

49.   Is The Only Reason to Perform Validation Because It Is Required by the Regulations?

Validation makes good business sense because it saves money by discovering system defects before production use. That alone should be reason enough to validate.

50.   How Long Should Validation Take?

Validation time for projects performed by users has been steadily decreasing over the past few years. COTS software, little or no software customizations, the use of the risk-based validation approach, and increased experience and knowledge have contributed to this reduction. Typical projects range from a few weeks to a few months. Gone are the days when projects took more than a year.

51.   Should Computer System Validation Be an IT Responsibility?

End users, usually a validation team, have the primary responsibility for performing validation. IT is often part of the validation team and provides assistance during the process but never has primary responsibility.

52.   When a Company Has Limited Resources, Can It Farm Out the Validation Responsibility to Temporary Staff or Consultants?

Companies can certainly retain experts to assist in the validation process, but only actual users have the experience necessary to perform effective computer system validation. Only they know the culture of the company and the record-keeping needs that the application will serve.

53.   Is There Anything that Companies Can Do to Make the Validation Process Less Laborious?

The reason why most validation projects take so long is a lack of project team understanding and experience. It's important to understand exactly what the requirements for the computer system are and to formulate a working plan that will drive the process forward. The use of document templates is highly advantageous.

54. What Is IQ/OQ/PQ?

IQ/OQ/PQ is a method of qualifying hardware, such as equipment and devices. The Q stands for qualification, which means to confirm. IQ is installation qualification, which confirms correct installation. OQ is operational qualification, which confirms the functionality related to the manufacturer's specifications. PQ is performance qualification, which confirms the functionality related to the user's needs. Often vendors perform IQ and OQ and PQ on equipment, devices, and other types of hardware. PQ is often performed in tandem with process validation (PV).

55. Is Qualification the Same as Validation?

Qualification and validation are similar but not the same. Qualification addresses individual components such as equipment and devices. Validation addresses the use of all components that comprise a process. Validation is a superset of qualification. Many validation methodologies incorporate the concepts of installation qualification (IQ), operational qualification (OQ), and performance qualification (PQ). However, validation must also address user requirements, process-based functional specifications, and, most important, process-based risk assessment.

56. Why Isn't IQ/OQ/PQ Sufficient for Software Validation?

Computer system validation (CSV) is better suited for validation of complex computer systems because it addresses the process involved with the use of complex software as well as the hardware it operates on. IQ/OQ/PQ is intended for qualification of both instruments and equipment. Today, most systems that have instruments also have interfaces to computers running complex software. In this case, IQ/OQ/PQ is performed for the instruments and CSV is performed for the entire system.

57. What's the Objective of Installation Qualification in Software Validation?

The objective is to document installation instructions and verify that the software is operational.

58. What Needs to Be Qualified?

Hardware components such as equipment and devices need to be qualified. Network cabling, firewalls, file servers, and other infrastructure components need to be qualified. Computers (both servers and workstations) that run software applications require qualification, and their software requires validation. Since validation is a superset of qualification, a separate computer qualification is not necessary.

59. How is Validating Software Different from Validating a Piece of Equipment with Computerized Controls?

Equipment is qualified. Computer systems are validated. Equipment that is connected to a computer is first qualified and then the entire system is validated. Equipment that contains computerized controls is often qualified unless the software is so complex that validation is more suitable.

60. How Are Validation and Verification Different?

Verification and validation differ in that validation includes a series of verifications. Verification is an activity that ensures the correctness of a document or event. Common verification activities include review and approval of documents, performing hazard analysis, and executing test cases. For example, verification confirms that each test you perform has the right outcome. Validation shows that a system is suitable for its intended use and that it will perform as it needs to. Validation starts with user requirements and ends with user acceptance of the computer system that implements those requirements. Validation is complete when the actual users, having been trained on approved instruction documents, have used the computer system in the production environment to do real work, thereby "proving" or "validating" the system works as intended.

61. Can the Validation Use Simulated Data to Test the System?

Yes. However, the validation environment should use a copy of production data whenever possible.

62. Do Scanners Require Validation?

Scanners usually don't require validation because they are used in manual processes. The most important thing is to be very thorough in the QC process of documents that are scanned.

63. If a Company Builds a Database Internally to Keep Laboratory Source Data, What Does It Need to Do to Make Sure It's Functional?

Any system that keeps source data requires validation, whether it is COTS or built in-house.

64. What Is Prospective Validation?

Prospective validation is validation of a computer system before production use of the system. After 1997, all new systems and upgrades to systems should be prospectively validated. Most systems are supplied by software vendors and are validated by the system users using the risk-based approach to computer system validation.

65.   What Is Retrospective Validation?

Retrospective validation is validation of a computer system after using the unvalidated system in production. Using an unvalidated computer system in a GXP production environment is a violation of 21 CFR Part 11 and GXP predicate rules. Retrospective validation follows the same methodology as prospective validation, and it usually requires the software to be cloned to a test environment in order to segregate test data from production data.

66.   What's the Difference Between Prospective Validation and Retrospective Validation?

Prospective validation occurs before production use. Companies sometimes think that prospective validation needs to occur only before an FDA inspection, but this is erroneous. Retrospective validation occurs after a system is in use. Retrospective validation should only apply to systems that were in place before the legislation requiring validation, and, as such, it should be phasing out.

67.   What Is Retrospective Evaluation?

A retrospective evaluation is a review of data outputs from a system to determine if failures have occurred. This exercise is very difficult because the data output, such as historical reports, may not provide enough context and detail to make a basis for the decision. Retrospective evaluation is often performed when failures are known to have occurred. Retrospective evaluation is not a substitute for prospective validation or retrospective validation.

68.   When Is Validation Required?

Validation is required for new computer systems, upgrades, and changes to existing systems. Complete revalidation usually occurs for major upgrades and widespread changes. Change control, meaning validation of only the changes that have occurred since the last validation activity, usually occurs for minor upgrades and small changes.

69.   What Is User Validation?

Validation performed by the users of the software is called computer system validation (CSV). CSV addresses the intended use of the software as it is used in business processes.

70.   What Types of Software Need Validation?

Much, but not all, software needs to undergo computer system validation. Examples of software that require validation are electronic document management systems (EDMS); laboratory information management systems (LIMS); spreadsheet applications made with Excel and used with regulated data; database applications made with FileMaker or Microsoft Access and used on regulated data; software that manages laboratory equipment; software that manages clinical trial data; software that is used in the manufacture of regulated products; and software that is contained in a medical device or is itself a medical device.

71. What Types of Software Do Not Need Validation?

Much, but not all, software needs to undergo computer system validation. Examples of software that do not require validation are operating systems; word processing programs; spreadsheet programs such as Microsoft Excel; database programs such as FileMaker; electronic mail; web browsers; and presentation programs such as Microsoft PowerPoint. Note that applications made with software programs that involved GXP data may need validation. For example, an Excel spreadsheet may need validation, while the Excel program itself does not need validation.

72. Who Is Responsible for Computer System Validation?

Users of computer systems are responsible for computer system validation on a continual basis as the system evolves. Often the users receive assistance from IT, QA, and software vendor resources, but users should not relinquish their responsibility to any other party.

73. Where Is COTS Computer System Validation Performed?

COTS computer system validation is most often performed in a test or validation software environment that closely resembles the production environment. In some cases, as in the first validation for a newly purchased system, the production environment may be used for all validation activities. In all cases, validation cannot be completed until the software is used in the production environment by the actual end users.

74. When Is COTS Computer System Validation Performed?

COTS computer system validation is always performed before production use of the system. Every change to the computer system, either as a result of upgrades from the software vendor or changes to the configuration or use by the users, must be validated. Validation ensures the changes are understood, and instruction procedures are updated to reflect proper use.

75. What Types of Changes Can Be Made to a Computer System Without Revalidation?

Major changes to a system of course require revalidation, the full COTS risk-based approach; and since there is existing validation documentation, the revalidation effort consists of editing copies of those documents to reflect the changes. Minor changes are usually handled via change control. Changes that have little or no impact on system functionality may be handled with a maintenance event. Changes to user accounts generally are handled by a specific procedure that has a verification step built into the process.

76. Why Is Computer System Validation Performed on COTS Software?

Users perform computer system validation on COTS software to learn the system, develop processes, assesses risk, and provide documented evidence that the system works as intended. During the process of validation, the users who are the project team members get an in-depth education about the workings of the software which makes them able to develop internal instructional procedures to train all the other users. When users have a good understanding of their system, they become more efficient and more effective in their jobs. This increases productivity, which is the primary reason the computer system was obtained in the first place.

77. What Is a Validation Document Template?

A validation document template is a word processing file that contains content labels followed by instructions and questions. The users on the project team follow the instruction and answer the questions. This fill-in-the-blank method ensures that all validation projects have documents with the same set of content. Content labels typically may not be deleted nor added. When all validation projects use the same templates, quality reviews are made efficient and adherence to the validation SOP becomes easy to verify.

78. How Is Validation Documentation Written?

The best way to write validation documentation is to have at least two authors, or even the entire team, working on the document at exactly the same time. It is important to get the validation project team members away from their normal work and get them into a quiet and relaxed area. A small conference room with a projector works well. If you need to see the software and author a document at the same time, then use two projectors.

79. How Are Large Software Systems Validated?

The best way to validate a large software system is to divide the system into small projects, typically on a module-by-module basis. It is easier and faster to validate several small projects in succession rather than try to validate a large system. Smaller documents are also easier to review and edit.

80. What Are the Validation Requirements When COTS Software Is Customized?

When COTS software is customized, meaning the vendor changes the source code, those customizations must undergo validation by both the software vendor and the users. Whenever possible, the user should request changes to the software that are then included in the software that the vendor releases to all customers. That turns the request into an enhancement rather than a customization. The user should not request customizations until the software has been fully validated. The process of validating the software educates the users and often teaches the users a better way of doing things, thereby eliminating or changing the customization desired. It is difficult for users to have the discipline to hold off on customization until after the validation is complete, but it is more beneficial in the long run.

REFERENCE

1. www.securesoftwarecertificate.com, accessed March 29, 2009.

CHAPTER 5

MAKING THE TRANSITION FROM HYBRID TO VALIDATED E-SYSTEM

INTRODUCTION

Making the transition to a workable electronic system or systems may seem like an overwhelming task, but it need not be. A systematic approach works best when implementing changes in how you create and manage documents and records. Far too many firms have spent needless hours validating systems when a simple, logical approach would have gotten them to the finish line in short order. Consider the firm that is engrossed in validating software and has devoted over six months to the task, only to find that the vendor has issued an upgrade to the software. Does the validation have to stop and take the new upgrade into account? Should it continue with the validation and then do a partial revalidation of the new features afterwards? How much easier would it be to validate in short order in the first place? This chapter provides guidance on transitioning to an e-system and avoiding the pitfalls that delay or derail validation efforts. It answers the following questions:

1. Does the FDA offer any help in validating systems?
2. What is the difference between validation and qualification of software?
3. Do we have to validate our network?
4. Should we validate firmware that controls automated instruments?
5. As a young company with no documentation system in place, can we go electronic from the very beginning?
6. Can we use our existing system to approve the validation documents for our new e-system?
7. Once we have purchased software, what should we do next?
8. Should an IT member be the validation lead?
9. Does the validation team need to have expertise in hardware and software to successfully validate?
10. What is IT's actual role in COTS validation?

Managing the Documentation Maze, By Janet Gough and David Nettleton
Copyright © 2010 John Wiley & Sons, Inc.

11. What is the role of Quality Assurance in COTS computer system validation?

12. What characteristics should a validation team have?

13. What characteristics should a validation team member have?

14. Should a validation team perform every system validation?

15. What are the duties of the validation team?

16. Is our QA manager correct in saying that since we are installing packaging software and three shifts will use it, we must validate three times?

17. What's the purpose of a validation packet, and does it have to be paper?

18. What are the deliverables for a COTS validation project?

19. How do you create validation documentation efficiently?

20. How are validation documents reviewed?

21. Who approves computer system validation documentation?

22. How do we determine what SOPs we need to have to support the system?

23. Can a system go into production if supporting documentation is drafted but not approved?

24. What should our validation documents look like?

25. Is it possible to purchase validation templates?

26. How should we number our validation documents?

27. What's the best way to prepare validation documentation?

28. How do you create validation documentation efficiently?

29. Our writing consultant told us never to use actual people's names in a document, but if we don't, how to we identify team members in the Project Plan?

30. What should the User Requirements document cover?

31. What should our Project Plan cover?

32. So what's the best way to install the new software into our system?

33. When we have installed software successfully, what is the next step?

34. How do the functional requirements and functional specifications relate to each other?

35. What should the Functional Specifications document include?

36. How do we assess hazards?

37. How can we best report on Hazard Analysis?

38. What is meant by severity?

39. What is meant by probability?

40. What is meant by risk level?

41. What should we report in our Hazard Analysis document?

42. After we've assessed the system for hazards and their likelihood of occurring, what should we do?

43. What's the most effective way to test a new system?

44. What is a Validation Error Control Plan?
45. What constitutes a test case?
46. What constitutes test results?
47. Isn't testing the same as validation?
48. Can a test script author execute the tests?
49. What do we do after executing the test cases?
50. What documentation is required to release a system for production use?
51. What documentation is required train users?
52. What happens if a user cannot perform a task on the new system?
53. What happens if we find that most users have difficulty completing a task?
54. Once we're live in production, can every user access the system?
55. Is validation complete when the system is released for production use?
56. What documentation is required to complete a validation project?
57. What do you do with the validation documentation after the project?
58. Should we write a summary for the validation packet?
59. What happens if the system isn't working after completion of the validation project?
60. How can we migrate data from one system to another?
61. Can we change our SOP format when we move to a new document management system?
62. How can we make sure the right people review and approve documents electronically?
63. If we want to bring manually signed documents into our system, what do we need to do?
64. What is a satisfactory QC check of a document?
65. What exactly is a legacy document?
66. How many signatures do we need to bring a legacy document into the system?
67. Who should bring legacy documents into the system?
68. Do we need to bring in every document that exists outside the system?
69. Which documents should we bring into the system first?
70. Can we load secure electronic files from our hybrid system into our new system, and, if so, how should we do it?
71. How can we bring in paper reports that our CRO has prepared for us and that bear numbers that are different from the ones in our system?

1. Does the FDA Offer Any Help in Validating Systems?

At present, the FDA has two guidance documents intended to help firms establish compliant electronic systems. After the publication of 21 CFR Part 11 Electronic

Records; Electronic Signatures in 1997, the FDA issued five guidance documents in an effort to help industry and the agency itself understand the application of the law. In August 2003, the FDA withdrew those Part 11 guidances and issued *Guidance for Industry Part 11, Electronic Records; Electronic Signatures—Scope and Application* (1). This guidance is still in effect. In May 2007, the FDA issued a new guidance directed at record keeping in clinical trials: *Guidance for Industry, Computerized Systems Used in Clinical Investigations* (2). Both guidances offer the agency's "current thinking" on the topic of Part 11.The agency is planning to issue additional guidances on the subject. Adherence to the guidances is not mandatory, but most companies comply with the intent. What the guidance documents do not do is identify how your actual system works, so validation must be geared to "intended performance." See the Appendix, pages 431 and 441, for the full text of these documents.

2.   What Is the Difference Between Validation and Qualification of Software?

Hardware is qualified, which means confirming that it is capable of the intended use as stated by the manufacturer. Software undergoes validation, which means to confirm the intended use of the process from the user's perspective. Verification is a component of both qualification and validation in which you confirm that something is correct. Verification includes documents reviews and executing test cases and is integral to both validation and qualification.

3.   Do We Have to Validate Our Network?

You don't validate a network; you qualify the components of the network. The network functionality as a whole is included as part of software validation that runs data over the network.

4.   Should We Validate Firmware that Controls Automated Instruments?

Firmware is software that is contained in hardware. It is validated by the software developer just like any software is validated by the developer. Since firmware is usually part of a device or instrument, the firmware is also qualified when the device or instrument is qualified.

5.   As a Young Company with No Documentation System in Place, Can We Go Electronic from the Very Beginning?

You will need at least a simple alternate system to prepare and approve documentation as you go through the process of validating your e-system. Before a system can go live, the validation documents must be in place, as must be SOPs to support the system.

6.   Can We Use Our Existing System to Approve the Validation Documents for Our New E-System?

Yes. You can keep the existing system operational, whether it is manual or another e-system, until you validate the new system and move your records and documents into the new system. This process means keeping both systems running concurrently until you have fully made the change.

7.   Once We Have Purchased Software, What Should We Do Next?

The next step is to assemble a validation team. If you had a team to determine initial user requirements, you probably already have a viable working team. The team should have a leader and a representative from every area of the company that will use the system, and at least one representative from Information Technology (IT). A member of the Quality Assurance unit should participate. Consultants may participate in the validation process as well, but they should not be the leader for the actual validation activities.

8.   Should an IT Member Be the Validation Lead?

It's usually best if the owner of the system is the validation lead. The owner is usually one of the people instrumental in bringing the system in-house and managing it. This person is a key player in the validation, because he or she knows what the system needs to do to satisfy the requirements of the end users. In this capacity, this team member is analogous to a congressman for a district who works on behalf of the constituents. In addition, an IT person is a member of every project team and provides the interface between the computer system and the existing infrastructure. The team member from IT is the lead for system installation.

9.   Does the Validation Team Need to Have Expertise in Hardware and Software to Successfully Validate?

Such expertise is not necessary for all validation team members. IT provides the expertise in getting the system up and running and maintaining it. The validation team must understand the system and what it will do, so initially there should be training for the team, so that all members know how it works. This knowledge is what will drive the validation, as well as provide know-how for future validation.

10.   What Is IT's Actual Role in COTS Validation?

The IT team member provides the interface between the computer software system and the existing infrastructure. The team member from IT is the lead for system installation.

11.   What Is the Role of Quality Assurance in COTS Computer System Validation?

In the risk-based approach to computer system validation, Quality Assurance (QA) approves every validation document. This QA approval is always the last approval because QA is verifying that the template has been filled out correctly and that all required approvals have been made. In this capacity, QA is an independent reviewer and is not part of the actual team. If QA will ultimately use the system as well as troubleshoot and maintain it, a QA person may be on the project team, but that person cannot also be the independent reviewer.

12.   What Characteristics Should a Validation Team Have?

A computer system validation team should consist of people that have been with the company long enough to gain experience with the existing systems and processes, know the other users, understand the culture of the company, and be available to perform future validations. The validation team needs a leader that will facilitate the activities and keep the momentum going.

13.   What Characteristics Should a Validation Team Member Have?

Validation team members should be team players, meaning they work well with the members of the group they represent and the other team members. Members should be logical thinkers able to troubleshoot problems, suggest solutions, develop new processes, and lead their respective groups to adopt new processes.

14.   Should a Validation Team Perform Every System Validation?

A team of users should validate every computer system. The team ensures that all user interests are considered. Synergy occurs when multiple people work together, and this helps ensure a positive and thorough validation effort. Validation teams should vary for each project so that more people gain exposure to and participation in validation projects and that each validation effort reflects actual system users.

15.   What Are the Duties of the Validation Team?

The validation team is responsible for all aspects of the project. Team members develop processes and scripts for testing, conduct training, compile the validation packet, write the validation documentation, and write new or revise existing SOPs in support of the system. They are the principal communicators of progress as the project moves forward. It's important that they demonstrate enthusiasm for the system, because it is common for people to fear change and be resistant to new systems.

16.   Is Our QA Manager Correct in Saying that Since We Are Installing Packaging Software and Three Shifts Will Use It We Must Validate Three Times?

No. This is another case where we make the validating process more onerous than it needs to be. The validation team needs to have a representative from each shift to ensure that all shifts are using the software the same way. If there are differences in the way the shifts use the software, the validation documentation will capture that, and when management and QA review the documents, they will learn what is actually happening. In no way should a company repeat validation per shift.

17.   What's the Purpose of a Validation Packet, and Does It Have to Be Paper?

A validation packet captures the history of the project, from User Requirements and vendor selection to validation completion. In a paper-based model, the packet is typically a binder, with tabs for each deliverable. If the packet is electronic, then the packet is typically an electronic folder system. First in should be copies of contracts, correspondence with the vendor, requests for information, references, marketing materials/vendor information, records of payments, and the initial User Requirements (all versions).

18.   What Are the Deliverables for a COTS Validation Project?

The risk-based commercial off-the-shelf validation approach uses a standard set of fill-in-the-blank templates that are organized chronologically. Below is a brief description of each document.

- **User Requirements.** This document describes what the users want and need and is the foundation for setting specifications for the system.
- **Project Plan.** This document identifies the members of the validation team and outlines their responsibility for implementing the required processes.
- **Installation Protocol.** This document describes how the hardware and software are to be installed and tested.
- **Installation Report.** This document describes the results of testing, deviations from the protocol, and the readiness of the system for further validation activities.
- **Functional Specifications.** This document describes the software functions in the context of business processes. Specifications are the basis for the risk assessment and trace to requirements. They define the functionalities to undergo testing.
- **Hazard Analysis.** This document determines the risk assessment related to the Functional Specifications. Hazards trace directly to specifications without needing a trace matrix.

- **User Testing Protocol.** This document describes the testing related to the Functional Specifications. All specifications are tested, but those with higher risk receive more testing. Test cases also trace directly to specification without needing a trace matrix.

- **User Testing Report.** This document describes the results of testing, deviations from the protocol, and the readiness of the system with respect to implementing the specifications.

- **System Release Report.** This document describes the plan for production use, identifies the Standard Operating Procedures and Work Instructions that detail how the users will accomplish the business processes, and releases the system for initial production use.

- **Validation Completion Report.** This document describes the results of the initial production use of the system and the readiness of the system for full production use. It completes the validation project.

19. How Do You Create Validation Documentation Efficiently?

Validation documentation is created efficiently when standard templates are used. Validation templates can work the way SOP templates work, with set sections for recording information as the validation progresses. One mistake that companies make when they validate software is to complete the entire validation packet and then put it into review. A better approach is to secure sign-off on each document that is completed during the validation process. That way, when the system goes live, the documents are in place. The validation concludes with the validation completion report.

20. How Are Validation Documents Reviewed?

Validation documents are reviewed by the project team, the user groups represented by the team members, Quality Assurance (QA), and others as needed. For COTS software, formal review documentation is not required but formal approval documentation is required.

21. Who Approves Computer System Validation Documentation?

Computer System validation documentation is approved by the project team members and a Quality Assurance (QA) person that is an independent reviewer. Users may elect to have an executive, such as a system owner, approve certain documents such as the (a) User Requirements that are used to make the purchasing decision, (b) the System Release Report that allows production use of the system to begin, and (c) any other validation document.

22.   How Do We Determine What SOPs We Need to Have To Support the System?

The place to start is by looking at your existing SOPs. Do you have any in place now to support your current system? Can they be modified to accommodate the new system, or do you need to create new procedures? You may also need to create SOPs to move you through a transition period as you build your new system. Chapter 7 gives more information about SOPs that support document systems. Most important, the Functional Specifications are the basis for the content of instruction documents used to educate and train the users.

23.   Can a System Go into Production If Supporting Documentation Is Drafted But Not Approved?

The validation process may uncover the need for supporting documents, such as an SOP for computer system validation. If such a document is drafted, but not approved and active, the system can go into production provided QA has given approval. However, users are responsible for ensuring procedures for system operation and maintenance are approved prior to production use of the system. Any documents that need to be in place should be propelled through the review and approval process concurrent with the actual validation, so that they are in place before the system is used in the work environment.

24.   What Should Our Validation Documents Look Like?

You can create templates that are much like SOPs in how they work; then draft the validation documents as you move through the process. A validation document template is a word processing file that contains content labels followed by instructions and questions. The users on the project team follow the instructions and answer the questions. This fill-in-the-blank method ensures all validation projects have documents with the same set of content. Content labels may not be deleted nor added. When all validation projects use the same templates, quality reviews are made efficient and adherence to the validation SOP becomes easy to verify.

25.   Is It Possible to Purchase Validation Templates?

Yes, but if you do so, make sure you can modify them as you need to, so that they are compatible with the conventions of your document system, such as numbering and type faces. There are published books that actually include templates on a disk that you can download to prepare your validation documents (3).

26.   How Should We Number Our Validation Documents?

Numbering should be consistent with your total document numbering system. Since validation is really a project with many documents, some companies number the entire validation and then have indicators for each document in the project. For instance, a company may have an alphanumeric system that identifies documents by category, then core number, with a subsequent number indicating the document sequence as follows: VAL 004-03. Thus, VAL is the category, 004 is the validation project, and 03 indicates the third document, the Installation Protocol, in the 10-step risk-based approach.

27.   What's the Best Way to Prepare Validation Documentation?

The best way to write validation documentation is to have at least two authors, or even the entire team, working on the document at exactly the same time. It is important to get the validation project team members away from their normal work and get them into a quiet and relaxed area. A small conference room with a projector works well. If you need to see the software and author a document at the same time then use two projectors.

28.   How Do You Create Validation Documentation Efficiently?

One mistake that companies make when they write validation documents is not to create a clear purpose statement for each document, so that reviewers fully understand what each document does. Another mistake is to complete the entire validation packet and then put it into review. A better approach is to review and secure sign-off on each document as it is completed during the validation process. By approving the validation documents in order, any issues can be address before they impact upcoming validation activities.

29.   Our Writing Consultant Told Us Never to Use Actual People's Names in a Document, But If We Don't, How to We Identify Team Members in the Project Plan?

Validation documents are not like SOPs. In the validation documents you should cite team members by name because they are participating in this project in real time. These documents are not subject to re-versioning over a long duration, only the project duration. If at a later point in the project a member leaves the team, for whatever reason, that event can be captured in a subsequent validation document and a new team member identified.

30.   What Should the User Requirements Document Cover?

Such a document typically covers the following:

- Purpose
- Justification—for the software
- Intended users

- Process overview
- Referenced documents
- Operating environment
- Software platform
- Sizing
- Interfaces
- Functional requirements
- Approval signatures

31.   What Should Our Project Plan Cover?

The project plan tells what you are going to do:

- Purpose
- Validation guidance
- Project deliverables
- History—of the project to this point
- Resources
- Project team
- Project scheduling
- Budget
- Approval signatures

The project plan is ideally in place before the software is actually installed into the company's system.

32.   So What's the Best Way to Install the New Software into Our System?

Since the system must have integrity, it's important that the installation is well thought out and planned. Does your contract specify who will install the software? The best way to do this is to capture everything in a protocol for installation that IT, the validation lead, and QA approve. An Installation Protocol covers the following:

- Purpose
- Impact analysis
- Installation liaison—by name
- IT representative—by name
- Installers—by name
- Hardware components
- Environmental conditions

- Software components, including server, clients, and licenses
- Design overview
- Backup plan
- Installation instructions
- Installation testing
- Testers—by name
- Media and materials
- Approval signatures

Once you have installed the software and tested it, you can write the installation report, which is the fourth validation deliverable. This document covers the following:

- Purpose
- Installation deviations (if there were none, say so)
- Environmental conditions
- Installation results
- Installation summary
- Approval signatures

33.   When We Have Installed Software Successfully, What Is the Next Step?

The next step is to write your functional specifications. Each specification is linked to a functional requirement. A specification is a paragraph or more that describes how a function in the computer system performs a piece of a process. The specification describes the process and software components according to how it will be used. Specifications are used to determine risk in the process. Specifications must be detailed so that test cases can be written to verify system operation. Specifications are also used as the basis for instructional procedures.

34.   How Do the Functional Requirements and Functional Specifications Relate to Each Other?

The user requirements describe what the users want the system to do on a high level. The functional specifications explain the details of how the system performs functions that implement the requirements. One user requirement may be related to many functional specifications. These are mapped out in a traceability matrix that is part of the functional specifications document. (See Table 5.1.)

35.   What Should the Functional Specifications Document Include?

Functional specifications should describe all functions used in the software, describe all processes that involve the software, and include the settings of all configuration

TABLE 5.1.   Traceability Matrix

Functional Requirement	Functional Specification
R1	S1, S3
R2	S2
R3	S14
R4	S4, S15, S16

options, user-defined parameters, and security permissions. It's important that you include all of the specifications for your system, since they will link to your hazard analysis. Once you have set your specifications, prepare the Functional Specifications document to include the following:

- Purpose
- System overview
- Functional specification
- Existing affected documents
- Traceability matrix
- Approvals signatures

36.   How Do We Assess Hazards?

Each functional specification is assessed for hazards by thinking what can go wrong in the process. This involves software functionality and processes related to the use of the software. Next think who and what are affected by that hazard occurring. Once the worst-case effects are known, the analysis requires the subjective measurement of severity and probability that are then used to determine the unmitigated risk level. This tells us our worst-case risk level independent of any and all precautions or preventative measure that are possible. The next step is to identify all mitigations for the hazard. What software functions and process steps will help to prevent the hazard from having the effect? Now considering the mitigations, reassess the severity and probability to determine the mitigated residual risk level. If the residual risk level is greater than can be tolerated, meaning greater than a moderate level of concern, the hazard analysis should be repeated and additional mitigations identified. (See Box 5.1.)

37.   How Can We Best Report on Hazard Analysis?

Hazard analysis is performed on each functional specification and is best performed in a sequential paragraph format. Many risk assessment approaches use a tabular format but this tends detracts from readability. A hazard analysis includes:

- Trace to functional specification
- Hazard
- Effect

BOX 5.1

HAZARDS

- People, including operators (users), field service, patients, subjects, animals, or others involved with the computerized system
- Hardware, software, and interfaces
- Automated or manual procedures, including development, assembly, calibration, and maintenance procedures as well as procedures for use
- Electrical power sources
- Packaging, shipping materials and transportation
- Environment, including temperature, humidity, dust, vibration, shock, etc.
- Sequences of events
- Possible errors in code
- Information or results that users rely on to make decisions

- Severity
- Probability
- Risk level
- Mitigations
- Residual severity
- Residual probability
- Residual risk level

38. What Is Meant by Severity?

Severity is a subjective measurement of the impact that results from the hazard causing the effect. It is looked up on a chart that gives a standard gradient. (See Box 5.2.)

39. What Is Meant by Probability?

Probability is a subjective measurement of how likely a hazard is to occur. It is based on volume of transactions and how probable a failure is bound to occur. It is looked up on a chart that gives a standard gradient. (See Box 5.3.)

40. What Is Meant by Risk Level?

Risk level is determined by looking up severity and probability on a standard chart. The ratings give a relative value that is used to compare the unmitigated risk level to the mitigated residual risk level. This allows the team to determine the power of the mitigations and therefore understand their importance on risk prevention and process controls. (See Table 5.2.)

BOX 5.2

PROBABILITY RATINGS

- *Frequent:* May occur with a 1 in 5 chance or once a week
- *Reasonably Probable:* May occur with a 1 in 10 chance or twice a month
- *Occasional:* May occur with a 1 in 100 chance or 12 times a year
- *Probability*
- *Remote:* May occur with a 1 in 1000 chance or 1 time a year
- *Unlikely:* May occur with a 1 in 10,000 chance or 1 time in the life of the system
- *Extremely Unlikely:* May occur with less than a 1 in 10,000 chance or 1 time in the life of 10 systems

BOX 5.3

SEVERITY RATINGS

- *Catastrophic:* Data may be incorrect and lead to loss or corruption of data. Incorrect results from the application may have significant financial ramifications.
- *Critical:* Data may be incorrect and lead to improper decision making which may cause severe problems including significant financial difficulties.
- *Moderate:* Data may be incorrect and lead to improper decision making which may cause minor or nonpermanent problems. Incorrect results from the application may cause moderate financial impact.
- *Marginal:* Data may be incorrect and lead to improper decision making which may cause inconvenience (e.g., delay). Incorrect results from the application may cause minor financial impact.
- *Minor:* Data may be incorrect but do not result in improper decision making or cause problems. Incorrect results from the application may cause minor financial impact.

41.   What Should We Report in Our Hazard Analysis Document?

A hazard analysis document covers the following:

- Purpose
- Hazard analysis
- Risk summary
- Approvals signatures

TABLE 5.2. Risk Levels

	Catastrophic	Critical	Moderate	Marginal	Minor
Frequent	Very high	Very high	Very high	Moderate	Moderate
Reasonably probable	Very high	Very high	Very high	Moderate	Moderate
Occasional	High	High	Moderate	Low	Low
Remote	High	High	Moderate	Low	Low
Unlikely	Moderate	Moderate	Low	Low	Low
Extremely unlikely	Moderate	Moderate	Low	Low	Low

42. After We've Assessed the System for Hazards and Their Likelihood of Occurring, What Should We Do?

The logical next step is to test the system, so you must train enough people to use the system so that you can test it. The project team members are the most knowledgeable users thus far. While every specification is tested, specifications for software mitigations are tested more thoroughly.

43. What's the Most Effective Way to Test a New System?

You will need to prepare a protocol for the testing and actual test scripts. The User Testing Protocol should cover the following:

- Purpose
- Testers—by name or group
- Testing approaches
- Test environment description
- Validation error control plan
- Test cases
- Test results
- Approval signatures

44. What Is a Validation Error Control Plan?

Issues such as bugs, anomalies, and errors in the test protocol need to be documented during formal testing of the software. This plan is part of the User Testing Protocol and describes the forms used to capture data, how forms are logged, and how issues are resolved.

45. What Constitutes a Test Case?

Test cases can be a single paragraph that explains the following:

- Test case number that traces to the function specification
- Description of the function being tested by the test case

- Test data
- Test execution instructions, including test results documentation
- Expected results (acceptance criteria)

46.   What Constitutes Test Results?

Test results need to include the following information:

- Test case number
- Pass/fail indicator
- Number of pages in the test results (usually print screens and reports)
- Test execution date
- Tester's initials
- Testing output identifying actual results
- Reviewer's initials
- Review date

Test results can be recorded on a Test Execution and Review Worksheet. Screens and reports may be printed to provide the testing output.

47.   Isn't Testing the Same as Validation?

Testing is not the same as validation. Testing is a form of verification, which is verification of the specifications. Each validation consists of a series of verifications. In total, verifications feed into validation and ensure that the user requirements are met in production use of the entire system. This is a much larger scope than testing.

48.   Can a Test Script Author Execute the Tests?

It is acceptable to have the test script author be the executor as long as the actual results of testing are approved by the tester and reviewed by another person. Both the approval and review need initials and dates.

49.   What Do We Do After Executing the Test Cases?

You should capture the results of the testing in a User Testing Report. Such a report has the following elements:

- Purpose
- Testing dates
- Configurations tested
- Validation error control log
- Test case review

- Testing conclusion
- Approvals signatures

Your user testing report confirms that the system is operating as it should or identifies problems that need further action or consideration.

50. What Documentation Is Required to Release a System for Production Use?

To release a system for production use, all validation documents up to the System Release Report need to be approved, instruction procedures need to be approved, training plans need to be approved, and a plan for determining how the system performs during the first few weeks of production needs to be approved. Once those are in place in the validation packet, you can write the System Release Report, which should include the following:

- Purpose
- Schedule for release
- List of completed documents: validation documents so far, SOPs, and training documents
- Release instructions
- User training plan
- Product change control
- Production review (your plan for assessing performance)
- Conclusion
- Approval signatures

Once this document is approved and in place, you can go ahead and release the system into production. This is the "go live."

51. What Documentation Is Required to Train Users?

Companies should have a training SOP in place that they can follow, and training should produce training records that trainees sign to verify their understanding of the training content. These records also serve as statements of accountability for using the system.

52. What Happens If a User Cannot Perform a Task on the New System?

If a user has difficulty performing a task, it usually stems from insufficient instructions or training. Retraining should occur first. If more than one user is having the same difficulty then the instruction document content should be revisited.

53.   What Happens If We Find that Most Users Have Difficulty Completing a Task?

If many users are having difficulty performing a task, then a review of the instructions, training, and functional specifications is in order. It may be that incomplete or inaccurate specifications resulted in poor instruction documents. The reason that production use occurs before the completion of the validation project is to allow updates to validation and instruction documents before full-scale production use.

54.   Once We're Live in Production, Can Every User Access the System?

Once a system is in production use, only users that have been formally trained may access the system. This doesn't mean that you have to train everyone before the system becomes fully operative, but that no one can use it who has not undergone training.

55.   Is Validation Complete When the System Is Released for Production Use?

The validation project does not end until the system has been assessed after production use, typically after 7–30 days. For some systems that have infrequent use, the assessment period may be longer, perhaps 90 days. Validation requires the actual users, trained on their instruction procedures, to use the system in real time and in the real production environment for a predetermined period of time.

56.   What Documentation Is Required to Complete a Validation Project?

The validation project completes once this production review period ends and the Validation Completion Report is approved. It is written after the production review period where actual system use occurred for a predetermined amount of time. It should include the following:

- Purpose
- Release date
- Production review results
- Project documentation archiving
- Conclusion
- Approval signatures

57.   What Do You Do with the Validation Documentation After the Project?

At the end of a project, all of the paper can be scanned to make a backup copy. The validation packet is ideal for quick reference and is the object to be audited by third parties and regulatory agencies.

58. Should We Write a Summary for the Validation Packet?

There is no requirement for doing so, but some auditors prefer to review a project summary at the beginning.

59. What Happens If the System Isn't Working After Completion of the Validation Project?

Once the Validation Completion Report is approved, no more changes to the validation documentation may occur. The system is in production use, so change control procedures determine how changes to the system are made.

60. How Can We Migrate Data from One System to Another?

Data migration from an old system to a new system is not required. Users may want to migrate a small portion of data or larger amounts of data. It is usually not advisable to migrate all data. Data migration is a difficult and error-prone activity. Data from one system most often does not perfectly match the requirements of the new system. Data migration involves moving data so that the user interface and, therefore, the data integrity protections are bypassed. This has considerable risk, and migration comes at considerable costs. Usually it is best to leave old data in the old system and refer to it as needed. The user must verify all data migrated to ensure quality and quantity integrity.

61. Can We Change Our SOP Format When We Move to a New Document Management System?

Yes. The first thing to do is to write an SOP for the transition period that spells out exactly how you are going to reformat and bring the documents into the system. Then look at the SOPs in your preexisting system. If they are ready for review (2-year cycle), you can bring them in as a new version of the existing version in the new format and then put them through the review and approval cycle electronically. You can maintain the dual systems until every SOP in the preexisting system has been either retired or moved into the new system. Of course, any new SOPs should generate initially electronically in the new system.

62. How Can We Make Sure the Right People Review and Approve Documents Electronically?

Review and approval of documents is the same regardless if this is done as a paper document or as an electronic document. Procedures that explain the review and approval process must address how document routing is performed and to whom based on the document scope.

63.   If We Want to Bring Manually Signed Documents into Our System, What Do We Need to Do?

The FDA recommends PDF files for scanning electronic documents. You may scan manually signed documents and import them into your system, provided that you do a QC check to show that all imaged text converted by software is accurate and an exact replica of the original. If you have the original electronic files (such as the Word files), used to make the printed documents, it is advisable to import them as well so that future revisions can be made easily.

64.   What Is a Satisfactory QC Check of a Document?

A 100% QC check verifies that the scan of a paper document has the correct number of pages, the overall scan is readable, and the first and last words of each page are identical to those in the original document. It's best when the person who does the scanning is different from the person who does the QC check.

65.   What Exactly Is a Legacy Document?

While the meaning of "legacy" has the connotation of being from the past, legacy refers to any document created outside a system and imported into it.

66.   How Many Signatures Do We Need to Bring a Legacy Document into the System?

This should be determined by your SOPs, but the industry standard is for at least two signatures. How the documents come into the system must first be thought out and documented in an SOP that covers the transitioning phase, and it can be retired when the system is fully loaded. The former system must stay up and running until all documents have either been retired or transitioned to the new system. Then the old system and its SOPs can be retired.

67.   Who Should Bring Legacy Documents into the System?

Usually a system administrator ensures that the legacy document coming into a new system is an exact replica, and a second person does a 100% QC check. Again, at least two signatures is the industry standard.

68.   Do We Need to Bring in Every Document that Exists Outside the System?

No. You may build your system so that it suits your current needs. If you have archived paper documents, there is no need to bring them in unless you want electronic searchability for those records.

69. Which Documents Should We Bring into the System First?

Typically, companies begin to build a document repository by first addressing their procedures; they handle documents such as work instructions and forms the same way. A reasonable place to start is by determining if any documents are ready for retirement. If they are, they can be put into review in the former system and simply retired within that system. They can also be brought in as any other SOPs that are ready to go into the review cycle, but then they must follow the SOP for importing documents to the new system.

Documents that are ready for review can also undergo reformatting, but not content changing. This is possible if the original system has secure electronic files of the final approved document. A reformatted document can then be imported into the system, checked for exact replication as the approved, signed document, and put into review. It is not the best idea to scan versioned documents because the text will have to be recreated anyway for updating during the review cycle. If the numbering system has changed for the new system, the document history in the electronic system should identify the previous number and version and link it to the new number. This history links the e-file in the new system to the document from the former system.

The next procedures to address are those that require updating before their periodic review cycle. Many companies set priorities and determine which of these documents should enter the system first. Bear in mind, however, that they are still active in the older system and will continue to be until they enter the new system. These can be imported the same way as other SOPs.

70. Can We Load Secure Electronic Files from Our Hybrid System into Our New System, and, If So, How Should We Do It?

Yes. Protocols, amendments, reports, and other documents and records that must be available for reuse can be handled the same way as versioned documents, since companies need to have electronic files for updates. Again, if the earlier system has a feature that ensures security of the electronic file, it can be imported with a QC check for exact replication.

71. How Can We Bring in Paper Reports that Our CRO Has Prepared for Us and that Bear Numbers that Are Different from the Ones in Our System?

A good way to do this is to assign the document a number from your system. A scan will create a PDF that shows the vendor's number, but the document in the system will be searchable by the number you give it. Make sure you have an SOP that spells out how you number such documents as they come into the system. (See Chapter 3 for acceptable scanning resolutions.)

REFERENCES

1. Guidance for Industry, Part 11, Electronic Records; Electronic Signatures—Scope and Application, www.fda.gov, accessed July 18, 2009.
2. Guidance for Industry, Computerized Systems Used in Clinical Investigations, www.fda.gov, accessed July 18, 2009.
3. Nettleton, David, and Gough, Janet. *Risk Based Software Validation: 10 Easy Steps*, 2nd edition, 2006. Baltimore, MD: Parenteral Drug Association; and River Grove, IL: Davis Horwood International.

CHAPTER 6

PART 11 COMPLIANCE

INTRODUCTION

Keeping abreast of evolving requirements for the management of documents and records means understanding the regulations and the current thinking on issues. 21 CRF Part 11, Electronic Records; Electronic Signatures is, by its very nature, a vague and indefinitely worded regulation. In 1997, when it was first issued, industry wasn't quite sure how it applied. Furthermore, fears of Y2000 issues with computer systems fueled the decision to remain paper-based, and the FDA itself was tasked with learning how exactly this law applied. In more than a decade, technology has improved, industry best practices have evolved to satisfy Part 11 requirements, and the agency has further defined its expectations for compliance with this law. This chapter addresses compliance with Part 11 and industry standards. It answers the following questions.

1. What documents does the FDA use to ensure compliance?
2. Does the government offer any advice for companies?
3. Does the FDA endorse international standards?
4. Does 21 CFR 11 apply to just those systems acquired after the final rule was made effective in 1997?
5. What does the regulation mean when it uses the term "grandfathering"?
6. What are computer system industry standards?
7. If a small company still uses paper records and does so compliantly, will the FDA ever mandate that it employ electronic record keeping?
8. Does Part 11 apply to systems that can print records from a portable storage medium such as a jump drive?
9. Does 21 CFR Part 11 apply to instruments not connected to computers but that have microprocessors inside?
10. What does "data integrity" mean?
11. Once an electronic record is printed, initialed, and dated, can the electronic record be discarded?
12. What responsibility does a company have for electronic records once they are submitted to the FDA?

Managing the Documentation Maze, By Janet Gough and David Nettleton
Copyright © 2010 John Wiley & Sons, Inc.

13. Does 21 CFR Part 11 permit companies, at their option, to submit to the FDA any required record in electronic format?

14. Does 21 CFR Part 11 apply only to those systems that employ electronic signatures?

15. Is there a deadline for implementing electronic signatures?

16. Should a company certify that every associate who signs electronically understands that the electronic signature is legally binding?

17. Can a single restricted login suffice as an electronic signature?

18. How do electronic signatures work?

19. What forms of electronic signature are available?

20. What is the primary difficulty with electronic signatures?

21. How does 21 CFR Part 11 relate to paper fax transmissions?

22. How does 21 CFR Part 11 relate to a fax transmission received electronically?

23. Is a scan of a paper document an electronic record?

24. Is a scanned paper containing a handwritten signature an electronic signature?

25. Can the technology used to capture a handwritten signature on a credit card receipt at a retail store be used as an electronic signature?

26. What data must be provided when an electronic signature is displayed?

27. Is an electronic signature the same as a digital signature?

28. Can a digital signature verify that the document hasn't been altered after signing?

29. When paper and electronic records are used, which is the "official" record?

30. Does outsourcing the hosting of a system make it an open system?

31. Does external access by a vendor for maintenance work (e.g., using a modem) to a computer system make that an open system?

32. What can a company do if it suspects it's not fully compliant?

33. What is data conversion?

34. What are some examples of audio data that may be captured in the pharmaceutical industry?

35. How is software categorized in the GAMP standard?

36. Is there a return on investment (ROI) for implementing a computer system?

37. What types of electronic records does 21 CFR Part 11 apply to?

38. How do you notify the FDA that you are using electronic signatures?

39. After a company notifies the FDA that it plans to use electronic signatures, does it have to wait until the FDA approves or acknowledges the notification before it can actually employ electronic signatures?

40. Do you have to notify the FDA for every system that uses signatures?

41. Does a company that uses electronic signatures have to inform the FDA of all users that have this capability?

42. Is an electronic signature made with a user name and password legally binding like a handwritten signature made with a pen?

43. What are the industry standards for electronic signatures?

44. What is a digital signature?

45. What can make a password more secure?

46. What is a biometric signature?

47. What are the advantages to biometric identification?

48. When many electronic signatures are required in a contiguous session, can one user have permission to copy the other signatures and paste them to a record?

49. How does logging on to a system differ from an electronic signature?

50. If someone leaves the company, can we reissue a user name to a new hire?

51. What are the industry standards related to user inactivity?

52. What does electronic security address?

53. What exactly are the industry standards for security for electronic systems?

54. What are threats to security?

55. What is role-based security?

56. When confidential data are copied to a laptop computer, how can security be ensured?

57. What is "hashing"?

58. What are the industry standards for passwords?

59. What are the industry standards for security logs?

60. What are the industry standards for data transferred to laptops and removable media?

61. What is the industry standard date format?

62. What are the industry standards for Audit Trails in electronic systems?

63. Must date and time on e-records be local time?

64. Why is an audit trail important?

65. What data should an audit trail support?

66. For clinical data management systems, where does the audit trail begin, after first entry or after the data have been verified?

67. What is an appropriate audit trail for an Excel Spreadsheet?

68. What data must an audit trail capture whenever a record is modified?

69. Does an audit trail always have to be electronic?

70. Are there any instances where it may be acceptable to have a paper audit trail for an electronic record?

71. Does the audit trail need to record all data entered by a user?

72. If an electronic record is annotated electronically, much like a sticky note on a paper record, is the annotation part of the record and subsequently subject to an audit trail?

73. What are the industry standards for data transfer?

74. How do we make sure all users know how to use a computer system?

75. If we have a hybrid system, how do we train users?

76. Does everyone have to be trained in all features of a system?

1. What Documents Does FDA Use to Ensure Compliance?

The FDA uses the Code of Federal Regulations (CFR), specifically GXP regulations that are referred to as predicate rules, and 21 CFR Part 11 Electronic Records; Electronic Signatures. Most important, compliance for 21 CFR Part 11 is measured against industry standard practices. These industry standards define the computer system security and operational norms that ensure data integrity. A comprehensive list of industry standards is not published in the CFR. Many industry standards are included in GXP, Health Insurance Portability and Accountability Act (HIPAA), and other regulations related to the many agencies of government.

2. Does the Government Offer Any Advice for Companies?

The government issues guidance documents on a full spectrum of industry activities and also removes them when the current thinking changes, or it issues updates. FDA Guidance documents fall into the following categories (1):

- **FDA Guidance Documents: General and Cross-Cutting Topics**
  General and Cross-Cutting Topics
- **Guidance for Industry**
  Animal and Veterinary
- **Guidance Documents**
  Cosmetics
- **Guidance, Compliance, and Regulatory Information**
  Drugs
- **Guidance Documents**
  Food
- **Guidance Documents (Medical Devices)**
  Medical Devices
- **Industry Guidance**
  Radiation-Emitting Products
- **Biologics Guidances**
  Vaccines, Blood, and Biologics

Guidance documents are not legally binding, and the FDA issues a disclaimer asserting this fact. (See Box 6.1.) The FDA is also accessible through its website (www.fda.gov). It's possible to query questions via e-mail, and the FDA will respond directly.

3. Does the FDA Endorse International Standards?

Yes. The FDA's website also makes the International Conference on Harmonization (ICH) guidance documents available. The FDA encourages compliance with ICH, but does not mandate it (2).

4. Does 21 CFR 11 Apply to Just Those Systems Acquired After the Final Rule Was Made Effective in 1997?

No. The regulation applies to all systems. However, the February 2003 guidance document indicates that the FDA is not enforcing Part 11 requirements on pre-1997 systems, provided that they don't have significant risk.

5. What Does the Regulation Mean When It Uses the Term "Grandfathering"?

"Grandfathering" refers to the possibility that Part 11 may not apply to a system in place before the rule became effective. Part 11 itself does not allow for grandfathering of legacy systems, so systems in place before August 20, 1997 must be made compliant.

6. What Are Computer System Industry Standards?

Computer system industry standards are best practices for software functionality and computer use. Industry standards relate to security, audit trails, and data transfer, and they address 21 CFR Part 11.

7. If a Small Company Still Uses Paper Records and Does So Compliantly, Will the FDA Ever Mandate that It Employ Electronic Record Keeping?

The FDA has already mandated that all documentation going to the government must be electronic. This means that any company in clinical testing or marketplace that files such documents as Investigational New Drug Applications (INDs), Common Technical Documents (CTDs), and annual reports must have electronic transmission capability. It is likely, as well, that companies that hold onto paper record keeping will fall behind and the inability to use electronic record keeping will impair their ability to maintain a viable business position. In reality, virtually all paper-based systems used today have electronic record components to make forms, compile data, report information, and create updates. This means the system is really hybrid.

8. 2. Does Part 11 Apply to Systems that Can Print Records from a Portable Storage Medium Such as a Jump Drive?

Part 11 does not apply to electronic systems that cannot be used to create an electronic record. If an electronic system can create an electronic record and send it somewhere else, then Part 11 is applicable.

9. Does 21 CFR Part 11 Apply to Instruments Not Connected to Computers but that Have Microprocessors Inside?

If such instruments do not generate electronic records or generate records that are not subject to the GXP regulations, then Part 11 does not apply.

10. What Does "Data Integrity" Mean?

Data integrity means that the electronic records generated and managed by the systems are trustworthy. The FDA looks at systems used for drug distribution, drug approval, manufacturing, and quality assurance to ascertain product quality and safety. What this means is that most major systems are critical: these include Laboratory Information Management (LIMS), Electronic Document Management System (EDMS), Clinical Data Management System (CDMS), and others used to generate and store information.

11. Once an Electronic Record Is Printed, Initialed, and Dated, Can the Electronic Record Be Discarded?

No. Electronic records must be maintained regardless of any printed records created from them. The electronic record is most likely the one to be updated, searched for, and used in business processes.

12.   What Responsibility Does a Company Have for Electronic Records Once They Are Submitted to the FDA?

The responsibilities are the same as for paper submissions. Once records are submitted to the FDA, the corresponding source records require control to ensure their reliability.

13.   Does 21 CFR Part 11 Permit Companies, at Their Option, to Submit to FDA Any Required Record in Electronic Format?

No. The types of records that the FDA will accept in electronic format are defined in regulations separate from Part 11.

14.   Does 21 CFR Part 11 Apply Only to Those Systems that Employ Electronic Signatures?

No. The regulation applies to all systems that manage electronic records, whether they use electronic signatures or not. Electronic signatures are optional under the regulation. Most companies use systems with electronic records because electronic records make updating and revisions to document easy and allow for the replacement of paper records.

15.   Is There a Deadline for Implementing Electronic Signatures?

No. Manual signatures on paper are acceptable if they link to electronic records so signers cannot repudiate records.

16.   Should a Company Certify that Every Associate Who Signs Electronically Understands that the Electronic Signature Is Legally Binding?

A company is only required to confirm the use e-signatures in the organization as a whole. As part of training for electronic record-keeping systems, trainees can be asked to sign a statement of accountability that attests to their understanding of the legally binding nature of an s-signature.

17.   Can a Single Restricted Login Suffice as an Electronic Signature?

Part 11 §11.50 says that (a) signed e-records must contain information associated with the signing that indicates the printed name of the signer, the date/time, and the intent and (b) this information must be included in any human readable form of the record (3).

18. How Do Electronic Signatures Work?

An electronic signature, which is most often created with a user name and password, may be used in place of a handwritten signature to make an approval. An electronic signature may also be a biometric signature, such as a finger print scan or a retinal scan. An electronic signature positively identifies the user, in a similar way as to when a user logs onto a computer system. In order to perform an electronic signature, the system must display the electronic record, the name of the intended user who is to perform the electronic signature, and a statement of testament that indicates the meaning of the electronic signature. Once the user enters their password or biometric, the system authenticates that the password is correct for that user name in order to confirm the user's identity, the electronic signature is permanently attached to the electronic record, and both the electronic record and the electronic signature are locked to prevent modification. The electronic signature is available for display or printing any time the electronic record is displayed or printed. The electronic signature contains the user name, date, time, and statement of testament.

19. What Forms of Electronic Signature Are Available?

Most electronic signatures use a public user name and private password. The password may be replaced with a biometric such as a fingerprint or iris scan. A token or keycard may be used to supplement or replace the user name but cannot be used in place of the password.

20. What Is the Primary Difficulty with Electronic Signatures?

Passwords are forgotten or stolen. In a single year, approximately 40% of help desk calls are related to forgotten passwords. Additionally, the lack of paper tends to make people feel insecure; but in reality, electronic signatures are more secure than handwritten signatures.

21. How Does 21 CFR Part 11 Relate to Paper Fax Transmissions?

If an electronic record is printed and then faxed to a second company that receives it on a standard paper fax machine, Part 11 does not apply. Part 11 specifically excludes the transmission of paper records. If the data received by the second company are used to support good practice (GXP) data or there is risk involved related to decision-making based on this information, predicate rules would apply to the paper record. Note that HIPAA regulations for privacy do apply to fax transmissions.

22. How Does 21 CFR Part 11 Relate to a Fax Transmission Received Electronically?

If a company receives a fax on a computer (e.g., fax server) and saves the data in a file, which is then printed, Part 11 does apply. If the data received are used to support

GXP data or there is risk involved related to decision-making based on this information, Part 11 would apply to the electronic record.

23. Is a Scan of a Paper Document an Electronic Record?

Once a scan of a paper document is made—for example, the scan makes a PDF or Tiff file—the file is an electronic record. Once the electronic record is verified to be a replica of the paper document, the electronic record is considered equal to the paper document.

24. Is a Scanned Paper Containing a Handwritten Signature an Electronic Signature?

No. Scanning does not perform authentication of the handwritten signature. The scanned image is an electronic record that is subject to Part 11.

25. Can the Technology Used to Capture a Handwritten Signature on a Credit Card Receipt at a Retail Store Be Used as an Electronic Signature?

No. Digital ink does not provide a way to authenticate the signer. However, there are systems in which handwriting captured with digital pens can be used as a means of document approval.

26. What Data Must Be Provided When an Electronic Signature Is Displayed?

The signer's printed name, date and time of signing, and meaning of the signature must be present.

27. Is an Electronic Signature the Same as a Digital Signature?

An electronic signature is not the same as a digital signature, but they both provide for user authentication. A digital signature involves encryption and certificates that are bound to the electronic record. A digital signature usually requires the user to enter a password or biometric in order to confirm the user's identity.

28. Can a Digital Signature Verify that the Document Hasn't Been Altered After Signing?

Yes. A digital signature is based on a set of rules and a mathematical algorithm so that the identity of the signatory and integrity of the data is verifiable. Signature generation uses a private key to generate a digital signature. Signature verification then uses a public key that corresponds to the private key. Each user has both a private and public key. The private key is necessary for signature generation.

29.   When Paper and Electronic Records Are Used, Which Is the "Official" Record?

When a system has both paper and electronic copies of a document, both records are official. However, the electronic records are more easily retrieved, searched, and updated, and the electronic records get more use. Since the electronic record doesn't contain approvals, the paper records provide the official support for the approval of the record. If an electronic record is updated without having an updated paper record, the result is regulatory and legal liability. Hybrid systems are thus more cumbersome than fully electronic systems that employ both electronic records and electronic signatures.

30.   Does Outsourcing the Hosting of a System Make It an Open System?

A closed system has restricted access that is controlled by people responsible and accountable for the content of the records it houses. If others can access the system to read, modify, or remove records, then it is an open system. If outsourcing a system, a company would need to take measures to ensure that records cannot be read, modified, destroyed, or otherwise compromised by people not authorized to access the system. Before establishing such a system, users must ensure that the hosting company has procedures in place to protect against security violations. Users will also have to access the system remotely. If this occurs over the Internet, the system is still an open system. Open systems require encryption for transport.

31.   Does External Access by a Vendor for Maintenance Work (e.g., using a modem) to a Computer System Make that an Open System?

It is an open system if the vendor can make changes to data in the system. However, if the users exercise appropriate controls so that no data change can occur by any unauthorized user, including the vendor, the system is closed.

32.   What Can a Company Do If It Suspects It's Not Fully Compliant?

First it needs to become informed; then it needs to take action. Company staff can become educated by attending conferences and seminars. A gap analysis, performed by a qualified employee or consultant, will also reveal where the company is deficient. Armed with the correct knowledge, the company can take measures to move into compliance.

33.   What Is Data Conversion?

Data conversion means changing the format or content of data to match a new requirement. This often occurs as part of data migration. A common example is converting two-digit year values to four-digit year values. Data conversion may involve changing text date values to real date values, or converting dates with TBD (to be determined) to real dates. Data conversion often deals with blank values, as well as with values outside of accepted ranges. Data conversion must occur prior to adding data to a new system.

34.   What Are Some Examples of Audio Data that May Be Captured in the Pharmaceutical Industry?

Audio recordings of patient information or experimental observations are not commonplace, but in some instances they can occur. But audio conferences about projects, reports, and data are common. If the data that such recordings capture are subject to predicate rules, then the audio file should be saved to durable media, and then Part 11 applies as well.

35.   How Is Software Categorized in the GAMP Standard?

The Good Automated Manufacturing Practices (GAMP) standard is a guideline maintained by a technical subgroup of the International Society of Professional Engineers (ISPE). GAMP divides software into five categories and provides validation requirements for each.

1. Operating systems do not require validation since they are validated indirectly when software applications are validated.

2. Standard Instruments, such as barcode scanners, do not require validation, but their configuration must be recorded and their operation verified.

3. Standard software packages are software programs that are used to create applications (Microsoft Excel and FileMaker are common examples); they do not require validation of the software itself, but applications made with the software do.

4. Configurable software packages such as Laboratory Information Management Systems (LIMS) and Document Control Systems (DCS) require validation of the user-configured application.

5. Custom Software requires validation from the developer's perspective, and users of the configured software application must perform validation as described in categories 3 or 4 (4).

36.   Is There a the Return on Investment (ROI) for Implementing a Computer System?

Computer systems have an excellent ROI. The cost of compliance is low compared to costs for potential losses. During validation, procedures for disaster recovery are verified. Validation reduces labor costs by increasing employee efficiency and effectiveness and educates the system users, which saves money by making the organization more productive. The use of electronic records is less expensive than the use of paper records with respect to creation, organization, searching, and retention.

Computer systems are essentially the tools for prime document management and compliance with Part 11.

37. What Types of Electronic Records Does 21 CFR Part 11 Apply to?

21 CFR Part 11 applies to electronic records in any format stored on any type of nonvolatile (semipermanent) media. Electronic records may be any form of text, graphic, audio, video, or other data types.

38. How Do You Notify the FDA that You Are Using Electronic Signatures?

21 CFR Part 11, §11.100(c) General Requirements states: "(c) Persons using electronic signatures shall, prior to or at the time of such use, certify to the agency that the electronic signatures in their system, used on or after August 20, 1997, are intended to be the legally binding equivalent of traditional handwritten signatures. The certification shall be submitted in paper form and signed with a traditional handwritten signature, to the Office of Regional Operations (HFC–100), 5600 Fishers Lane, Rockville, MD 20857" (3).

39. After a Company Notifies the FDA that It Plans to Use Electronic Signatures, Does It Have to Wait Until the FDA Approves or Acknowledges the Notification Before It Can Actually Employ Electronic Signatures?

No. It is sufficient to give notification "prior to or at the time of such use" or typically within 30 days after a system goes into production The certification, either by an individual or by a business establishment, certifies in writing that electronic signatures are the legally binding equivalent of handwritten signatures. If the certification is somehow deficient, the FDA will contact the submitter and instruct on how to correct the deficiency. Many companies use the language in the Part 11 to attest to their intent to use electronic signatures.

40. Do You Have to Notify FDA for Every System that Uses Signatures?

A company that uses electronic signatures only has to notify the FDA that it is using electronic signatures once. The company must keep records that show each workforce member that uses electronic signatures understand that their electronic signature is the legally binding equivalent of their handwritten signature.

41. Does a Company that Uses Electronic Signatures Have to Inform the FDA of All Users that Have this Capability?

No. It is sufficient to inform the FDA that the company is using electronic signatures. It is unreasonable to expect notification every time the company has a new hire or transfers an employee who will access electronic records and sign them electronically. Regulated companies have to keep internal records of all people who use electronic signatures.

42. Is an Electronic Signature Made with a User Name and Password Legally Binding Like a Handwritten Signature Made with a Pen?

Yes. They are equally binding.

43. What Are the Industry Standards for Electronic Signatures?

Electronic signatures use the same user names and passwords to log onto a system. While a biometric identification may be used in place of the password, biometrics have not been widely adopted and have not become an industry standard. Both the user name and password components should follow the industry standards when used for login or for electronic signature.

44. What Is a Digital Signature?

A digital signature involves encryption and certificates that are bound to the electronic record. A digital signature usually requires the user to enter a password to confirm the user's identity. A digital signature may be used in place of an electronic signature, especially when approvals are required from persons that are not system users. In that case, a trusted third party generates certificates that are authenticated by the system.

45. What Can Make a Password More Secure?

Passwords are the weakest component in security. A user may have many passwords to remember; passwords change frequently, and their makeup is not standard language. The best way to make passwords more secure is to use biometrics in their place. A biometric may replace the password as the method of authentication, or the biometric may activate a keychain utility that securely enters a password into a standard password entry field. Longer passwords tend to be more secure, and pass phrases consisting of 15 characters or more may evolve as an option. Passwords should not be written as plain text and should not be sent by e-mail. Users may write down passwords, provided that they are encrypted. This is usually performed with character substitution based on a set of rules devised and known only by the user.

46. What Is a Biometric Signature?

A biometric signature is a form of authentication that relies on the unique physical characteristics of a person. The most common forms of biometric identification are fingerprint and retinal scan. Fingerprint scans do not record the user's fingerprint pattern but instead take a statistical sampling of the whorl patterns of the finger.

While fingerprint scanners are not possible for users who are wearing gloves, iris scanners use a laser beam to penetrate goggles, glasses, and the eye to read the rod and cone patterns in the back of the eye. Biometrics using voice recognition, facial recognition, and DNA have not evolved to the point of being reliable and low cost.

47. What Are the Advantages to Biometric Identification?

Biometric identification is currently the best form of identification because it doesn't rely on the user remembering anything. It is very difficult to fake without the user's knowledge. Biometrics have not become an industry standard, and this may be due to concerns over civil rights and the use of the personal data collected by them. Biometric signatures are unique to every individual; they can't be shared; and they don't have to be changed periodically. Furthermore, users don't need to remember them because they can't forget them; they can't lose them and others can't find them, steal them, or guess them.

48. When Many Electronic Signatures Are Required in a Contiguous Session, Can One User Have Permission to Copy the Other Signatures and Paste Them to a Record?

No. Electronic signatures may not be copied by anyone. In the event that a person who must sign a document is unavailable, that person can assign a designee. Most companies give more than one person authority to sign a record and maintain a log or reference document that indicates who can sign for which records and who are alternate signatories.

49. How Does Logging on to a System Differ from an Electronic Signature?

Logging into a computer system and performing an electronic signature have common elements but are very different. Both use a user name and password to authenticate a user. An electronic signature permanently links to a electronic record, requires the user to agree to a statement of testament that describes the meaning of the electronic signature, and locks both the electronic record and the electronic signature from further modification.

50. If Someone Leaves the Company, Can We Reissue a User Name to a New Hire?

User names must uniquely identify an individual. They should not represent a group of users, or a generic user such as an administrator. If a person leaves a company, the user name may not be reissued to another person of the same name. So if Mary_Doe leaves the company, and another Mary Doe later joins it, the second Mary Doe must have a different user name, perhaps Mary2_Doe. This convention ensures that all activity is traceable back to whoever performed an action in the system. Each user's signature is linked to the records he or she signed electronically. This does not mean, however, that an employee who leaves the company remains a user of the system. Once a person leaves the company, access to the system should be disabled.

51. What Are the Industry Standards Related to User Inactivity?

There are several industry standards related to user inactivity. Each software application should require the user to enter a password, such as login, after a period of inactivity, 20 minutes or less. The operating system should blank the screen and require the user to enter a password after a period of inactivity, usually 20 minutes or less. Additionally, if a user does not access a system for a predetermined period, usually 30 days, the system locks out the user and requires and administrator to unlock the account.

52. What Does Electronic Security Address?

Electronic security addresses four main areas.

1. Authentication validates that a user should have access to a system based on the user entering a private password.
2. Authorization restricts the user to predefined areas of the software based on user roles or groups.
3. Privacy means precautions intended to prevent observation and snooping.
4. Data integrity relies on software functions such as encryption, audit trails, and screen indication of data changes.

53. What Exactly Are the Industry Standards for Security for Electronic Systems?

As of 2008, security means the following:

- Access limited to authorized individuals (roles and privileges defined by data owners).
- No users with "God" role, no IT people with user system administrator role.
- Password minimum length (8 characters).
- Password makeup requirements (no words in dictionary, alphanumeric).
- Password change frequency (90 days).
- Password reuse frequency (1 year).
- Passwords are not displayed when entered.
- Passwords are not remembered by browsers and applications.
- Password is known only by individual user.
- Password encryption (upon entry, storage).
- Password cannot be copy and pasted.

- Passwords are not e-mailed or written down.
- Temporary passwords are unique.
- Temporary passwords must be changed at next logon.
- Temporary password expires (24 hours).
- User name appears on screen.
- User name is unique.
- User name is not deleted, just inactivated. Therefore, it cannot be reused.
- Automatic log off after period of inactivity (10–20 minutes).
- OS screen saver with password (10–20 minutes).
- Auto lockout after too many failed log on attempts; e-mail notification to system administrator/security staff (3–5 attempts).
- Logging of all user activity. When logging onto a system from a second location, the user is notified that they are logged in another location.
- Auto lockout of inactive accounts (30 days).
- Last log on displayed when logging on.
- The network is secure with respect to user access, Internet access, virus protection, and physical security.
- Removable media, including laptops and PDAs, have confidential data encrypted.
- Device checks confirm that once data starts from a device, another device doesn't take over.

54. What Are Threats to Security?

Security threats primarily result from system users, meaning "insiders" and "crackers" who are outsiders that break into systems via the Internet. Security threats include physical threats such as theft of backup media, removable media, and laptop computers.

55. What Is Role-Based Security?

Role-based security requires users to be assigned to roles or groups, and then permissions to access functions are granted to the role. When multiple users have the same security, users are more readily able to detect incorrect security settings.

56. When Confidential Data Are Copied to a Laptop Computer, How Can Security Be Ensured?

Confidential data should be encrypted before being transferred to the mobile media (laptop computer, CD, memory stick, etc.). A password, known only to the intended user, allows that user to reverse the encryption and access the data. In this manner, if the mobile media is lost or stolen, there is no way for other people to access the encrypted data. Therefore, the data remain confidential even if lost or stolen. IT assists the users with security by providing tools and policies, but the users are ultimately responsible for security.

57. What Is "Hashing"?

Hashing is a way of accessing data or for maintaining data security. A hash is a number generated from a string of text. The hash is substantially smaller than the text itself, and it is generated by a formula in such a way that makes it unlikely that some other text will produce the same value. Hashes play a role in security systems when they're used to ensure that transmitted messages have not been subject to tampering. A sender generates a hash of the message, encrypts it, and sends it with the message itself. A designated recipient then decrypts both the message and the hash, produces another hash from the same message, and compares the two hashes. If they're the same, there is a very high probability that the message transmitted intact.

58. What Are the Industry Standards for Passwords?

The current standard is 8 characters alphanumeric. A password must not be a dictionary word. And to further distinguish a password as such, many companies are including a symbol as well as letters and numbers.

59. What Are the Industry Standards for Security Logs?

Security logs are to record login, logout, password changes, lockouts, and changes to security settings. As of press time, there isn't a standard for security log retention, but security logs are usually kept for a minimum of one year.

60. What Are the Industry Standards for Data Transferred to Laptops and Removable Media?

Data transferred outside of the company's physical environment must be encrypted. Operating systems and software utilities provide a means to encrypt a folder so that any contents are automatically encrypted. It is not advisable to encrypt entire hard drive partitions, especially those containing the operating system.

61. What Is the Industry Standard Date Format?

The worldwide date format has the month identified by letters, usually three letters, such as Jan for January. The order of the day, month, and year is different throughout the world, but with the year unambiguous this does not present a problem. All year indications must be four digits.

62. What Are the Industry Standards for Audit Trails in Electronic Systems?

As of 2008, Audit Trails means the following:

- Audit trail records the creation, modification, or deletion of electronic records.
- Audit trail records user name, date, time, previous data, new data, and reason for change (if required by predicate rules).
- All computers must be synchronized to a standard time source.
- The indication of changed data is known to the user by on-screen indication, not just in audit trail.
- Application is aware if data integrity has been compromised, database encryption.
- Some audit trails also record the reason for the change when this is a requirement associated with predicate rules.
- Most important, the audit trail is read-only and can be viewed by the users.

63. Must Date and Time on E-Records Be Local Time?

The FDA says "You should implement time stamps with a clear understanding of what time zone reference you use. Systems documentation should explain time zone references as well as zone acronyms or other naming conventions." Firms need to implement procedures and controls to ensure that time stamps are both accurate and reliable and based on computer system clocks (5).

64. Why Is an Audit Trail Important?

Audit trails are required to record the creation, modification, and deletion of electronic records. Audit trails provide a history of a record, from conception through archiving. It allows the electronic record to be recreated for any point in time.

65. What Data Should an Audit Trail Support?

Part 11 §11.10(e) says that audit trails must be secure, computer-generated, and time-stamped to independently record the date and time of user entries that create, modify, or delete electronic records. Audit trail documentation must be accessible and retrievable. Audit trails should say who did what to a record and when they did it. GLP records also require an explanation of why. Part 11 does not specify the format for audit trails (3).

66. For Clinical Data Management Systems, Where Does the Audit Trail Begin; After First Entry or After the Data Has Been Verified?

It may be both. Clinical research organizations are mandated to comply with 21 CFR Part 11, which requires tracking the activity and ownership of electronic clinical data in audit trails. If you are using Remote Data Entry (RDE) software for data entry, or especially a web-based RDE, you need to exercise due diligence to protect your data from inadvertent or malicious changes.

67. What Is an Appropriate Audit Trail for an Excel Spreadsheet?

Excel has a built-in audit trail that tracks all changes to all cells. Once GXP data are entered into a spreadsheet, all data changes must be tracked using the audit trail, and all spreadsheet versions must be tracked using validation documentation.

68. What Data Must an Audit Trail Capture Whenever a Record Is Modified?

The audit trail must include the date, time, user, old data, and new data values.

69. Does an Audit Trail Always Have to Be Electronic?

The audit trail has to be electronic when the related record is electronic. 21 CFR Part 11 11.10(e) calls for a computer-generated audit trail (1).

70. Are There Any Instances Where It May Be Acceptable to Have a Paper Audit Trail for an Electronic Record?

No. Audit trails must be automatically generated by the computer, so there is no chance of user introduced fraud.

71. Does the Audit Trail Need to Record All Data Entered by a User?

No. Only data that are saved to the computer system need to be recorded in the audit trail. If data are entered, changed, or deleted prior to saving, they should not be recorded in the audit trail.

72. If an Electronic Record Is Annotated Electronically, Much Like a Sticky Note on a Paper Record, Is the Annotation Part of the Record and Subsequently Subject to an Audit Trail?

Electronic annotations are part of the electronic record. And like the record to which they are affixed, they need to be trustworthy and reliable. Often annotations carry significant information such as reviewer agreement or instructions about the record. While this information is part of the electronic record, they are generally not included in the audit trail because they did not actually change the data.

73. What Are the Industry Standards for Data Transfer?

Data transfer means the following:

- Limited and controlled delete capabilities.
- Data transferred outside of the intranet firewall is encrypted.
- Data taken offsite is encrypted (laptops, removable media).
- The system must include operational system checks to enforce correct sequencing of events and validity of input data.
- Date format dd-MMM-yyyy (10-Jan-2003).

74. How Do We Make Sure All Users Know How to Use a Computer System?

All users of a computer system must be trained to perform their specific functions according to written procedures. When a computer system is validated, a project team consisting of representatives of each user group is selected. The project team often obtains training from the software vendor and then develops new processes, which are incorporated into new procedures. These project team members are best suited to train the user groups. This is why most vendors state that they will "train the trainers." Training is necessary when new users must access the system, and when there are changes to a system.

75. If We Have a Hybrid System, How Do We Train Users?

A hybrid system has the same training requirements as a fully electronic system: Users must undergo training to perform their specific functions according to written procedures for the system.

76. Does Everyone Have to Be Trained in All Features of a System?

No. People must be trained to perform the activities for which they have privileges, regardless of the type of system. For instance, a document manager may have the authority to post documents to an electronic document repository or an intranet server as a PDF. Another system user that does not have access to perform this function would not have to be trained in this function.

REFERENCES

1. www.fda.gov/regulatory/Guidances, accessed July 19, 2009.
2. www.fda.gov/regulatory/Guidances, International Conference on Harmonization (ICH) Guidance Documents, accessed July 19, 2009.
3. 21 CFR Part 11 Electronic Records; Electronic Signatures, Final Rule, 1997.
4. Nettleton, David, and Gough, Janet. *Risk Based Software Validation: 10 Easy Steps*, 2nd edition, 2006. Baltimore, MD: Parenteral Drug Association; and River Grove, IL: Davis Horwood International.
5. Guidance for Industry, 21 CRF Part 11; Electronic Records; Electronic Signatures, Time Stamps (draft), February 2002.

CHAPTER 7

STANDARD OPERATING PROCEDURES

INTRODUCTION

The regulations tell companies what they must do, but they don't say how to do it. In short, they are prescriptive, not descriptive. The regulations all call for procedures, so companies are compelled to have them in place. Standard Operating Procedures (SOPs) show how companies apply what the regulations say to their specific operations. As such, they must be clear, they must be true, and they must work together. SOPs also reflect good business because they are safeguards for ensuring that processes and activities occur as they should, so that they yield the same results every time. Surely then, having sound procedures in place makes for good compliance; moreover, and perhaps equally important, they help keep companies on track and functioning optimally. This chapter addresses the procedural infrastructures that companies must have. It answers the following questions:

1. Why is there so much emphasis on SOPs and where can we find the requirements?

2. How binding are SOPs for our company?

3. How can we determine which SOPs we need?

4. How many SOPs should a company have?

5. At what point do we need to put SOPs in place?

6. If we follow the regulations, and do quality assurance audits, will we be compliant?

7. What are global SOPs?

8. What are quality manuals, and why do companies have them?

9. What's the difference between a work instruction, an SOP, and a policy?

10. Do work aides, which are documents that help interpret our SOPs, need controls?

11. Is it a good practice to maintain SOPs and IOPs?

12. Is it acceptable to have SOPs but not to follow them?

Managing the Documentation Maze, By Janet Gough and David Nettleton
Copyright © 2010 John Wiley & Sons, Inc.

13. If our equipment already has a manual that tells how to use it, why should we write an SOP?

14. Must we have an SOP on SOPs?

15. Should every area of a company have its own physical set of SOPs?

16. How do we prevent duplicate or overlapping SOPs?

17. Shouldn't an SOP be a perfect document before it's put in place?

18. Is it a good practice to post our SOPs to our company's intranet so that anyone can access them?

19. If we are rolling out a new process that has four phases, and we have only implemented phase 1, can we write an SOP to cover all four phases that will eventually be in place?

20. Since we have a multicultural, global workforce, should our SOPs all be in English, as the universal business language?

21. Can a form serve as an SOP?

22. Can we deviate from our SOPs?

23. Is it okay to have pilot SOPs that only a few people follow, so that we can test the procedure?

24. How should we control reference documents that are instructional?

25. Does a sponsor need SOPs in place if CROs are managing clinical activities?

26. If a CRO relies on our SOPs, is it necessary that a CRO has its own SOPs?

27. What's the best way to control procedures?

28. Why do companies follow a document control process?

29. What's the difference between superseding and obsoleting an SOP?

30. What do we do if we want to permanently obsolete an SOP and it's cross-referenced in other SOPs?

31. Can we make our SOPs effective on the day they are signed off?

32. How much time should we allow between approval of an SOP and issuing an effective date?

33. Can we get rid of training forms if we scan them into our document management system?

34. How often do SOPs require review?

35. Does the FDA expect us to keep comments on SOPs that were generated during the review process?

36. How can you paginate attachments to SOPs?

37. Should forms be contained within SOPs?

38. How can we link who completed an electronic form to the form itself?

39. Is it okay to annotate a working copy of an SOP to reflect an improvement until we can get the new version through the system?

40. What should we do when we recognize the need for a new procedure and need it faster than our system permits?

41. How do you handle the situation where there is a new procedure that doesn't reflect what we actually do?

42. If you make changes to an SOP, do you need to look at any other documents?

43. Should the version number change when an SOP undergoes review and there are no changes to it?

44. How can we know what changes we've made to SOPs in sum since the first version?

45. Should SOPs have a table of contents?

46. How long should SOPs be?

47. What happens if we identify a need for an SOP and then decide we don't need it?

48. Can I revise an SOP from the copy I have on my workstation, since I am the author?

49. If a person who has authored SOPs leaves a company, what happens to those documents in the next review cycle?

50. Do the regulators define any specific format for SOPs?

51. What are the conventions for SOP formats?

52. Should we simply put N/A if an element in an SOP doesn't apply?

53. Is it good to use a template that has a series of tables with item and action?

54. Should we update an SOP if we find a spelling or mechanical error?

55. What documentation practices should SOPs cover?

56. What SOPs do we need for our document and record-keeping systems?

57. What SOPs should you have to support document and record-keeping systems?

58. Must we include training requirements in every SOP?

59. What happens if people undergo training, but fail to follow a procedure?

1. Why Is There So Much Emphasis on SOPs and Where Can We Find the Requirements?

Written procedures are a requirement of all the regulations that drive the therapeutic product industry. The International Organization for Standardisation (ISO) sums it up well: "Say what you do, and do what you say" (1). The predicate rules in the United States Code of Federal Regulations (CFR) all call for procedures, as do the International Conference on Harmonization (ICH) guidelines (2) and the regulations for most companies involved in developing, manufacturing, or distributing products in the health-care industry.

2.   How Binding Are SOPs for Our Company?

If you put an SOP in place, you must adhere to it. SOPs are legally binding for the company. They are based on the requirements in all the binding regulations for therapeutic products that say there "shall be procedures." The Office of the Inspector General also gives this message: "Procedures are really the law for the company. Companies are very tightly bound by their procedures, which are required to be established. By regulation, such procedures must be written, trained to, followed, documented, and supervised in execution. They must be approved by the quality unit of a company. The agency (FDA) holds a company accountable for this compliance" (3).

3.   How Can We Determine Which SOPs We Need?

There is no magic formula. SOPs must reflect what you do relative to your company. In sum, procedural documents form the backbone of the operation, whether you are in preclinical, clinical, manufacturing, distribution, or tracking products in the marketplace. If you are a contractor providing services along this continuum, you are also subject to the regulations that call for SOPs. These documents tell "how it happens here" (4).

4.   How Many SOPs Should a Company Have?

There is no correct answer, because each company should have SOPs that address all areas of their business. SOPs, work instructions, and other informative documents work together to promote a uniform working platform and provide a basis for consistent training.

5.   At What Point Do We Need to Put SOPs in Place?

It depends on your level of comfort with risk. SOPs provide the basis for doing work in a uniform manner. If your work is not uniform, how good are the results? If you are doing straight research, procedures are not a requirement, except for those that support chemical hygiene, and safety, and companies typically put documents in place for compliance with 29 CRF 1910. However, once you have proof of concept, it's time to start thinking about standardizing what you do. Good controls actually help the business function, they help ensure that your development activities will be compliant, and thus they support your product from early development and beyond.

6.   If We Follow the Regulations and Do Quality Assurance Audits, Will We Be Compliant?

You must follow the regulations, to be sure. But while the regulations tell you what you must do; they don't tell you how to do it. Furthermore, they are explicit in calling for written procedures that are followed. Even though you may follow the regulations, you must have SOPs that explain how you do so. The Office of the Inspector General issued a guidance on Developing a Compliance Program for

Pharmaceutical Manufacturers which called for these elements for an effective compliance program:

- Implementing written policies and procedures
- Designating a compliance officer and compliance committee
- Conducting effective training and education
- Developing effective lines of communication
- Conducting internal monitoring and auditing
- Enforcing standards through well-publicized disciplinary guidelines
- Responding promptly to detected problems and undertaking corrective action

Note that the very first item is written policies and procedures. The additional elements, in turn, should be reflected in written policies and procedures (5).

7. What Are Global SOPs?

These are documents that address the big picture. Often a parent company will issue global SOPs to its sites, and the sites will then, in turn, create SOPs that define how things happen there. (See Figure 7.1.) Another way to look at the structure of the SOP system is as a pyramid: The top-level or global documents give the "big picture," and SOPs show the backbone of what actually happens across the company, while work instructions tell how to perform specific tasks that have defined start and stop points and one or two people can carry them out.

8. What Are Quality Manuals, and Why Do Companies Have Them?

Quality manuals are comprehensive, big picture documents that show how quality drives the operations, from the CEO down. They tell what regulations a company adheres to, and they indicate how quality is communicated throughout the company. They address quality records as well. They are required for device companies and companies adhering to ISO standards. Right now there is no dictate that pharmaceutical companies put them in place, but that may change as GMPs evolve. Required or not, many pharmaceutical and biotech companies have quality manuals because they make sense, particularly if the companies supply product or services to a client base.

9. What's the Difference Between a Work Instruction, an SOP, and a Policy?

Companies often make distinctions as to document types. Typically, a policy is a top-level document that describes what is to be done but now how it is to be done. An SOP is a high-level document that describes how a process occurs. A work instruction provides details that describe a process in detail, typically with defined start and stop points. Regardless of what a company calls its documents, they require tight controls.

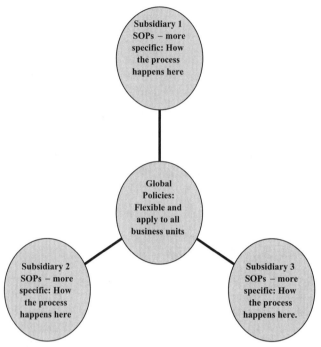

Figure 7.1.   Organization of SOPs for multiple sites.

10.   Do Work Aides, which Are Documents that Help Interpret Our SOPs, Need Controls?

Yes, any "how to" document that gives instructions needs to be controlled. "How to" documents include work instructions, instructions, global policies, franchise documents, and more. It doesn't matter what you call them; they are subject to the same processes as your formal SOPs.

11.   Is It a Good Practice to Maintain SOPs and IOPs?

Some companies maintain two sets of procedures: Standard Operating Procedures and Internal Operating Procedures. The difference is that SOPs are what are made available during inspections and audits, and IOPs reflect what is actually done in more detail. There's no need to cover the same processes in two sets of documents, however. A danger is that you will update one document, but fail to adjust its coordinate document. Auditors may and often will ask to review both SOP and IOPs because all instruction documents are relevant to the processes under inspection. Such redundancy also implies that the company does not want to reveal what it actually does, and that is a red flag for investigators.

12.   Is It Acceptable to Have SOPs but Not to Follow Them?

No. If a procedure is in place, regardless of whether the regulations call for it, it is binding, and you must follow it. If you have procedures that are superfluous, the best thing to do is retire them and create SOPs that you want to follow because it makes good business sense.

13. If Our Equipment Already Has a Manual that Tells How to Use It, Why Should We Write an SOP?

The regulations for all therapeutic products are specific in requiring that companies have SOPs that tell how activities occur, so companies need to comply with the law. But more importantly, SOPs make sense. A manual may cover more than your process calls for; it will not detail scope and responsibilities as they apply to your company; and you have no guarantee that the manual will stay current with your actual practices.

14. Must We Have an SOP on SOPs?

Some industry experts believe that all companies should have an SOP on SOPs, a document that explains the system for creating, reviewing, approval, and controlling SOPs. But there is no specific dictate for such a document. What is more important is that SOPs cover all facets of documentation, so a company may have a document management SOP that includes SOPs. This model is particularly effective when there is an e-system in place and all documents are contained within it.

15. Should Every Area of a Company Have Its Own Physical Set of SOPs?

It's not necessary today with hybrid and fully electronic systems. It used to be convenient to have a full manual of SOPs for each area when systems were purely paper-based. Now, because electronics have made controlled accessibility possible, all areas of a company can adhere to common SOPs. Since it's unlikely that every SOP will apply to every area, SOPs can tell where a procedure applies in the Scope section. Such an approach to procedures prevents overlaps where more than one SOP addresses the same process. In the past, such a practice has proved to be troublesome, since through the revision process such replicate SOPs often become modified and eventually may be significantly different and even contradictory.

16. How Do We Prevent Duplicate or Overlapping SOPs?

The key to ensuring that SOPs do not duplicate information is to have each SOP address a process or a part of a process. SOPs must also identify where the SOP applies. Furthermore, if an area of a company determines that it needs an SOP to cover a certain process, the first thing to do is search the system to see if there is an SOP that covers related information that can be modified to incorporate the new information and the scope of application.

17. Shouldn't an SOP Be a Perfect Document Before It's Put in Place?

It's better to put an SOP in place even if you don't think it's 100% perfect. An SOP in place is better than a planned SOP, and a planned SOP is better than no concept of an SOP. Putting an SOP in place lets you perform the process in real time, and you'll see what needs correcting pretty quickly. As soon as you do, you can put it into review for revision, while you still have the original in place. The next version will be better. That's the process of SOPs—they are really organic documents that grow and change as the company's processes do. With that said, you should not put an SOP in place that you know is incorrect. It is one thing to lack detail, but is quite another to provide incorrect information.

18.   Is It a Good Practice to Post Our SOPs to Our Company's Intranet So that Anyone Can Access Them?

It can be if you exercise some controls. Posted documents must be exact replicas of the official document, and they must be posted in a form that doesn't allow for alteration of the document (read only). It they are printed, it's possible to include a time and date stamp with the words "Valid only on print date." It's never a good idea to have uncontrolled copies of SOPs because you run the risk of people not following the correct SOP, and that is definitely something that inspectors look for.

19.   If We Are Rolling Out a New Process that Has Four Phases, and We Have Only Implemented Phase 1, Can We Write an SOP to Cover All Four Phases that Will Eventually Be in Place?

No SOP should be approved and made effective if the process has not been verified or the users of the SOP cannot perform the procedure. It is better to issue an SOP for the current phase and then update it as the subsequent phases roll out.

20.   Since We Have a Multicultural, Global Workforce, Should Our SOPs All Be in English, as the Universal Business Language?

This is a matter of quality. Which language will best ensure compliance with the SOPs? SOPs need to be in the language that the users understand.

21.   Can a Form Serve as an SOP?

Many companies are moving toward this model, where forms are instructional. Such companies build directions into the forms themselves, so that the forms become the documentation for the procedure. It's important that the procedures themselves be identified in a master list or log, and that training on the forms is the same as for any other procedure.

22.   Can We Deviate from Our SOPs?

Deviations are either planned or unplanned. A company may have a special project that requires a somewhat different process than the SOP in place specifies. In such an instance, for the special project, the company can write a planned deviation in advance of the project. Deviations that are unplanned—say a failure to do a QA check on gowning—require documentation of the deviation, investigation, and resolution of the outcome.

23.   Is It Okay to Have Pilot SOPs that Only a Few People follow, So that We Can Test the Procedure?

Yes, provided that your SOP on SOPs explains how the pilot SOP system works. A company can approve a pilot SOP and train a limited number of operators to test the new process. The existing, effective SOP remains in place, and the rest of the workforce follows it until the pilot either becomes effective and supersedes it or there is a decision not to implement the pilot process.

24.   How Should We Control Reference Documents that Are Instructional?

Reference documents can provide important information. But if they are not binding, you can say so, much the way that the FDA does in its guidance documents. In short, your formal procedures are binding; reference documents do not have to be. Reference documents are a good place to put information that is nice to have, but does not affect the outcome of a process. For instance, preferences for speaking about products can go into a reference document such as a style guide. As another example, if you say "put one space between sentences" in an SOP, you make the directive binding, and that's just silly. Such information is best reserved for reference documents.

25.   Does a Sponsor Need SOPs in Place If CROs Are Managing Clinical Activities?

Yes. While there is no requirement in the regulations, guidance documents clearly establish this expectation. But if you think about it, a sponsor needs to ensure that its products, in any phase of testing, are subject to correct controls so that the outcome is valid. A sponsor can rely on a CRO's SOPs, but only if the SOPs are approved equivalents of the sponsor's own procedures.

26.   If a CRO Relies on Our SOPs, Is It Necessary that a CRO Has Its Own SOPs?

Yes. As a service organization, a CRO must be compliant with the regulations that govern its activities. The regulations for this industry require SOPs.

27.   What's the Best Way to Control Procedures?

Procedures are documents that have a life cycle that includes review and reversioning. Before anyone writes a procedure, there should be a search of a log or electronic system to make sure that no one else is writing a similar document, or that the new document will be redundant or contradictory to one that's already in place. For revisions, the search should provide evidence that no one else is currently revising the document. So the first step is approval of the concept for creating a new or revising an existing document. A new document is drafted in a blank template, and a revision is to a controlled copy of the electronic file of the previous version. Review by designated reviewers and revision are part of the cycle. When there is concurrence, designated approvers sign off, and the document history log gets updated. Next there should be a training period, followed by release of the new document or new version of a document. Finally, the document is issued as "effective," and any previous versions are archived.

28.   Why Do Companies Follow a Document Control Process?

The regulations are quite clear on what has to happen. Even though a company may not be a "manufacturer," the standards for products, regardless of the stage of development, need to be consistent. Here's what the regulations say:

Sec. 820. 40: Document Controls.   Each manufacturer shall establish and maintain procedures to control all documents that are required by this part. The procedures shall provide for the following:

(a)   Document Approval and Distribution.   Each manufacturer shall designate an individual(s) to review for adequacy and approve prior to issuance all documents established to meet the requirements of this part. The approval, including the date and signature of the individual(s) approving the document, shall be documented. Documents established to meet the requirements of this part shall be available at all locations for which they are designated, used, or otherwise necessary, and all obsolete documents shall be promptly removed from all points of use or otherwise prevented from unintended use.

(b)   Document Changes.   Changes to documents shall be reviewed and approved by an individual(s) in the same function or organization that performed the original review and approval, unless specifically designated otherwise. Approved changes shall be communicated to the appropriate personnel in a timely manner. Each manufacturer shall maintain records of changes to documents. Change records shall include a description of the change, identification of the affected documents, the signature of the approving individual(s), the approval date, and when the change becomes effective (6).

29.   What's the Difference Between Superseding and Obsoleting an SOP?

Supersede usually mean that a new version of the document is effective and the previous version is no longer effective. Obsolete usually means that no new version of the document will be made and the current version is no longer effective.

30.   What Do We Do If We Want to Permanently Obsolete an SOP and It's Cross-Referenced in Other SOPs?

Before an SOP can be made obsolete, all SOPs that reference it must be updated to remove the reference. The history of the obsolete document must also indicate why it is no longer active.

31.   Can We Make Our SOPs Effective on the Day They Are Signed Off?

It's not the best practice. There must be sufficient time to make sure everyone impacted by a new or revised SOP knows about it and undergoes training as appropriate.

32.   How Much Time Should We Allow Between Approval of an SOP and Issuing an Effective Date?

It depends on how quickly you can train the people who will use the SOP. You may be able to train the users the same day the SOP is made effective, or it may take a week or more. The key is to train the users before they use the SOP. This doesn't that mean all users of the SOP have to be trained by the effective date.

33.   Can We Get Rid of Training Forms If We Scan them into Our Document Management System?

As with all paper records, once they are scanned to make electronic records and those records are backed up, the paper record can be destroyed. Most companies keep paper records for a short time period such as when they are actively being accessed or are subject to audit. Most companies also realize the cost savings associated with only keeping electronic records and destroy paper records at regular intervals.

34.   How Often Do SOPs Require Review?

The regulations call for "periodic review." Industry in the United States has determined that a 2-year cycle is appropriate for SOPs, but some companies adhere to a yearly or 18-month cycle. In some countries, an initial SOP is reviewed after one year, and then the SOP goes into a 3-year cycle. More important, SOPs require review any time the processes that they control are modified.

35.   Does FDA Expect Us to Keep Comments on SOPs that Were Generated During the Review Process?

No. The regulations are clear in the requirement for review, and accordingly you must show that there is a review component to your SOP system. If you have an e-system, the review process is documented in the metadata. Some companies export a document for review, and they route a paper copy with a signature form. Others make a PDF copy and route it through the company's intranet, so reviewers can post comments. Either way, based on the review comments, the SOP draft changes until there is concurrence and a version that is approvable. The final version then receives approval. You don't need to keep reviewer's comments.

36. How Can You Paginate Attachments to SOPs?

This is a matter of preference and precedent. Most documents are created from MS Word. You can either continue numbering or restart numbering for the attachment. In some cases, such as a single page form, you may not want a page number on an attachment. As long as the SOP identifies the attachment and it is clear where the attachment begins and ends, you are free to paginate as you like.

37. Should Forms Be Contained Within SOPs?

Whether forms should be included in SOPs is a matter of opinion. Forms should have a unique document number and version number and should be controlled, and SOPs should reference those used in a process. Forms typically undergo revision faster than SOPs do. That's one reason companies keep forms separate, because revising the form could mean revising the SOP. Some companies that do handle forms this way state that "Revision to the form does not require revision to the SOP." If a form is in an SOP, then the SOP and form have to be updated together. If a form is separate from the SOP, care must be taken to update the SOP if updates to the form impact the instructions. In no cases should a form be in an SOP and also be separate.

38. How Can We Link Who Completed an Electronic Form to the Form Itself?

Depending on the electronic form used, you may be able to capture the user logged on to the operating system. You may have to prompt the user to enter their name or other identifier. Some forms are part of computer systems that require the user to log in, so the user is automatically captured.

39. Is It Okay to Annotate a Working Copy of an SOP to Reflect an Improvement Until We Can Get the New Version Through the System?

It's best not to. SOPs are binding for companies, and annotations violate the process for review and approval of changes. An option is to prioritize the review process for a new version, so that it goes through your system more quickly. If you violate your process for SOP generation, review, and approval in the process, however, it is a deviation, and you must document it.

40.   What Should We Do When We Recognize the Need for a New Procedure and Need It Faster than Our System Permits?

This is a deviation from your standard SOP on SOPs, and you must document the deviation. Make sure all the appropriate people have buy in for the new procedure, then write, approve, train, and implement as quickly as possible.

41.   How Do You Handle the Situation Where There Is a New Procedure that Doesn't Reflect What We Actually Do?

Suppose your boss pushes an SOP through the system for a process that you are not aware of, and then tells you to train your staff, but it's not workable for what you actually do. Several issues are at play here. If you have a procedure for generating an SOP, has the boss followed it? If so, why don't you know about it in advance? You may have a faulty SOP on SOPs because impacted areas need to know about planned procedures and ideally participate in the generation and review. You can then suggest revisiting the SOP on SOPs. But your problem may reflect personnel issues if the boss believes that he or she is immune from adhering to established procedures. In such a case, you may need to enlist (tactfully) the assistance of your QA group to stress that failure to adhere to standard procedures poses a compliance risk to the company.

42.   If You Make Changes to an SOP, Do You Need to Look at Any Other Documents?

Yes. Any documents referenced or affected by an SOP should be considered when making changes to an SOP. These include forms, templates, or other documents that are part of the process related to the SOP being updated. Documents rarely stand alone. For instance, a quality manual may talk about quality in general terms, but an SOP may give an overview of calibration within the company. A next-level document might be instructions for calibrating specific equipment, such as a scale. A calibration sticker is also part of the documentation.

43.   Should the Version Number Change When an SOP Undergoes Review and There Are No Changes to It?

Companies do not need to change a version number if no changes are made to the SOP. The review is recorded in the document's history. Not changing the version ensures that the training records always link to the correct document. That said, changing the version number for every review cycle is optional, and many companies build that feature into their systems to verify regular review cycles. Regardless, the SOP that addresses the handling of SOPs must tell how the company handles reversioning and training.

44.   How Can We Know What Changes We've Made to SOPs in Sum Since the First Version?

SOPs should contain a version history that details the types of changes made to each version. This document history is a roadmap that shows the evolution of the procedure.

45.   Should SOPs Have a Table of Contents?

Typically, SOPs have standard sections and therefore a table of contents is not necessary.

46.   How Long Should SOPs Be?

SOPs should be a short as possible. Most companies prefer to have SOPs that are less than 10 pages; but outside the United States longer SOPs are common. It's best to remember, however, that SOPs are working documents, not manuals.

47.   What Happens If We Identify a Need for an SOP and Then Decide We Don't Need It?

Once an SOP has been assigned a document number, that number may not be reused. The number and title and status remain in the log as a history of what you actually planned but then decided against. Your log should also explain the decision not to issue an SOP to justify why you identified an SOP but have no actual document.

48.   Can I Revise an SOP from the Copy I Have on My Workstation, Since I Am the Author?

Doing so is a weak practice on two counts. One, the document is an electronic file, even if you maintain an official paper copy; and as such, it requires Part 11 controls as they apply. The secure electronic file is the one that should be released for revision. Two, the copy you have on your workstation may not reflect every change that was made to the document before approval. For instance, does your system allow for mechanical changes such as for typos or spacing? Such changes may have been made to the official previous version, and you may not have them in your version, and are thus not revising the actual document in place.

49.   If a Person Who Has Authored SOPs Leaves a Company, What Happens to Those Documents in the Next Review Cycle?

Documents should not be owned and authored by a single person. SOPs describe a process, and maintenance of the SOP is the responsibility of all people that are affected by the document. When a document has to be reviewed, the people with the most process experience participate in the revision process.

50. Do the Regulators Define Any Specific Format for SOPs?

No. But while procedures may look somewhat different from company to company, they mostly adhere to the same conventions.

51. What Are the Conventions for SOP Formats?

SOPs have headers and footers that identify the type of document, the title, the document ID or number and version, pagination, and, for hybrid systems, an effective date. Common elements such as the following are arrangement with military numbering (1.1, 1.1.1, 1.1.1).

Objective or Purpose

Scope

Acronyms and Definitions

Materials and Equipment

Responsibilities

References/Related Documents

Procedure

Warnings, Cautions, and Notes

Documentation

Appendices

Approvals (for paper and hybrid systems)

Distribution (for paper and hybrid systems)

(See Box 7.1.) Each company must choose the elements their SOPs need. Small companies may have as few as three elements, while a larger organization may have eight or more elements.

52. Should We Simply Put N/A If an Element in an SOP Doesn't Apply?

Many companies do that, since not every procedure calls for materials and equipment, for example. Other companies create templates that are more flexible, so that if a section doesn't apply to a certain process, an author can remove it. Once deleted, the template numbering adjusts accordingly. In such a system, not every SOP has identical elements, but rather elements that are appropriate to the type of process the documents addresses.

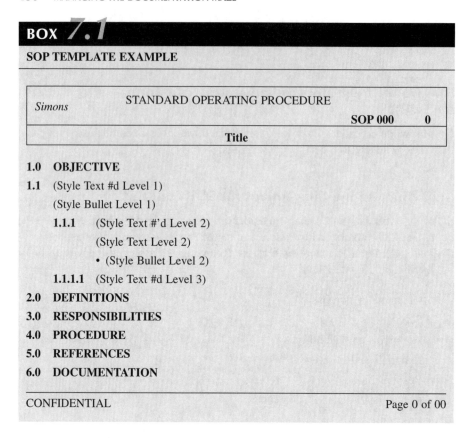

BOX *7.1*

SOP TEMPLATE EXAMPLE

Simons	STANDARD OPERATING PROCEDURE		
		SOP 000	**0**
	Title		

1.0 OBJECTIVE

1.1 (Style Text #d Level 1)

(Style Bullet Level 1)

1.1.1 (Style Text #'d Level 2)

(Style Text Level 2)

• (Style Bullet Level 2)

1.1.1.1 (Style Text #d Level 3)

2.0 DEFINITIONS

3.0 RESPONSIBILITIES

4.0 PROCEDURE

5.0 REFERENCES

6.0 DOCUMENTATION

CONFIDENTIAL   Page 0 of 00

53. Is It Good to Use a Template that Has a Series of Tables with Item and Action?

SOPs that use tables for each section are based on information mapping. A strong argument for this sort of organization is that it may be easier to read. Information mapping, however, requires a lot more paper if documents are printed.

54. Should We Update an SOP If We Find a Spelling or Mechanical Error?

There's no need to reversion an SOP for a minor error that has no bearing on the content. Some companies keep a log of changes that they will make to SOPs in the next review cycle. Such a log is a place to record document imperfections—such as a numbering gap or inaccurate punctuation mark that wasn't caught in the last review cycle. It's a vehicle that helps ensure that the review cycle will catch and repair these minor issues. The SOP on SOPs or document management should address this practice.

55.   What Documentation Practices Should SOPs Cover?

SOPs should drive all documentation practices. So if a company prepares GLP study reports, for instance, an SOP should delineate the "how to" for that document type.

56.   What SOPs Do We Need for Our Document and Record-Keeping Systems?

Each system needs to have an SOP that explains how to use it. Administration of the system may be an additional SOP. For instance, if a LIMS system allows input from scientists and technicians in a laboratory, they will need to have instructions for inputting and accessing data. An administrator, on the other hand, must know how to structure security access and archive data. Similarly, users of an electronic document system must know how to access the system and use it to initiate, review, or approve documents.

57.   What SOPs Should You Have to Support Document and Record-Keeping Systems?

Since systems these days are either fully electronic or hybrid electronic and paper, there must be SOPs that support how systems function. Here are some SOPs to think about:

- **Facilities Security.** This is the first line of defense for your company records and activities. Such an SOP covers facility access control accounts, visitor policy, and loss management.
- **Network Security.** This is the second line of defense for your company records and activities. Such an SOP covers network user access, password requirements, mandatory password change intervals, screen saver password use, remote access, Internet access, and virus protection.
- **System Backup.** Such an SOP tells how records are backed up on networked servers, individual workstations, and parts of systems. It also covers media labeling, reuse, and destruction.
- **Data Archiving.** Such an SOP tells how to archive data to make more storage space available and how to make archived data retrievable for review and inspections.
- **Computer System Event Recording.** Such an SOP identifies logs for recording modifications to a system. Logs create an audit trail for the system by capturing the date of the event, description of the event, testing explanation, tester, and date of completion.
- **Computer System Change Control.** One change control process may be used for all systems or each system can have a system-specific change control procedure. Such SOPs identify logs to record modifications.

- **Computer System Disaster Recovery.** Such an SOP tells how systems can be brought back online following the loss of facility, hardware, software, connectivity, or data.

- **Electronic Signature Policy.** If a system employs electronic signatures, an SOP delineates how users are held accountable for actions initiated under electronic signature and that e-signatures are legally binding equivalents of wet signatures.

- **Record Retention.** Such an SOP tells how long the company retains records and how data can migrate from a system. The legal requirements for e-records are the same as for paper.

- **Scanning.** Such an SOP tells how to scan documents and verify replication.

- **Computer System Validation.** Such an SOP tells how to validate a computer system, usually for COTS software applications.

- **E-Mail Policy.** Such an SOP tells how e-mail is backed up, archived, and retained, and it may include deletion policies.

- **Time Management.** If the company has sites in different time zones, such an SOP tells how to handle time and date stamps across the organization.

- **Software/Hardware Procurement.** `Such an SOP tells how to evaluate potential vendors in relation to user requirements.

- **Software Vendor Auditing.** Such an SOP tells how a company evaluates the software development life cycle (SDLC), regulatory experience, and vendor and product history.

- **Systems Inventory.** Such an SOP tells how systems are inventoried, ranked according to risk, and analyzed for gaps with the industry standards. The inventory includes plans for upgrades or retirement.

- **Software Development.** If a company creates its own software, such an SOP tells how it the software development life cycle is performed and documented.

- **Internal Auditing.** Such an SOP tells how a company audits its own systems and provides an auditing schedule.

- **Hosting a Compliance Inspection.** Such an SOP tells the protocol for receiving investigators, what records the company makes available, who serves as a host, and how investigators/auditors access records and documents,

- **Central Records Management.** Such an SOP tells how the company stores, protects, and accesses paper records.

- **Transitioning to an E-System.** Such an SOP tells how documents will move into a new, validated system. Typically, an SOP like this retires after the process is done.

- **Part 11 Committee.** Companies with substantial resources often form Part 11 Committees to make decisions about systems. If they do, they should put a Part 11 committee SOP in place to explain the purpose, responsibilities, and protocol for action.

58.   Must We Include Training Requirements in Every SOP?

No. Training is mandatory, and firms need to have an SOP that addresses training overall. To reiterate this requirement in each SOP is redundant.

59.   What Happens If People Undergo Training, But Fail to Follow a Procedure?

Retraining is in order. If failure to follow a procedure is consistent, the company needs to look at the procedure itself to see if it reflects an efficient and workable process. It may be that the procedure requires revision. If the procedure is sound, the company may have a personnel problem, perhaps an employee who is unable to perform the task or willfully disregards the established procedure. In that case, reassignment of the employee to another work area or disciplinary action may be in order. If many people fail to follow a procedure, management must investigate possible language and prerequisite skill sets deficiencies.

REFERENCES

1. www.iso.org, accessed July 18, 2009.
2. www.fda.gov/regulations/guidance, International Conference on Harmonization Guidance.
3. Food and Drug Administration, Drug Information Comment FDA/CDER from Office of Compliance, e-mail to author, February 13, 2009.
4. Gough, Janet, and Hamrell, Michael. Standard Operating Procedures (SOPs): Why Companies Must Have Them, and Why They Need Them. *Drug Inform J.* **43**:69–74, 2009.
5. Developing the compliance program guidance for pharmaceutical manufacturers, *Federal Register* **68**(86), Monday, May 5, 2003/Notices.
6. 21 CFR Part 820.40 Document Controls.

CHAPTER 8

NONCLINICAL RECORDS

Nonclinical documentation has a wide scope. What you must manage depends in total on where you are in the life cycle of a therapeutic product. If you are just beginning to develop a product and have proof of concept, you may be ready to move into Good Laboratory Practices (GLPs) to test for safety in the animal model. GLP regulations are clear in what you must do. If you are further along in development, you are amassing documentation that supports proof of concept for testing in humans. If you have product in the marketplace, you must maintain manufacturing and distribution records and track products in the market, including handling complaints and managing field alerts and recalls. This chapter provides answers to these frequently asked questions about document controls.

1. As a young company that has proof of concept for a new compound, at what point should we put document controls in place?

2. Why do so many products fail during development?

3. What controls should we think about first in the laboratory research setting?

4. What's the best way to manage our laboratory notebooks during research?

5. What's the best way to ensure that information goes into the notebooks correctly?

6. Is a notebook the property of the scientist or engineer to whom it is issued?

7. How do electronic notebooks work, and what are the advantages?

8. What do electronic notebooks require?

9. What exactly are Good Laboratory Practices (GLPs)?

10. Where should we keep amendments to GLP protocols?

11. When a contractor performs GLP studies, who should keep the original protocols and reports?

12. Should we allow our contractor to give us copies of the reports or should we demand the original reports?

13. What records do companies developing devices need to control?

14. What documents do companies manufacturing drugs need to control?

15. What kind of controls should manufacturing facilities have in place for documentation and records?

Managing the Documentation Maze, By Janet Gough and David Nettleton
Copyright © 2010 John Wiley & Sons, Inc.

16. What's the difference between QA and QC, and what documentation does each require?

17. Our engineers use the terms "design verification" and "design validation" interchangeably, but should they?

18. Can we document our production and process control functions as soon as we complete them?

19. Does maintenance require planning?

20. How should maintenance activities be tracked?

21. Who keeps maintenance records?

22. Where should we keep housekeeping records?

23. What is manufacturing change control, and what kinds of records does it require?

24. Why do companies have asset numbers on their equipment, and are they part of documentation?

25. What is the difference between an asset number and a serial number?

26. What is equipment IQ/OQ/PQ, and what documentation and controls do they require?

27. What is process validation, and what documentation and controls does it require?

28. Is barcoding documentation, and how should we control it?

29. What's a workable process for initiating, completing, and archiving batch records?

30. Is it possible to handle shipping and receiving records electronically, and, if so, how do we go about it?

31. What documents should we generate as we look for a new piece of manufacturing equipment, and where should we keep it?

32. If we follow SOPs for manufacturing exactly, how important is a corrective and preventive action (CAPA) program?

33. What's the difference between "correction" and "corrective action" and "preventive action"?

34. Who should manage manufacturing records?

1. As a Young Company that Has Proof of Concept for a New Compound, at What Point Should We Put Document Controls in Place?

If you are doing straight research, there's no requirement for controls, and the point at which you put controls in place really depends on your comfort with risk. Many young companies find that controls help their intellectual property and investor positions, and therefore the controls pay for themselves. If you have a product that you want to bring through nonclinical and into clinical testing, you must be prepared for rigorous development activities going forward. The earlier you put controls in place, the better able you will be to trace your product through the development process and thus build from one study to the next. Controls will also help ensure that you are developing and testing appropriately—for instance, with enough replicates to get solid data. Many a young company has been hurt by doing inadequate or undocumented studies early on, because they had to repeat them at a later point, which proved costly.

2. Why Do So Many Products Fail During Development?

According to FDA's Critical Path Initiative, many failures are "attributable to problems relating to the transition from laboratory prototype to industrial product" (1). In other words, the ability to take "proof of concept" from laboratory research and move it into product development can be problematic. Very often the problems are related to controls and verification of those controls through documentation.

3. What Controls Should We Think About First in the Laboratory Research Setting?

The place to start is with your SOPs and a chemical hygiene plan. At a minimum, you should control your laboratory notebooks, laboratory practices, inventory from incoming materials to reconciliation, and how you capture data from early testing. Documentation of your product development needs to be traceable from where you are now to the first proof of concept.

4. What's the Best Way to Manage Our Laboratory Notebooks During Research?

Laboratory notebooks require tight controls, because in research they are often the repository for intellectual property. Controls for notebooks involve two groups of activities: (a) issuance, retrieval, and archiving and (b) compiling data within the individual notebooks. Notebooks should be numbered and tracked, so that there is a record of where each notebook is at all times. Usually, companies assign control of notebooks to Legal or Quality. The controlling department maintains a log that shows where each notebook is, its identifying number, when it was issued, who it was issued to, and confirmation of receipt. its The controlling department also retrieves notebooks, typically upon completion, and makes a verified copy of the content for reference, so that scientists or technicians researching previous work can access them in a controlled manner. These copies are also tightly controlled. Notebook management practices for the company needs to be covered in an SOP.

5. What's the Best Way to Ensure that Information Goes into the Notebooks Correctly?

An SOP on notebook compilation should spell out how data are entered into a notebook and how verification by a second analyst or supervisor occurs. Such an

SOP would cover making entries, handling prints such as chromatograms, and making corrections to an entry.

6. Is a Notebook the Property of the Scientist or Engineer to Whom It Is Issued?

No. Notebooks are the property of the company and represent critical documentation integral to the company's research and development. Notebooks may not be removed from the company premises.

7. How Do Electronic Notebooks Work, and What Are the Advantages?

Electronic notebooks are not paper books; they are electronic documents. Entries can be text, graphics, and scans of paper records. All entries are automatically date and time stamped along with a link to the user that made the entry. Major advantages include the ability to immediately distribute information entered into a notebook, guarantee of proper date and time stamping, cost reduction due to elimination of paper, enhanced security access, and fast search capabilities.

8. What Do Electronic Notebooks Require?

Electronic notebook systems require a qualified network infrastructure, validation of the software system, administration for secure user access, and training of the users. Usually there are significant initial implementation challenges that are resolved with experience and acceptance. Just about everything related to an electronic notebook is superior to a paper notebook.

9. What Exactly Are Good Laboratory Practices (GLPs)?

GLPs are Part 58 of the federal regulations and address collection, storage, retrieval, and retention of laboratory data and testing. The regulation addresses controls for assays, test articles, and reagent standards. GLPs also address the conduct of nonclinical, nonhuman animal studies, and animal care facilities. GLPs demand extensive documentation including SOPs, protocols, amendments to protocols, if any, and final reports. Data that nonclinical studies generate also require careful management.

10. Where Should We Keep Amendments to GLP Protocols?

Amendments to GLP protocols are handled just like any other amendments to protocols. The amendments are filed along with the protocol. Controls for the approval of protocol amendments follow the same rigor as approval for the original protocol. All copies of the protocol that have been distributed must be updated to include amendments.

11.   When a Contractor Performs GLP Studies, Who Should Keep the Original Protocols and Reports?

Companies determine who keeps the original reports when they establish agreements with contractors. What is important for companies to understand is that the documentation a contractor produces remains the responsibility of the sponsor. Sponsors can delegate, but never abrogate, this responsibility. If contractors retain original documents, they should make copies that are exact replicas available to the sponsor. Furthermore, to ensure that contractors adhere to acceptable documentation practices, sponsors need to audit source documentation.

12.   Should We Allow Our Contractor to Give Us Copies of the Reports or Should We Demand the Original Reports?

It is acceptable for the sponsor to have verified copies of the original reports. However, the sponsor should audit the contractor to determine that the reports they receive are indeed exact copies of the original reports.

13.   What Records Do Companies Developing Devices Need to Control?

21 CFR Part 820 the Quality System Regulations (QSRs) are clear in what device developers need to control as they develop a device and ultimately manufacture for testing in a clinical setting, as well as for distribution.

Section 820.181 device master record (DMR) is a compilation of records that include device specifications, production process specifications, quality assurance procedures, packaging and labeling specifications, installation, maintenance, and servicing specifications.

Section 820.184 device history record (DHR) is a compilation of records that ensure that the device is manufactured according the device master record and address the batch, lot, date, quantity, acceptance, and labeling for devices.

- Approval of design plans (820.30(b))
- Approval of design input (820.30(c))
- Approval of design output (820.30(d))
- Result of design review (820.30(e))
- Results of design validation (820.30(g)
- Approval of the device master record (DMR) or changes to it (820.40)
- Equipment maintenance and inspection activities (820.40)
- Calibration records (820.72(b))
- Approval of process validation (820.75(a))
- Performance of validated processes (820.75(b) (1), (2))
- Release of finished devices (820.80(d)
- Device acceptance activities records (820.80(e))

- Authorization to use nonconforming product (820.90(b))
- Labeling inspection (820.120(b))
- Audit certification (820.180(c))
- Complaints (820.198(b))

In addition, device developers need to have quality manuals, standard operating procedures, and work instructions to cover these activities (820.40(a)). Device developers also need to have procedures in place to tell how change is controlled for the following:

- Product requirements
- Product specifications
- Engineering specifications
- Components
- Verification, validation, and testing
- Design processes
- Manufacturing processes
- Quality control processes
- Quality assurance processes
- Distribution and support processes
- Investigational device exemption (IDE) and updates during clinical testing (citation)

14. What Documents Do Companies Manufacturing Drugs Need to Control?

21 CFR Parts 210 and 211 address Good Manufacturing Practices for drugs. Part 210 addresses manufacturing, processing, packaging, and holding of drugs. Part 211 addresses finished pharmaceutical products.

15. What Kind of Controls Should Manufacturing Facilities Have in Place for Documentation and Records?

The controls for all documentation systems are basically the same. Chapter 3 explains standard controls. (See also Box 8.1.)

16. What's the Difference Between QA and QC, and What Documentation Does Each Require?

Quality Control is concerned with making sure products and components meet specification as they are manufactured, whether they are incoming, in-process, or finished goods. QC documentation is typically verification of product suitability and may include laboratory notebook entries, out-of-specification (OOS) result

BOX 8.1

FEATURES OF MANUFACTURING DOCUMENT CONTROLS

According to the FDA, manufacturing document control systems typically include the following:

- Documents named and numbered in accord with a logical scheme that links the documents to the product or component they describe or depict and illuminates a drawing hierarchy.
- A master list or index of documents that presents a comprehensive overview of the documentation which collectively defines product and/or process.
- Approval procedures that govern entry of documents into the document control system.
- A history of document revisions.
- Procedures for distributing copies of controlled documents and tracking their location.
- Files of controlled documents periodically inventoried to ensure that they are up to date.
- A person or persons assigned specific responsibility to oversee and carry out these procedures.
- A process for removal and deletion of obsolete documents

Source: Adapted from FDA's Design Control Guidance for Medical Device Manufacturers (2).

resolution, and reporting. QC frequently uses Laboratory Information Management Systems (LIMS) for routine reporting. Quality Assurance, on the other hand, verifies activities after they occur, and it audits systems, processes, and units to ensure compliance, so the documentation QA produces is typically sign-off and verification, logs of activities, and written reports.

17.   Our Engineers Use the Terms "Design Verification" and "Design Validation" Interchangeably, But Should They?

No. Design verification confirms that the design output or specifications meet the design requirements. Design validation determines whether the device has been made correctly and that it meets its intended uses. Using these terms interchangeably may lead to confusion in the review of documentation.

18.   Can We Document Our Production and Process Control Functions as Soon as We Complete Them?

No. These functions require documentation at the time of performance. Documentation at the time of performance verifies each step in a process *as it occurs*. These controls confirm that the products comply with the specificaions from the start of a batch to final product.

19. Does Maintenance Require Planning?

All maintenance activities require planning, but planning takes different forms. Preventative and scheduled maintenance requires planning and documenting the schedule and activities to take place. Unscheduled maintenance such as repairs requires plans for how to make and verify the changes to the system. Planning provides a means to think out a problem before acting on the problem. Different levels of documentation are necessary, but they must detail the plan to be followed and the activities that actually take place.

20. How Should Maintenance Activities Be Tracked?

Most companies today use electronic records, such as work-order systems to track maintenance activities, but many companies still use paper records. Regardless of how the records are retained, the records indicate the plans for making changes, authorization, actual work performed, and means for verifying proper operation.

21. Who Keeps Maintenance Records?

Companies decide which group is responsible for keeping maintenance records, and they detail how they do it in their SOPs. Often it is the maintenance group, operations group, or the Quality Assurance group that assumes responsibility.

22. Where Should We Keep Housekeeping Records?

Housekeeping records may involve many types of regulated and nonregulated records. It is usually best to keep all records in the standard systems. There is no problem including nonregulated data with regulated data. Care must be taken to ensure that system access limits users to only the data appropriate for their job duties.

23. What Is Manufacturing Change Control, and What Kinds of Records Does It Require?

When a manufacturing system is changed due to preventative maintenance, repair, or redesign, care must be taken to ensure that it is operating as intended. For changes that don't alter the design, verification activities must ensure that the system is operating in the same manner as when the system was validated. For changes that alter the design, the system must be revalidated. Revalidation is considerably more involved because it addresses the operation of the entire system with respect to predefined specifications and acceptance criteria.

24. Why Do Companies Have Asset Numbers on Their Equipment, and Are They Part of Documentation?

Asset numbers uniquely identify equipment. When multiple pieces of identical or similar equipment are in use, it is easy to get them confused, so a unique asset tag helps to differentiate them. Asset identification is also useful for accounting controls so that proper depreciation and business requirements can be followed.

25. What Is the Difference Between an Asset Number and a Serial Number?

A serial number is assigned by the manufacturer and uniquely identifies a piece of equipment. The company that uses the equipment assigns an asset number. Since serial numbers vary greatly between manufacturers and are often placed in a difficult-to-access location, they are less popular for unique equipment tracking. Uniform asset numbers prominently displayed are easier to work with and are less prone to error.

26. What Is Equipment IQ/OQ/PQ, and What Documentation and Controls Do They Require?

Installation qualification (IQ) addresses installation of equipment according to the recommendations and specification of the manufacturer. Operational qualification (OQ) addresses proper operation of the equipment in accordance with manufacturer specifications after installation. Ideally, the equipment should function the way it did before it was shipped and installed. Performance qualification (PQ) addresses the functionality of the equipment according the needs of the company that is using the equipment. All qualifications consist of a preapproved protocol that contains instructions and verification activities, along with a report that contains a record of the actual activities performed and actual results which is subsequently approved. PQ addresses how the equipment functions in manufacturing the actual product. PQ is often performed in conjunction with process validation (PV).

27. What Is Process Validation, and What Documentation and Controls Does It Require?

Process validation (PV) is confirmation that an entire process, a series of steps such as a manufacturing procedure, is working as intended. It utilizes equipment that has already been qualified, but is often performed in tandem with PQ, the last step in qualification. Process validation documentation may take many forms and is usually outlined in a process validation plan or an SOP.

28. Is Barcoding Documentation, and How Should We Control It?

Barcodes are labeling, and labeling is documentation. Strict controls must be applied to barcodes to ensure that there are no duplicates or missing identifiers. Barcoding systems, because they produce electronic records, require validation.

29. What's a Workable Process for Initiating, Completing, and Archiving Batch Records?

Batch records are usually paper records, but many companies are transitioning to electronic batch records. Controls must be in place to ensure that correct batch instructions are present on the record and it is approved prior to initiation. When batch records are filled out, care must be taken to ensure that they are accurate and traceable to both equipment and operator. Rigid documentation standards are prudent. Batch records must go through a series of quality control checks before product may be released. Completed batch records must be stored with strict access controls. It is advisable to make a backup copy, usually electronic, to aid in their retrieval at a future date. Older batch records may be archived to make room for new ones. Archived batch records must adhere to the same access restrictions that applied before archiving.

30. Is It Possible to Handle Shipping and Receiving Records Electronically, and, If So, How Do We Go About It?

Yes, many companies today have shipping and receiving records managed by the same systems that manage inventory. As with any electronic record system, the key is to have the electronic records contain all of the same information as the paper records. Many shipping companies themselves have gone paperless. Any system that manages electronic records related to a regulated product must be validated per 21 CFR Part 11.

31. What Documents Should We Generate as We Look for a New Piece of Manufacturing Equipment, and Where Should We Keep It?

As with any purchase, you must start out with your requirements. These requirements are often included in a Request for Information (RFI). The responses allow you to uniformly compare vendors and products. As you learn more about what is available, be sure to update your requirements documentation. When you have narrowed the list of vendors, you can place a Request for Proposal (RFP). This is a formal request for information that includes specifications and pricing. From this information you will be able to develop a purchasing contract. Some companies choose to collect these documents and store them in purchasing files, while others store them in equipment files. Companies that have electronic document management systems, often store them electronically.

32. If We Follow SOPs for Manufacturing Exactly, How Important Is a Corrective and Preventive Action (CAPA) Program?

It is very important. The expectation of the agency is that manufacturers will have CAPA programs in place. A CAPA program focuses on investigating and correcting discrepancies and preventing recurrences. CAPA is really three concepts: remedial corrections as something goes awry; root cause analysis with corrective action to prevent recurrence; and preventive action to prevent initial occurrence. Documentation for such systems require controls spelled out in an SOP. Documentation that a system produces includes identification of a problem, analysis, and remediation action and then verification of corrective/preventive action, communication of action, and managerial review. Documentation produced in the CAPA program must also be available, since it becomes essential for tracking and trending.

33. What's the Difference Between "Correction" and "Corrective Action" and "Preventive Action"?

Correction addresses repairs or rework and the disposition of an existing nonconformity, while "corrective action" addresses the elimination of causes of nonconformities. Preventive action addresses the elimination of possible nonconformities. All three require controlled documentation.

34. Who Should Manage Manufacturing Records?

FDA recommends that a person who is not directly involved with developing or using the documents administer the system. If a company is small, perhaps a librarian or full-time clerical or paraprofessional employee could manage the system (2).

REFERENCES

1. www.fda.gov/oc/initiatives/critical path, accessed May 5, 2009.
2. Design Control Guidance for Medical Device Manufacturers, March 1997, www.fda.gov/cdrh, accessed Jul 19, 2009

CHAPTER 9

CLINICAL AND SUBMISSION RECORDS

INTRODUCTION

Maintaining clinical and regulatory data and documents is critical to bringing a therapeutic product to market and keeping it there. Failure to adequately produce and manage clinical records can be very costly. Of 5000 to 10,000 new chemical entities, five will go to the clinic, but just one will gain approval. And the cost of bringing a single drug to market is estimated to be well over $1 billion dollars US for a process that typically takes 10 to 15 years. Documents that go to the regulators, either in the United States or abroad, require tight controls so that submissions over time tell a complete story of a product slated for the marketplace. Similarly, once a product gains approval, the sponsoring company must report on it annually to confirm its continued suitability. This chapter answers the following questions.

1. What is a regulatory record versus a regulatory document?
2. We provide records to the FDA, so it is really a one-way street for information, isn't it?
3. Where can a company find assistance in understanding clinical documentation?
4. What documentation does a company need to have to begin a clinical trial in the United States?
5. How can we keep track of our submission documents and supporting records?
6. Does regulatory "own" the submission documents?
7. What are the typical submission documents for a product in development?
8. When there are many versions of submitted documents, how is the "official" copy determined?
9. Should we use the number the FDA assigns to our project as our official number for our submissions?
10. Is there a legal requirement that says submission documents themselves must be signed?
11. What is a CTD?

Managing the Documentation Maze, By Janet Gough and David Nettleton
Copyright © 2010 John Wiley & Sons, Inc.

12. What is the advantage of a CTD?

13. What exactly does publishing mean?

14. Now that we have to file electronically, do we need to have an electronic publishing system in house?

15. If we decide to outsource the publishing function, what should we look for?

16. How should we go about filing our own e-submission?

17. Can we file submissions for one product electronically and file paper for another product?

18. Once we file an IND or IDE, do we have to wait for FDA to give approval before we can begin our clinical trials?

19. What is "clinical hold"?

20. What's the real purpose of submission document review?

21. Is there any industry standard for referring to people who participate in clinical trails?

22. How can we make the process for generating submission documents more efficient?

23. How many reviews should our documents go through, and how many people should see them?

24. Why are document review cycles so long?

25. What is a call out?

26. Does everyone have to agree to the changes in a submission document?

27. Do we have to review a submission document in our document management system, or can we export it for this purpose?

28. How can we get people to stop the endless wordsmithing?

29. Who has the final say about document content?

30. Is someone who edits a clinical document an author?

31. Do you need to keep a history of every draft used in developing a submission document?

32. If our CRO prepares our submission documents, how much input should we have?

33. What are QA checks on submissions, and why do we need them?

34. After a QA check, does the document require an additional full review?

35. What's the best way to hold a concurrence meeting so that we can reach consensus on a document?

36. How do we make updates to a template used for an IND?

37. What is an electronic submission, and how does it differ from a paper submission?

38. Can we file submissions for one product electronically and file paper for another product?

39. What is XML language, and do we have to learn it?

40. How can we make the transition from paper to eCTD submissions?
41. What does the guidance "Computerized Systems Used in Clinical Trials" apply to?
42. What documentation does a 483 or warning letter require, and how should we control it?
43. What is an annual report?
44. What exactly is a clinical trial versus a clinical study?
45. What is clinical documentation, and what controls does it require?
46. ICH E6 tells us what documents a clinical study should have and that both site and sponsor should have copies, but who is responsible for making those copies?
47. What are clinical source data?
48. Who keeps clinical trial source data?
49. Who is responsible for clinical trial source data?
50. Must documentation/records be original copies?
51. Who is responsible for standard operating procedures during a clinical trial?
52. Is there a specific industry preference for organizing and maintaining clinical trial data?
53. How long should a case report form be?
54. What is a follow-up form?
55. Do ICH GCP guidelines outline specific standards for archiving documents?
56. Can we outsource clinical document management?
57. Can we use software to capture e-mail from our clinical trial sites?
58. Who owns source data, and who keeps it?
59. What should go into the case report forms?
60. Can study staff make corrections/edits to original records?
61. Should events that are not related to the therapeutic treatment be included in case report forms?
62. Can we keep information not collected on CRFs in a memo to file?
63. What are the uses of memos to file?
64. When our regulatory director refers to "case histories," what does she mean?
65. Can we still use paper Case Report Forms and be compliant?
66. After a clinical trial, how are data kept?
67. Does a sponsor have to worry about patient records generated during a clinical trial, in terms of retention after the trial is over?
68. What's the best way to access information in a patient diary and bring it into the clinical trial database?
69. If a site uses electronic patient diaries, how can a sponsor ensure data integrity?
70. If data are entered directly into a validated e-system, what is the source document?

71. Can we scan CRFs into our study files?

72. If a sponsor discontinues a study, how long must records remain available?

73. What is a certified copy?

74. What is the difference between clinical trial monitoring and auditing, and where do the records reside?

75. How can sponsors make sure that clinical trial staff, such as PIs, do not publish (disclose) information about the trial?

1. What Is a Regulatory Record Versus a Regulatory Document?

Any information that supports an FDA-regulated activity or submission is a regulatory record. Documents required by the regulations call upon your records in support of what your documents say. Furthermore, any records that you keep on the development, use, or maintenance of information systems to perform a regulated activity (a filing) constitute regulatory records.

2. We Provide Records to the FDA, So It Is Really a One-Way Street for Information, Isn't It?

No it's a two-way path. All submission documents are the result of dialogue between the regulator and the company. Therefore, it's critically important that incoming information from the regulator is captured and tracked appropriately. A clinical trial has a document life cycle, and it's important for a company to know where documents, records, and data reside, where they come from, and who manages them. The way to look at it is as data flows. An application is part of the clinical data flow to the regulator.

3. Where Can A Company Find Assistance in Understanding Clinical Documentation?

A good source is The Clinical Data Interchange Consortium (CDISC). CDISC is a global nonprofit organization dedicated to supporting clinical trials. Since 1997, CDISC has developed a group of standards for language, electronic data, and regulatory submissions. CDISC has worked extensively with the FDA and industry and has evolved from a purely submission focus to the development of standards for end-to-end data flow in clinical trials, as well as from source data to database to analysis and, finally, to regulatory submissions (1).

4. What Documentation Does a Company Need to Have to Begin a Clinical Trial in the United States?

A clinical trial can only occur if extensive preclinical animal testing is adequate to show that a product is safe. The company must have controlled records of all activities up to this point. Regulatory people then contact the agency, and typically they set up a meeting prior to filing an Investigational New Drug application (IND) or

Investigational Device Exemption (IDE) in the United States. The purpose of such a meeting is to discuss proof of concept and scientific data as well as the regulatory requirements for filing. Such discussions determine what the deliverables will be for the initial filing to test in humans. A submission then draws upon the relative documentation to date and summarizes it as proof of a successful drug candidate for testing in humans. Thus, it is prudent to have good documentation controls in place before a company goes into clinical testing.

5. How Can We Keep Track of Our Submission Documents and Supporting Records?

The best place to keep submission documents is in the company's document management system. If the company has a secure system with taxonomy that categorizes documents by type and has determined privileges to access documents, then a single document system is effective. If all documents reside in the same system, it is much easier to maintain control and ensure consistency.

6. Does Regulatory "Own" the Submission Documents?

Regulatory may be the custodian of the regulated information at key points during the information life cycle, but the company owns the content, process, and outcomes. Regulatory's role is to support the coordination, compilation, analysis, review, and submission of the data and documentation a submission requires.

7. What Are the Typical Submission Documents for a Product in Development?

Submission documents are those that a company files with a regulator once it has determined to test a product in humans. The typical submission documents begin with an application to test in humans, an investigational new drug application (IND) for drugs and biologics, or an investigational device exemption (IDE) for devices. This application contains a general investigation plan, the protocol for the study, the investigator brochure (which is the information for the investigator), and chemistry, manufacturing, and controls and/or design controls. Once a trial is underway, the protocol may be amended. The final outcome for each clinical trial is the clinical trial report. When clinical testing is complete, the company files an application for permission to bring the drug or device to market, a New Drug Application (NDA) for a drug, a Biologics Licensing Agreement (BLA) for a biologic, or a Pre-Market Approval (PMA) for a device. Because of the volume of documentation that each therapeutic product requires, tight controls are essential. In essence, each document after the first filing builds from that filing, so there is continuity and progression from document to document.

8. When There Are Many Versions of Submitted Documents, How Is the "Official" Copy Determined?

Once a document is finalized and sent to a regulator, the electronic version of the same document must be housed in a secure folder, with limited access, so that it is available for reuse in the next filing. The document, as an electronic file, is subject to 21 CFR Part 11 requirements and must be controlled. But it's more than a matter of complying with the regulations; tight controls ensure consistency and a document trail that shows exactly what was submitted and when. Subsequent version of the document must be maintained and kept separate from the versions of documents submitted. In this manner you have a history of submissions and can identify documents that need to be resubmitted.

9. Should We Use the Number the FDA Assigns to Our Project as Our Official Number for Our Submissions?

Your submission documents should be consistent with your internal document numbering system and should be linked to the project number assigned by the regulator. Remember that you own your documents and your system must work to house all of your documents and records, so that they are accessible through your system. In addition, a document may be submitted to multiple regulatory agencies, so a system of cross-referencing is essential.

10. Is There a Legal Requirement that Says that Submission Documents Themselves Must Be Signed?

Not directly, but FDA regulations require a review and approval process, and the approval of a document is by signature, either manual or electronic. A document copy that comes from a system that uses electronic signatures often has the e-signature as an additional page. Typically, a PDF of the document is included in a submission and it has the e-signature page included. Thus, a 10-page original document would be 11 pages in the PDF. A submission that is entirely electronic and comes from a validated e-system has been reviewed and approved with e-signatures, and this process is inherent in the system.

11. What Is a CTD?

CTD as an acronym could be clinical trial data, but most commonly it means "common technical document." The CTD is a format for presenting uniform information to multiple regulatory agencies. The structure and organization are thus the same for all submission, regardless of the agency, with the exception of regional information. It is a document of five sections: regional information; table of contents; chemistry, manufacturing, and controls (CMC) or quality; nonclinical; and clinical. The CMC section requires a summary, while the nonclinical and clinical sections require overviews and summaries (2). It is important to note, however, that the CTD, while recommended by the agency, does not replace the requirement for a New Drug Application (NDA), Pre-market approval (PMA), or Biologic Licensing Agreement (BLA). The predicate rules are still in effect, and the CTD must satisfy them as well.

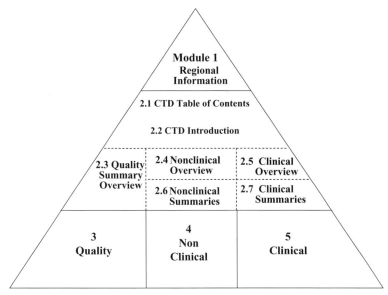

Figure 9.1.   The CTD model.

12.   What Is the Advantage of a CTD?

The CTD allows the harmonization of drug applications and provides standards for preparing documents for the submission phases. The strength of the CTD is that once a company is in the clinic, it can start building the CTD with pivotal CMC information and pivotal nonclinical reports and continue to build until all the data are in place. Then it's a matter of compiling the summaries. (See Figure 9.1.). The CTD Model. The CTD is acceptable in the United States for Investigational New Drug Applications (INDs), New Drug Applications (NDAs), Abbreviated New Drug Applications (ANDAs), and Biologic Licensing Applications (BLAs); in the European Union for the Investigational Medicinal Product Dossier (IMPD) and the Marketing Authorization Application (MAA); in Japan for New Pharmaceutical Applications (NDas); and in Canada for Clinical Trial Applications (CTAs), New Drug Submissions (NDs), and Abbreviated New Drug Submissions (ANDS).

13.   What Exactly Does Publishing Mean?

Publishing is the final step in the preparation of a submission document. Publishing ensures that the document meets all the requirements of a submission and that the mechanics are clear and consistent. For instance, in a submission document, each page must be accounted for; publishing does that. When submissions were paper-based, publishing involved such activities as compiling reports and documents into volumes, page numbering, indexing, photocopying, and binding. With electronic submissions, a more sophisticated skill level is necessary. Electronic submissions are electronic records, so 21 CRF Part 11 applies to the system used for publishing.

14. Now that We Have to File Electronically, Do We Need to Have an Electronic Publishing System in House?

Not necessarily. The determiner is usually cost. Larger companies with high submission volumes may opt to implement such systems and ensure that the operations team knows how to use them and submit electronically. Small to mid-size companies with just one or two filings may decide to outsource this function. Many software vendors and consultants provide turnkey publishing services.

15. If We Decide to Outsource the Publishing Function, What Should We Look For?

Many contractors offer this service. These companies have the advantage of optimal practices gained through experience. To find the best contractor for your company's needs, start with an internal discussion as to the precise requirements for your submission. If you are doing multiple submissions to different countries, you will need a vendor that understands how the eCTD will adapt for another region. Next is a search of contractors. Once you have identified a few as potential service providers, query each about the track record and ask for verification of quality of service. In other words, evaluate the contractors as you would any vendor before making a decision.

16. How Should We Go About Filing Our Own E-Submission?

If you want to make an eCTD submission to the FDA you must make a sample submission to show you can produce an eCTD of appropriate technical quality (2). This submission is strictly for evaluation of compliance with FDA and ICH eCTD specifications. Such a submission should contain sample data that the FDA will not evaluate for scientific or technical content. In the EU, submissions require certification of every individual that contributes to it.

17. Can We File Submissions for One Product Electronically and File Paper for Another Product?

For many therapeutic products such as devices, the FDA determines the method for filing on a per-product basis, so the only way to know what the agency will accept is to ask. Filings for drug products are not as flexible. The mandate has been for electronic filing since January 1, 2008. Many drug companies since then have asked for and received waivers allowing them to file paper submissions, but the agency has announced that after June of 2009, they would no longer accept paper.

18. Once We File an IND or IDE, Do We Have to Wait for FDA to Give Approval Before We Can Begin Our Clinical Trials?

No. If FDA does not respond within 30 days, you may begin your trial. That is not to say that the agency may not stop a trial once it is underway.

19.   What Is "Clinical Hold"?

Clinical hold means that a trial may not start or must stop when it is underway for a number of reasons. If an application to test in humans is incomplete, the agency may ask for more data, for instance, or put the trial on clinical hold. A company can also put its own trial on clinical hold for a number of reasons. For example, if a company is having trouble with a formulation, it may decide to suspend trial activities until it resolves its difficulties with a compound.

20.   What's the Real Purpose of Submission Document Review?

Revision and change are integral parts of the writing process. Review ensures that a document meets its goals from a scientific, medical, legal, and regulatory perspective. Review verifies the accuracy of the document, identifies any potential ambiguities and misleading statements, checks for consistency, and ensures cohesive messages from document to document. In the review process, others will see what authors may not. Authors, when they read their own writing, see what they expect to see, not always what's actually there.

21.   Is There Any Industry Standard for Referring to People Who Participate in Clinical Trials?

Some companies refer to people who participate in clinical trials as patients, others refer to them as subjects, and yet others call them volunteers. It's best to be consistent. The regulations consistently call people who participate in clinical trials "subjects," because subjects can be both healthy volunteers or patient volunteers.

22.   How Can We Make the Process for Generating Submission Documents More Efficient?

Many companies have successfully implemented a system of metrics for the process of generating, reviewing, approving, and quality checking documents. Metrics are based on the concept that all processes have steps, and steps have timelines. For instance, a company can go from database lock to a submission in two months as follows:

Day 0	Database lock
Day 1	Final statistical output issued
Day 28	Report draft
Day 35	Comments received
Day 42	Report draft 2
Day 45	Comments received
Day 55	Report finalized and signed
Day 60	Report published and QA checked

Of course, the timeframe depends on the resources a company has. Some companies can go from database lock to final report in a much shorter timeframe. If you adhere to metrics, make sure that everyone on the team is on board with the timelines. If timelines require adjustment during the process, one person, perhaps a project manager, should assume that responsibility. Make sure, as well, not to lock the timelines into an SOP, because you may need some flexibility, and if you violate the timelines, you don't want to have an SOP deviation.

23. How Many Reviews Should Our Documents Go Through, and How Many People Should See Them?

In an ideal world, a first draft is nearly perfect. In the real world, however, the review process usually brings important insights and adjustments to a document that improve it. To prevent the endless cycle of comments and revision, a good rule of thumb is to have no more than two review cycles. By the time the first review cycle is over and comments are incorporated, the document should be nearly complete. The second review process should affirm that the revisions are valid. If you cannot reach agreement after the second review cycle, hold a concurrence meeting to discuss the issues that are problematic and resolve them.

24. Why Are Document Review Cycles So Long?

Sometimes the review cycle is too long because too many people have input. Get the correct input from the departments that hold responsibility for the end document in their areas of expertise. Remember, nothing is gained by having a regulatory person move commas or rephrase an acceptable and clear sentence based on preference. So you want to make sure that medical staff review for scientific accuracy, legal for intellectual property, engineering for design (for devices or combination products), regulatory for correct deliverables to the agency, and medical writers/editors for language. A checklist for reviewers may be helpful in expediting the process. (See Table 9.1.)

25. What Is a Call Out?

A call out is simply an identification of a disconnect in a document. Do data in the charts reflect the text accurately for instance? Do data in text match source data? Has the document made a statement that is unsupported or contradicted? "Call out" is simply jargon for bringing attention to an issue found during the review process.

26. Does Everyone Have to Agree to the Changes in a Submission Document?

No. It is not mandatory that an author incorporate everyone's comments in a revision. But all comments deserve consideration. Perhaps a comment is the result of misreading; in such a case, a revision to the text for purposes of clarity may be the outcome. Or a comment may simply reflect the preference of the reviewer; in that case, discussion of the merits of change should take place.

TABLE 9.1. Document Review Checklist

Category	Item	Verify
Formatting	Margins are correct.	√
	Pages are numbered correctly.	√
	Page and section breaks are appropriate.	√
	Headers and footers are acceptable.	√
	All page contents are visible on the printed page.	√
	All other aspects of the document layout are acceptable.	√
	All hidden text, tracked changes, comments, and highlighted material are removed or otherwise made invisible.	√
	Consistent table and figure formats are used throughout document.	√
Administrative	Document version information is correct.	√
	All appendices are available and correct.	√
Scientific, Statistical, and Medical Validity	Document is consistent with the goals and objectives of the General Development Plan.	√
	Information is consistent with related documents for the same product, from a scientific and medical perspective.	√
	All methods and procedures are described accurately.	√
	Data are presented accurately, from a scientific and medical perspective.	√
	Table and figure titles are an accurate reflection of the information presented.	√
	Information in tables and figures is presented in a clear, relevant, and understandable fashion.	√
	All relevant scientific and medical issues are addressed.	√
	All data interpretations, discussions, and conclusions are medically and scientifically accurate.	√
	All terminology used in the document is clear and scientifically and medically accurate.	√
	All statements of fact agree with referenced publication, study, or other source of information.	√
	Information is presented logically, from a scientific and medical standpoint.	√
	Appendix information is correct.	√
Consistency / Quality Control	In-text tables and numbers match source data.	√
	All statements of fact agree with referenced publication, study, or other source of information.	√
	Document and data are consistent throughout, across all sections including synopsis and summary sections.	√
	Correct numbers and titles are used for all tables and figures.	√
	Source information for tables and figures is provided and is correct.	√
	All figure axes are labeled.	√
	Symbols used in figures are labeled and defined clearly and correctly.	√
	All references to section and appendix numbers are correct.	√
	All references to table and figure numbers are correct.	√
	All other references (e.g., to publications) are correct.	√
	Table of contents is correct and updated.	√
	List of abbreviations is correct; all abbreviations used in the document are defined, and all abbreviations in the list are used in the document.	√
	All references and bibliography information have been checked against the source and is presented in the appropriate style (e.g., Vancouver style).	√
	All appendices are complete and correct.	√
	Document adheres to company template and style guide.	√
Regulatory Writing/ Editorial	All sections and topics required by regulation are present.	√
	Grammar, punctuation, and spelling are correct.	√

27. Do We Have to Review a Submission Document in Our Document Management System, or Can We Export It for This Purpose?

In an e-system, everything you do to a document is captured in the audit trail. Some companies find it easier to export a document from the system for initial review, then import it for final review and sign off. What is important is that reviewers and signatories are identified in this final review process. When a document leaves an e-system the controls are eliminated, and therefore care must be taken to identify changes between the exported version and the imported version.

28. How Can We Get People to Stop the Endless Wordsmithing?

There's more than one way to say just about anything in a language, and say it well. Reviewers need to understand that first. Sometimes a minor change makes a world of difference, but just as often a change may be the result of an application of ego. The final document should always be the primary focus, and the final document should reflect wording in any previous submissions for the product. In essence, documents should deliver information about the product in consistent terminology. For issues that may require change, reviewers should ask themselves this: What can I live with, and what will keep me up at night? The issues that deprive sleep are the one to address; let the others go.

29. Who Has the Final Say About Document Content?

The final outcome of a document is the result of a give and take between author, reviewers, and final signatories. The final signatory has legal responsibility for the content of the document.

30. Is Someone Who Edits a Clinical Document an Author?

No. Submission documents are usually the product of multiple authors who contribute to the content. Editing does not constitute authoring, even though editing may substantially improve or change a document.

31. Do You Need to Keep a History of Every Draft Used In Developing a Submission Document?

No. You must have a process in place that shows that you review and edit appropriately—that is, that you circulate, collaborate, comment, and revise in a controlled manner. The final document is the official one, and it is the one linked to the approval final signature.

32. If Our CRO Prepares Our Submission Documents, How Much Input Should We Have?

Never lose sight of the fact that it is your document, not the CRO's. Even if the CRO compiles the document for you, you must have input during the preparation and review process, and you must have the final say in the end product.

33.   What Are QA Checks on Submissions, and Why Do We Need Them?

A quality check and sign off on the final document ensures that there are no disconnects or contradictions that may have been overlooked during the formal review process. QA provides an impartial look at the document by people who have not had direct input into the trial or the creation of the document itself.

34.   After a QA Check, Does the Document Require an Additional Full Review?

No, but any suggestions a QA check delivers must be reviewed and approved by the appropriate parties. If changes are major, then key reviewers should have input.

35.   What's the Best Way to Hold a Concurrence Meeting so that We Can Reach Consensus on a Document?

If a submission document has outstanding issues that have not been resolved in the review cycles, you can plan a concurrence meeting to specifically address these issues. The best meetings occur when participants know in advance what the purpose of the meeting is and come prepared to resolve them. Good meetings have agendas, and they stick to them. You can bring the document in question to a meeting in electronic form, and then you can project the areas in dispute so that all meeting participants see the issue in context and reach a decision about the final wording or display. It's also a good idea to indicate that those who do not attend the concurrence meeting lose the right to have input into the final decisions about a document.

36.   How Do We Make Updates to a Template Used for an IND?

It's probably best to stick with the template you already have in place, because you have already established the "look" for the company's submissions. However, if you feel your template is flawed, it's best to retain the elements that will remain constant in the format they're already in and then adjust the weak areas.

37.   What Is an Electronic Submission, and How Does It Differ from a Paper Submission?

A paper submission may include actual paper documents and electronic documents that are delivered on media such as CDs or DVDs. An electronic submission is one in which the entire delivery of the data is performed electronically. With the mandate to file electronically, companies have to change how they submit documents. E-submissions are those that go to the regulator electronically through the regulator's Internet portal.

38.   Can We File Submissions for One Product Electronically and File Paper for Another Product?

For many therapeutic products such as devices, the FDA determines the method for filing on a per-product basis, so the only way to know what the agency will accept is to ask. Filings for drugs products are not as flexible. The mandate has been for electronic filing since January 1, 2008. Many drug companies since then have asked for and received waivers allowing them to file paper submissions, but the agency has announced that after June of 2009, they would no longer accept paper.

39. What Is XML Language, and Do We Have to Learn It?

XML is extensible markup language. It is the basis for many data retention and transmission standards. XML provides data and a description of the data, such as field names, in a highly structured format. eCTD uses XML to provide highly uniform and structured data that are compatible with many computer systems. Since publishing systems do all of the XML work for you, there is no need to learn it. XML is so popular as a data exchange method that Microsoft Office uses it as part of their file formatting. Most computer systems are using XML for exporting and importing data because it eliminates proprietary file formats.

40. How Can We Make the Transition from Paper to eCTD Submissions?

The firsts step is to make each paper document available as an electronic source file. This means having the original file that made the paper document. It does not mean that you should scan the paper documents to make graphic files. The original content of the paper files needs to be actual data that can be manipulated and searched. Next, organize the electronic files so their content matches the requirements of the CTD. You can use the CTD table of contents to create a folders structure to store the files. This electronic data can then be used to fill in templates that match the eCTD format. Once all of the data are electronic and organized, it can be used in publishing systems that perform the eCTD submission.

41. What Does the Guidance "Computerized Systems Used in Clinical Trials" Apply to?

The guidelines apply to clinical trials for human and animal drugs, biologics, medical devices, and certain food/color additives (3).

42. What Documentation Does a 483 or Warning Letter Require, and How Should We Control It?

Any citation by the agency requires a response that indicates corrective action and a timeframe for such action. Responses are typically in letter form, and they require collaboration between regulatory and other areas of the company to determine if the corrective action is feasible. Regulatory typically sends the response to the agency, and then it keeps the letter on file. Correspondence with the agency belongs in the document management system for the company, where it is secure and searchable.

43.   What Is an Annual Report?

Companies file annual reports for products that they have in the marketplace. Such reports tell how many lots of a product were manufactured and distributed in the past year and what the status of these products is. In a sense, an annual report is simply an extension of an IND or IDE after approval. These reports should be retained in the document system of the company.

44.   What Exactly Is a Clinical Trial Versus a Clinical Study?

The terms are interchangeable. ICH defines a clinical trial this way: "Any investigation in human subjects intended to discover or verify the clinical, pharmacological, and/or other pharmacodynamic effects of an investigation product(s), and/or to identify any adverse reactions to an investigation product(s), and/or to study absorption, distribution, metabolism, and excretion of an investigation product(s) with the object of ascertaining its safety and/or efficacy. The terms clinical trial and clinical study are synonymous (4).

45.   What Is Clinical Documentation, and What Controls Does It Require?

This documentation reflects testing in humans. In sum, it tells the story of what you planned to do, what you did, and what the outcome was in clinical testing. Clinical documentation can be looked at as a three-phase process, with documentation in place before a trial begins, documentation accrued during the trial, and documentation that reports on the trial once it is over. Both sponsor and site must have the appropriate documents to support the trial. A clinical trial must be traceable back to its source data, so a sponsor must exercise tight controls on all documents and records it generates as part of the process.

46.   ICH E6 Tells Us What Documents a Clinical Study Should Have and that Both Site and Sponsor Should Have Copies, But Who Is Responsible for Making Those Copies?

If a sponsor uses a contract research organization (CRO) to conduct a trial, the sponsor may allow the CRO to make the copies of records generated during the study according SOPs that the sponsor has approved. The SOP must show that the copies are exact replicas of the originals and that they are closely controlled. If the sponsor conducts the study, the sponsor must have procedures in place that tell how the sites make copies and confirm replication to the original. The CRO/site monitors, while the sponsor audits, to affirm good documentation practices. (See Table 9.2.)

47.   What Are Clinical Source Data?

Source data, often called raw data, are initial recordings of data that support a study's findings. Source data may be hand annotations on a paper checklist, a medical record, a computer printout, a patient diary, or even a note to file. Source data must

TABLE 9.2. ICH Recommended Clinical Trial Documentation

			In the Files of	
	Title of Document	Purpose	PI or Institution	Sponsor
The following are documents that ICH recommends be in place *prior to the start of a trial.*				
1	Investigator brochure	To document that relevant and current scientific information about the investigational product has been provided to the investigator	X	X
2	Signed protocol and amendments, if any Sample case report form (CRF)	To document Principal Investigator (PI) and sponsor agreement to the protocol/amendment(s) and CRF	X	X
3	Information given to trial subject • Informed consent form (including applicable translations)	To document informed consent	X	X
	• Any other written information	To document that subjects will be given appropriate written information (content and wording) to support their ability to give informed consent		
	• Advertisement for subject recruitment	To document that recruitment measures are appropriate and not coercive		
4	Financial aspects of the trial	To document the financial agreement between the investigator/institution and the sponsor of the trial	X	X
5	Insurance statement (as required)	To document that compensation to subject(s) for trial-related injury will be available	X	X
6	Signed agreement(s) between parties, for example,	To document agreements		
	• PI/institution/sponsor		X	X
	• PI/institution/CRO		X	X (where required)
	• Sponsor and CRO		X	X
	• PI/institution/ authority(ies) where required		X	X

188

TABLE 9.2. (*Continued*)

| | | | In the Files of | |
---	Title of Document	Purpose	PI or Institution	Sponsor
7	Dated, documented approval/favorable opinion of an IRB or IEC of the following: • Protocol and any amendments • CRF (if applicable) • Informed consent forms • Any other written information provided to subjects • Advertisement for subject recruitment (if used) • Any other documents given approval/ favorable opinion	To document that the trial has been subject to IRB/ IED review and given approval/favorable opinion. To identify the version number and date of the document(s)	X	X
8	Institutional review board/ independent ethics committee composition	To document that the IRB/ IEC is constituted in agreement with GCP	X	X (where required)
9	Regulatory authority(ies) authorization/ approval/notification of protocol (where required)	To document appropriate authorization/approval/ notification by the regulatory authority(ies) has been obtained prior to initiation of the trial in compliance with the applicable regulatory requirements(s)	X (where required)	X (where required)
10	Curriculum vitae and/or other relevant documents evidencing qualifications of investigator(s) and subinvestigators	To document qualifications and eligibility to conduct trial and/or provide medical supervision of subjects	X	X
11	Normal value(s)/ range(s) for medical/ laboratory/technical procedure(s) and/or test(s) included in the protocol	To document normal values and/or ranges of tests	X	X

(*Continued*)

TABLE 9.2. (*Continued*)

	Title of Document	Purpose	In the Files of	
			PI or Institution	Sponsor
12	Medical/laboratory/ technical procedures/ tests. Certification of Accreditation or Established quality control and/or external quality assessment or Other validation (where required)	To document competence of facility to perform required test(s), and support reliability of results	X (where required)	X
13	Sample of label(s) attached to investigational product container(s)	To document compliance with applicable labeling regulations and appropriateness of instructions provided to the subjects		X
14	Instructions for handling of investigational product(s) and trial-related materials (if not included in protocol or IB)	To document instructions to ensure proper storage, packaging, dispensing, and disposition of investigational products and trial-related materials.	X	X
15	Shipping records for investigational product(s) and trial-related materials	To document shipment dates, batch numbers, and method of shipment of investigational product(s) and trial-related materials	X	X
16	Certificate(s) of analysis of investigational product(s) shipped	To document identity, purity, and strength of the investigational products to be used in the trial		X

TABLE 9.2.   (*Continued*)

	Title of Document	Purpose	In the Files of PI or Institution	In the Files of Sponsor
17	Decoding procedures for blinded trials	To document how, in case of an emergency, identity of blinded investigational product can be revealed without breaking the blind for the remaining subjects' treatment	X	X (third party if applicable)
18	Master randomization list	To document method for randomization of trial population		X (third party if applicable)
19	Pretrial monitoring report	To document that the site is suitable for the trail (may be combined with #20)		X
20	Trial initiation monitoring report	To document that trial procedures were reviewed with the investigator and investigator's trial staff (may be combined with #19)	X	X

The following are documents that ICH recommends be in place *as the trial progresses.*

	Title of Document	Purpose	In the Files of PI or Institution	In the Files of Sponsor
1	Investigator brochure updates	To document that the PI is informed in a timely manner of relevant information as it becomes available	X	X
2	Any revisions to: Protocol/amendment(s) and CRF Informed consent form Any other written information provided to subjects Advertisement for subject recruitment	To document revisions of these trial-related documents that take effect during the trial	X	X

(*Continued*)

TABLE 9.2. (*Continued*)

	Title of Document	Purpose	In the Files of PI or Institution	Sponsor
3	Dated, documented approval/favorable opinion of institutional review board (IRB)/ independent ethics committee (IEC) of the following: Protocol amendment(s) Revision of informed consent form Any other written information to be provided to the subject Advertisement for subject recruitment (if used) Any other documents given approval/ favorable opinion Continuing review of trial	To document that the amendment(s) and/or revision(s) have been subject to IRB/IEC review and were given approval/ favorable opinion. To identify the version number and date of the document(s)	X	X
4	Regulatory authority(ies) authorizations/ approvals/ notifications where required for protocol amendments and other documents	To document compliance with applicable regulatory requirements	X (where required)	X
5	Curriculum vitae for new investigator(s) and/or subinvestigators	[To verify credentials]	X	X
6	Updates to normal value(s)/range(s) for medical laboratory / technical procedures(s)/test(s) included in the protocol	To document normal values and ranges that are revised during the trial	X	XS

TABLE 9.2.   (*Continued*)

	Title of Document	Purpose	In the Files of PI or Institution	Sponsor
7	Updates of medical/ laboratory/technical procedures/tests Certification or Accreditation or Established quality control and/or external quality assessment or Other validation (where required)	To document that tests remain adequate throughout the trial period	X (where required)	X
8	Documentation of investigational product(s) and trial-related materials shipment	[To document shipment dates, batch numbers, and method of shipment of investigational product(s) and trial-related materials]	X	X
9	Certificates of analysis for new batches of investigational products	[To document identity, purity, and strength of the investigational products to be used in the trial]		X
10	Monitoring visit reports	To document site visits by, and findings of, the monitor		X
11	Relevant communications other than site visits: • Letters • Meeting notes • Notes of phone calls	To document any agreements or significant discussions regarding trial administration, protocol violations, trial conduct, adverse event (AE) reporting	X	X
12	Signed informed consent forms	To document that consent is obtained in accordance with GCP and protocol and dated prior to participation of each subject in the trial and to document direct access permission	X	

(*Continued*)

TABLE 9.2. (*Continued*)

	Title of Document	Purpose	PI or Institution	Sponsor
			In the Files of	
13	Source documents	To document the existence of subject and substantiate integrity of trial data collected. To include original documents related to the trial, to medical treatment, and to history of subject	X	
14	Signed, dated, and completed case report forms (CRFs)	To document that the investigator of authorized member of the investigator's staff confirms the observations recorded	X (copy)	X (original)
15	Documentation of CRF corrections	To document all changes/ additions or corrections made to CRFs after initial data were recorded	X (copy)	X (original)
16	Notification by originating investigator to sponsor of serious adverse events and related reports	Notification by originating investigator to sponsor of serious adverse events and related reports	X	X
17	Notification by sponsor and/or investigator, where applicable, to regulatory authority(ies) and IRB(s)/IEC(s) of unexpected serious adverse drug reactions and of other safety information	Notification by sponsor and/ or investigator, where applicable, to regulatory authority(ies) and IRB(s)/ IEC(s) of unexpected serious adverse drug reactions and of other safety information	X (where required)	X
18	Notification by sponsor to investigators of safety information	Notification by sponsor to investigators of safety information	X	X
19	Interim or annual reports to IRB/IEC and authority(ies)	Interim or annual reports to IRB/IEC and authority(ies)	X	X (where required)
20	Subject screening log	To document identification of subjects who entered pretrial screening	X	X (where required)

TABLE 9.2.   (*Continued*)

	Title of Document	Purpose	In the Files of PI or Institution	Sponsor
21	Subject identification code list	To document that investigator/institution keeps a confidential list of names of all subjects allocated to trial numbers on enrolling in the trial. Allows investigator/institution to reveal identify of any subject	X	
22	Subject enrollment log	To document chronological enrollment of subjects by trial number	X	
23	Investigational product(s) accountability at the site	To document that investigational product(s) have been used according to the protocol	X	X
24	Signature sheet	To document signatures and initials of all persons authorized to make entries and/or corrections on CRFs	X	X
25	Record of retained body fluids/tissue samples (if any)	To document location and identification of retained samples if assays need to be repeated	X	X

The following are documents that ICH recommends be in place *after completion or termination of the trial.*

	Title of Document	Purpose	In the Files of PI or Institution	Sponsor
1	Investigational product(s) accountability at site	To document that the investigational product(s) have been used according to the protocol. To document the final accounting of investigational product(s) received at the site, dispensed to subjects, returned by the subjects, and returned to sponsor	X	X
2	Documentation of investigational product(s) destruction	To document destruction of unused investigational product(s) by sponsor or at site	X (if destroyed at site)	X

(*Continued*)

TABLE 9.2. (*Continued*)

	Title of Document	Purpose	PI or Institution	Sponsor
			In the files of	
3	Completed subject identification code list	To permit identification of all subjects enrolled in the trial in case follow-up is required. List should be kept in a confidential manner and for agreed upon time	X	
4	Audit certificate (if required)	To document that audit was performed 9if required)		X
5	Final close-out monitoring report	To document that all activities required for trial close-out are completed, and copies of essential documents are held in the appropriate files		X
6	Treatment allocation and decoding documentation	Returned to sponsor to document any decoding that may have occurred		X
7	Final report by investigator/ institution to IRB/ IEC where required, and where applicable, to the regulatory authority(ies)	To document completion of the trial	X	
8	Clinical study report	To document results and interpretation of trial	X (if applicable)	X

Source: Adapted from Guidance for Industry: E6 Good Clinical Practices Consolidated Guidance (4).

be clear and readable, because it provides the tools for determining the progress and outcome of a trial. In essence, source data are the audit trail for the trial. The International Conference on Harmonization (ICH) was the first organization to define source data and source documentation.

Source Data. "All information in original records and certified copies of original records of clinical findings, observations, or other activities in a clinical trial necessary for the reconstruction and evaluation of the trial. Source data are contained in source documents (original records or certified copies)."

Source Documentation. "Original documents, data, and records (e.g., hospital records, clinical and office charts, laboratory notes, memoranda, subjects' diaries or evaluation checklists, pharmacy dispensing records, recorded data from automated instruments, copies or transcriptions certified after verification as being accurate and complete, microfiches, photographic negatives, microfilm or magnetic media, X rays, subject files, and records kept at the pharmacy, at the laboratories, and at medico-technical departments involved in the clinical trial)" (4).

48. Who Keeps Clinical Trial Source Data?

Usually a clinical trial site keeps source data, and the sponsor gets case report forms based on the source data. However, it's important to note that the sponsor "owns" the data collected at a site, and, for that reason, sponsors spell out contractually how the data are to be kept.

49. Who Is Responsible for Clinical Trial Source Data?

Ultimately, the sponsor is responsible for the integrity of the data. The outcome of the trial needs to be traceable back to the source data. Also, all data collected during a medical procedure are subject to Health Insurance, Portability and Accountability Act (HIPAA) regulations, so the clinical site also has responsibility for source data (4).

50. Must Documentation/Records Be Original Copies?

There is no regulatory requirement that an original record must be kept. But the regulations do require that the investigator prepare case histories, which may include case report forms and supporting data, including signed and dated consent forms and medical records, such as progress notes of the physician, the individual's hospital charts, and nurses notes and other such records.

51. Who Is Responsible for Standard Operating Procedures During a Clinical Trial?

The sponsor is ultimately responsible for the processes of the trial and, as such, must either provide SOPs to the site or ensure that the site's SOPs are equivalent to the sponsor's own procedures. Furthermore, the sponsor needs to make sure the sites have training records that verify clinical trial staff have undergone training and understand the procedures. The sponsor can delegate, but not abrogate, responsibility for the conduct of the trial.

52. Is There a Specific Industry Preference for Organizing and Maintaining Clinical Trial Data?

Each company must establish its own model for maintaining clinical trial data. Because business models and product in development can be extremely diverse, each company must assess what data its trials generate and then orchestrate the method for capturing, storing, and accessing it. There are many commercially available software products that provide the means to collect clinical trial data, manage clinical trials, and perform data manipulation and statistics. In addition, CROs can provide services and computer systems to perform part or all of the clinical trial requirements.

53. How Long Should a Case Report Form Be?

Data collected during a clinical study are usually documented on case report forms (CRFs). Each form usually consists one or two pages but may be longer. One or more forms are used for data collection at each visit. The entire set of forms for a subject may range in length from just a few pages to more than 100, depending on the data collection requirements and term of the trial.

54. What Is a Follow-Up Form?

Some trials use follow-up forms as separate from the CRF to record patient data at specific times after study enrollment. Other trials include these data as part of the CRF.

55. Do ICH GCP Guidelines Outline Specific Standards for Archiving Documents?

No, but the expectation is that all documents are easily accessible and clearly identified with the trial name, reference number, and trial site numbers as applicable, the name of the PI, and the date of archiving.

56. Can We Outsource Clinical Document Management?

Yes, but be careful that your contractor is following practices that you endorse and that are compatible with your business model. Evaluate the contractor for documentation practices and compliance with the regulations; spell everything out contractually; then audit at regular intervals to make sure that documents and records are complete, have integrity, and are securely maintained. Be sure to audit computer systems to ensure compliance with 21 CFR Part 11 industry standards and system validation requirements.

57. Can We Use Software to Capture E-Mail From Our Clinical Trial Sites?

Yes. There are software program expressly for those purposes. Such software is configurable so that it catches all messages from identified senders and positions them relative to the project. If you do elect to use software for this purpose, it is important that the software undergoes validation for its intended performance, that it is secure with limited access, and that there are backup and disaster recovery capabilities.

58. Who Owns Source Data, and Who Keeps It?

Usually a clinical trial site maintains the source data, and it issues case report forms to the sponsor. The sponsor relies on the data it receives in its CRFs to assess the trial's progress and outcome. The sponsor performs audits that compare the source data to the data in the case report forms.

59. What Should Go into the Case Report Forms?

Case report forms must match the requirements of the clinical study protocol and source data. The case report forms should reflect the deliverables promised to the regulatory agency. The FDA's guidance on "Computerized Systems Used in Clinical Investigations" says this: "FDA's acceptance of data from clinical trials for decision-making purposes depends on FDA's ability to verify the quality and integrity of the data during FDA on-site inspections and audits" (4).

60. Can Study Staff Make Corrections/Edits to Original Records?

Yes, they can. But source data must be secure and edits to documents must be made according to an authorized procedure and good documentation practices. Any correction must not obscure the original entry (a simple strike through, for example); and any such adjustment should have a date and time of change, along with the initials of the person making the change. Many procedures required a reason for the change to be indicated as well.

61. Should Events that Are Not Related to the Therapeutic Treatment Be Included in Case Report Forms?

Certainly all adverse events require reporting in accord with the regulations. However, it can happen that a therapeutic product shows an effect that was unanticipated, such as a decrease in blood pressure with a product undergoing testing for reducing cholesterol. The best way to track such information is to include it on the case report form (CRF) and enter it into the database. That way it can be accessed in house, if needed. The statistical analysis plan addresses which variables and fields will undergo analysis. You can design your CRF so that it has a section for "additional events" to further distinguish this information from the information subject to analysis for the trial.

62. Can We Keep Information Not Collected on CRFs in a Memo to File?

If an event occurs during a trial, it needs to be captured in the medical record. An adverse event must be reviewed, so it should also be recorded on a CRF. Other events that do not relate to a specific CRF should be recorded so that they are included with the clinical study results. For an isolated event this may be a memo to file, but for reoccurring events it is often wise to amend the CRFs to include a form for this type of data collection.

63.   What Are the Uses of Memos to File?

Memos to file can be useful for clarifying information or adding information—for instance, if there is a discrepancy, in explaining how a discrepancy was resolved. Note however, that memos to file should never be substitutes for complete source documents. Too many memos to file may indicate issues with procedures or study documentation. A memo may repeat what has already been recorded elsewhere—or, worse, contradict it.

64.   When Our Regulatory Director Refers to "Case Histories," What Does She Mean?

A case history is the sum of source documents and information reported on the case report form for each participant in a clinical trial, either on investigational drug or on control. Case histories include the case report forms and other documentation relative to the participant, such as signed and dated consent forms, medical records, physician/nurse annotations, hospital charts, laboratory test results, and so forth.

65.   Can We Still Use Paper Case Report Forms and Be Compliant?

Yes, but great care must be taken when the data are transcribed into a database because there is a great opportunity to make an error. Most clinical trial systems that involve transcription of paper CRFs require double data entry, meaning that two different people in a blinded fashion enter the data into the system. The system compares the two data entry values, and any discrepancies are noted for correction. Another common method uses double data entry with the second person having veto power over the first.

66.   After a Clinical Trial, How Are Data Kept?

After a clinical trial has ended and the data have been submitted, all of the data need to retained for an extended period of time. The industry standard is a minimum of 15 years and may be extended much longer if regulatory approval is obtained. Typically, the papers records are scanned to make electronic copies and the electronic files are archived on removable media.

67.   Does a Sponsor Have to Worry About Patient Records Generated During a Clinical Trial, in Terms of Retention After the Trial Is Over?

Typically, the sponsor is only responsible for the CRFs after the trial is complete. However, during the trial the sponsor is actively working to ensure that the source data at the clinical trial site matches the data on the CRFs. By contractual agreement the clinical trial site retains patient data related to a clinical study. The clinical trial site must retain patient data per HIPAA and other medical records regulations.

68. What's the Best Way to Access Information in a Patient Diary and Bring It into the Clinical Trial Database?

Patient diaries may include paper documentation, or electronic records such as voice or device recordings. Typically, the summary or meaning of each entry is included in the clinical trial documentation, often on CRFs, and the diaries are retained as supporting references.

69. If a Site Uses Electronic Patient Diaries, How Can a Sponsor Ensure Data Integrity?

Any electronic data included in a clinical study must comply with 21 CFR Part 11. This means that the software used for data collection must be validated, and controls for security and audit trails are in place.

70. If Data Are Entered Directly into a Validated E-System, What Is the Source Document?

Source documentation is the first record, regardless of format, obtained for a patient in a clinical study. Usually the source documentation is the clinical trial site documentation and the data transcribed into the eCRF is not the source documentation. However, data may be entered in the eCRF for which there is no clinical trial site documentation. In this case the eCRF data in the e-system is the source documentation.

71. Can We Scan CRFs into our Study Files?

A CRF may be a single paper document or a multipart form. One copy of the CFR must remain on the clinical trail site. The sponsor or CRO may receive copies of the paper document by mail, fax, or scan. In all cases the sponsor must ensure that the CRFs at the clinical trial site match the CRFs received. This includes the total volume of CRFs and versions of CRFs that result from updates to them at the clinical trial sites.

72. If a Sponsor Discontinues a Study, How Long Must Records Remain Available?

Data collected during a clinical study must be retained by the clinical trial site in accordance with HIPAA and other regulatory regulations. While there is no exact specification for the data retention period, the sponsor needs to retain the clinical trial data as a reference for future clinical studies and development work. Most companies elect to keep this type of data indefinitely.

73. What Is a Certified Copy?

A certified copy is often called an "official copy." This means that after the copy was made, a verification process took place to ensure that the copy matches the original with respect to the number of pages and the content of those pages. The copy is marked to indicate that this process has been completed. Typically, a log of certified copies is kept to indicate the person performing the verification, date, time, and recipient of the copy.

74. What Is the Difference Between Clinical Trial Monitoring and Auditing, and Where Do the Records Reside?

Monitoring is an ongoing activity during a trial, and the study team is responsible for ensuring, and documenting, that the trial is proceeding as planned. Auditing, on the other hand, is a snapshot in time, and the persons responsible for the auditing do not have direct responsibility for the matters being audited. The sponsor must retain its monitoring records as part of the trial documentation. It must keep audit reports and responses to audit findings as well.

75. How Can Sponsors Make Sure that Clinical Trial Staff, Such as PIs, Do Not Publish (Disclose) Information About the Trial?

The contractual agreement between the principal Investigator and the sponsor should spell out the company's position on publishing. It has happened that some study staff believe that the data they are accruing as a trial progresses belong to them, and they are entitled to draw conclusions and publish what they have observed. A contractual agreement can make it clear that any data and results/conclusions resulting from them belong to the sponsor and staff may not publish without the express consent of the sponsor. The sponsor should also have an SOP in place that explains the process for disclosing information via journal article or abstract or another form of publishing. Any publication prepared during a trial must be subject to appropriate scrutiny and approval prior to release.

REFERENCES

1. www.cdisc.org/standards, accessed March 30, 2009.
2. The eCTD Backbone File Specification for Study Tagging Files V2.6. International Conference on Harmonisation of Technical Requirements for Registration of Pharmaceuticals for Human Use. http://estri.ich.org/STF/STFv2-6doc, accessed October 2008.
3. Guidance for Industry: Computerized Systems Used In Clinical Trials, www.fda.gov/cder/guidance, accessed March 29, 2009.
4. Guidance for Industry: E6 Good Clinical Practice Consolidated Guidance, www.fda.gov/cder/guidance/959fnl.pdf, accessed October 31, 2008.

CHAPTER 10

CONSISTENCY AND READABILITY IN DOCUMENTS

INTRODUCTION

An integral part of document management is the generation of the documents themselves. Because the therapeutic products industry must produce massive amounts of documentation, people tasked with writing may feel pressured because they do not have a grasp of what goes into good writing and how to control the language itself. The industry is truly multicultural and global, so for many people English is a second language. For many others, English may be the first language, but they don't like to write, or lack the skills to do so effectively. Many native English speakers chose science, medicine, engineering, or even business as career paths because they didn't want to be writers. And here they are, tasked with writing, which is ironically a critical skill for this industry. This chapter addresses writing challenges in the industry and answers the following questions.

1. Why is there so much emphasis on documentation?
2. Why is English the preferred language for the health-care industry?
3. What kind of language is English?
4. Why are so many of the words in English the same as the words in the European languages?
5. What is the correct English form for submissions to the United States versus Europe?
6. Why does European English often use different spelling than the US English?
7. Since there are many words in American English that other forms of English don't use, how can we make sure we are understood?
8. How can we make the translation process easier?
9. What should we do if we can't translate a word precisely?
10. Why do we have so many problems making our translated text fit our labeling specifications?
11. What does the term "red thread" refer to?

---

12. Does the government offer any advice about writing?

13. Is it acceptable to use "I" or "we" in journal articles?

14. Doesn't the active voice require including a person?

15. What is meant by "person" in writing?

16. What is the imperative voice?

17. Is it okay to mix voice in a document?

18. Why would a trainer onsite advise us not to cut and paste directly from a protocol into a final report, which is our normal practice?

19. Is it good advice not to shift tenses when writing?

20. Our publications department always writes press releases about our studies in the past tense, but shouldn't the information about what the results show be in the present tense?

21. When we write letters to the agency, is there a way to identify that the signature represents the position of the company?

22. Is our project manager correct in saying wording in IND updates should be the same from document to document?

23. Why do we have so many acronyms, and what's the best way to handle them?

24. If a document such as a submission has an acronym and definition list, should we still spell out the acronym in the text?

25. How do we clarify when different writers use the same terms to mean different things?

26. When preparing submissions, how do we present our product in the best light?

27. Is our consultant correct in telling us to write SOPs like the regulations?

28. Is our boss correct when he maintains that we can give procedures flexibility by using the word "should" in our process steps?

29. In procedure writing, is it better to say "Start the packaging run when QA has cleared the line" or "When QA has cleared the line, start the packing run"?

30. How should we structure the titles to our SOPs?

31. Is it okay to include peoples' names in SOPs?

32. In what documents can you include actual peoples' names?

33. When we number sections of documents, do we have to indent for each level of numbering?

34. When my boss tells me I am too "wordy," what does she mean?

35. I learned never to split infinitives in school, but now I see that rule violated regularly, so has it become acceptable?

36. Is it acceptable to start sentences with "and, but," or "because"?

37. Is there any rule about not ending a sentence with a preposition?

38. What's the difference between a phrase and a clause?

39. Why do our European colleagues use quotation marks differently than we do in the United States?

40. What's the proper way to punctuate bullets?

41. Which is the correct title: "Physicians Conference" or "Physicians' Conference"?

42. Word underlines "which" if I don't put a comma before it; is it correct?

43. Our Australian colleagues insist on the "Oxford" comma, but what exactly is it?

44. How can we clarify a series where one unit has two parts?

45. How do you use semicolons?

46. Should "in vivo" and "in vitro" and other Latinate terms be in italics?

47. Why do our Eastern European and Asian writers leave out articles?

48. In the review cycle, our QA people always change the article "an" to "a" before acronyms like FDA and SOP, but it doesn't sound right. Are they correct?

49. How can writers know when to use the infinitive or gerund ("to see" or "seeing") in a sentence?

50. How should we comment during the review cycle?

51. Does an author have to incorporate every suggestion that a reviewer makes?

52. How many review cycles should a document go through?

53. What should I do when my boss changes whatever I write during the review cycle?

54. How can I suggest changes to a document that I know someone has worked hard on before it goes into review?

55. What's the best way to review documents of non-native writers?

56. How can we speed up the review process when the regulatory manager moves commas and restructures sentences?

57. We hear a lot about metrics, but what exactly are they?

58. How can we avoid wasting time deciding mechanical issues such as how to abbreviate "standard deviation" as "s.d." or SD?

1. Why Is There So Much Emphasis on Documentation?

Documentation provides proof of your activities. The regulations are clear about what companies need to document in relation to their business model, and documentation must be ongoing and constant. Furthermore, it must be clear, and it must be true. Documentation, in sum, provides the big picture of a company's operations and products. Well-written documents make inspections flow smoothly, help prevent regulatory citations, and speed the review process for new products.

2. Why Is English the Preferred Language for the Health-Care Industry?

English is less difficult to learn than most of the other major languages. It has a simpler tense system and a fixed word order. You can make yourself understood in

English, even if the conventions are not perfect. English offers the ability to be concise—it uses approximately 20% fewer words to convey the same message as, for instance, French or Italian. Only Mandarin Chinese is spoken by more people, but English is more widely spoken around the globe and has wider dispersion than any other language. English is the official language of England, Ireland, the United States, Canada, Australia, and New Zealand. It is also the official language of Ghana, Liberia, Nigeria, Uganda, and Zimbabwe in Africa; Jamaica, the Bahamas, the Dominican Republic, and Barbados in the Caribbean; Vanuatu, Fiji, and the Solomon Islands in the Pacific; and a dozen other nations and territories. In more than 20 nations, English shares official status with another language. Some of these nations are Singapore, the Philippines, India, and Pakistan. In still other nations, English holds no official status but is widely spoken, particularly in the business sector. English is also the official language of the United Nations. It widely used in science and other technical arenas. And more Nobel Prizes in literature have been awarded to writers using English than any other language.

3.   What Kind of Language Is English?

English belongs to the Germanic group of languages. Other Germanic languages include Norwegian, Swedish, Danish, Icelandic, Dutch, Afrikaans, and, of course, German, although German has had some linguistic shifts that distance it somewhat from the other languages in this group.

4.   Why Are So Many of the Words in English the Same as the Words in the European Languages?

Three-fifths of the words in English come from French, Latin, and Greek, and one-fifth are Anglo-Saxon in origin, so the words have common roots with European languages. Another one-fifth come from languages around the word. English is quick to embrace new words; and there are no restrictions on adding to vocabulary. Historically, English has well over one million words; of these, three-quarters are technical.

5.   What Is the Correct English Form for Submissions to the United States Versus Europe?

For submissions in the United States, it's best to use traditional American English. In Europe, British English is the standard. That said, there is little danger of being misunderstood if you use the conventions of either American or British English. Note also that countries that have been British territories generally adhere to the conventions of British English because that is what was taught in those areas.

6.   Why Does European English Often Use Different Spelling than the US English?

Spelling is a variable, not just in American and British English, but in the forms of English worldwide. In British English, for instance, "harmonization" is spelled with an "s" instead of the "z." British English also commonly uses a "u" in words like color, so that the acceptable spelling is "colour." The word "judgment" in British English has an "e," making it "judgement." American dictionaries show two spellings, with "judgment" being the first choice. In England, the word "whilst" is still in use, but in America, "while" is the standard.

7. Since There Are Many Words in American English that Other Forms of English Don't Use, How Can We Make Sure We Are Understood?

English—in all its forms—is highly idiomatic. In America there is no official organization to keep the language pure, unlike other countries such as France or Spain. For writing in this industry, it's best to use standard, traditional terminology because the documents that companies produce provide a history of their products over the life cycle of those products. The language must be clear and comprehensible throughout the entire life cycle.

8. How Can We Make the Translation Process Easier?

Translation continues to be problematic because word-for-word translation usually yields inaccurate results. Translating context rather than specific wording is best. Many companies use a process called "back translation" where one person translates from English to another language, and a second person translates it back to English for a comparison to the original.

9. What Should We Do If We Can't Translate a Word Precisely?

It is the context that you need to concentrate on, not the words. English has the largest vocabulary of any language ever, which means that it has more than one word to convey a message. Other languages can convey the same meaning, but do not necessarily have equivalent words.

10. Why Do We Have So Many Problems Making Our Translated Text Fit Our Labeling Specifications?

English allows for precision and requires fewer words to get a message across. If an English language label is translated to Italian, for instance, the Italian text will be about 20% longer. It's thus best to consider the languages the label will ultimately bear before setting the specifications.

11. What Does the Term "Red Thread" Refer to?

The red thread is an excellent concept for the type of writing we do in the therapeutic product industry. The red thread means that each document you write links to the others that came before it, so, for instance, if you are writing a validation report for a device, you identify the product, the indication, and where you are in development, so that readers know the exact status of the device and can find the history leading up to this point easily. In sum, the protocols and reports for a single product can show continuity and progression from first proof of concept in a laboratory notebook to a submission for approval and beyond.

12. Does the Government Offer Any Advice About Writing?

The US government endorses the plain language movement (1). A directive for plain language has also been signed by US presidents and is supported by the FDA.

13. Is It Acceptable to Use "I" or "We" in Journal Articles?

Yes, but many journals still prefer the passive voice. That preference seems to be changing, possibly because of the global nature of English. "We found differences in the reactivity in the ferret trachea" is the active voice. In the passive voice the sentence becomes "Differences in the reactivity in the ferret trachea were found." The latter construction has been termed the "scientific voice," but it is harder to read and relegates the verb to the past participle, with the tense held in the "to be" verb.

14. Doesn't the Active Voice Require Including a Person?

Not necessarily, but it does require an agent of action. Consider this sentence: "The compound is being developed in the serotonin laboratories." The verb here is "is being developed," a present progressive passive construction. "Serotonin laboratories" sits in a prepositional phrase. A rewrite can make the laboratories the subject: "The serotonin laboratories are developing the compound."

15. What is Meant by "Person" in Writing?

Person refers to agents of action in a sentence. First person is "I" or "we"; second person is "you," and third person is "he," "she," or "it." It is common to see "we" in writing such as company mission statements or websites. "You" works well in procedures with defined start and stop points that one or two people can carry out. Third person is appropriate for most of the writing that occurs in this industry. "QA initials the batch record" is an example of third person. So is "Equipment undergoes qualification."

16. What Is the Imperative Voice?

The second person "you" is the only subject you can actually remove in English and still have a complete sentence. When writers do this, they create the imperative voice, as in "Remove the beaker from the heat." The imperative voice is excellent for giving instructions, particularly in procedures that have defined start and stop points.

17. Is It Okay to Mix Voice in a Document?

Yes, provided that you control it. Protocols, amendments, and reports are usually best in the third person, but sometimes writers refer to the company using "we," which is known as the "corporate 'we.'" Records such as investigations typically use the third person "The operator failed to notify QA" is an example. Documents such as chemical hygiene plans and SOPs often mix voice. A chemical hygiene plan may talk about the overall operations of a laboratory, but then give directions in the imperative voice. Similarly, a procedure may speak directly to the key person(s) carrying it out and refer to other persons in the third voice, as in "When QA clears the line, begin packaging." Begin packaging" is the imperative voice.

18. Why Would a Trainer Onsite Advise Us Not to Cut and Paste Directly from a Protocol into a Final Report, Which Is Our Normal Practice?

Evidently, this is a common practice, but it can pose some problems. If a final report, say for a good laboratory practice (GLP) study, includes wording from the protocol that says what the study will do, it will be very confusing to the reader, since the report tells what has happened and what the outcome is. If you do cut and paste, make sure you change the tenses to indicate what the study did.

19. Is It Good Advice Not to Shift Tenses When Writing?

Better advice would be to control your tenses when writing. Consider this passage, which tells a little story: "We were working on a particularly sensitive process, when the power failed." The first verb, "were working" is the past progressive tense; the verb "failed" is the simple past. Records like investigations often identify a procedure in the present tense, tell what happened in the past tense, explain the outcome in the present tense, and tell action that is going to happen as a result in the future tense.

20. Our Publications Department Always Writes Press Releases About Our Studies in the Past Tense, But Shouldn't the Information About What the Results Show Be in the Present Tense?

You are correct. When a company releases information about its activities, it tells what it is doing, what it has accomplished, and what it knows as a result. So in essence, a press release announcing a completed clinical trial may state what the company is developing (present tense), what the specific trial did (past tense), what the outcome was (past tense), and what the company knows as a result (present tense). It may even talk about what the company will do next (future tense).

21. When We Write Letters to the Agency, Is There a Way to Identify that the Signature Represents the Position of the Company?

Yes. The letter, by content, will already have established the position of the company, but to indicate legal responsibility for the message on behalf of the company, according to the industry book *Write It Down* (2), the following format is standard:

Sincerely yours,

RONWAY PHARMACEUTICAL COMPANY

(signature)

Erin Andreas

Vice President, Regulatory Affairs

22. Is Our Project Manager Correct in Saying Wording in IND Updates Should Be the Same from Document to Document?

Yes. It's good advice. Presenting information about your product in development the same way every time you submit to the agency allows the message to be consistent and uses wording that the reviewers are familiar with. Changes in an update to an IND reflect what is different from the last filing; in other words, what do you know now about your product that you didn't know at the last filing?

23. Why Do We Have So Many Acronyms, and What's the Best Way to Handle Them?

Like many other industries, this one has a multitude of acronyms. Acronyms are abbreviations formed with the first letters or group of letters in a phrase. These components may be individual letters, as in PI for Principal Investigator, and/or parts of words, as in MedDRA for medical dictionary of Regulatory Affairs. Writers create confusion when they use the same acronyms to mean different things. For instance, cGMP actually has at least two meanings: current good manufacturing practices and cyclic guanosine monophosphate. The industry standard right now for writing acronyms is, in the first citation, to spell out the words in their entirety and put the acronym in parentheses immediately thereafter. After that, using the acronym is fine, as in the following sentences: Standard Operating Procedures (SOPs) are a requirement for the industry. SOPs work in tandem with instructions. The FDA also posts a list of preferred acronyms on its website (3).

24. If a Document Such as a Submission Has an Acronym and Definition List, Should We Still Spell Out the Acronym in the Text?

Absolutely. While the regulations call for a list of terms, it's still best to follow the convention of spelling out the words in their entirety in the text and then repeating the process in major sections of a lengthy document. Since submission documents typically have several readers, following this practice helps ensure that each reader fully understands what the acronyms mean. (See Figure 10.1)

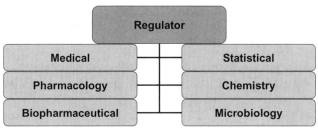

Figure 10.1.   Readers at the FDA.

25.   How Do We Clarify When Different Writers Use the Same Terms to Mean Different Things?

It's good practice to establish a terminology list that includes definitions as well as acronyms and other abbreviations. For example, a company had several reports that used the following terms interchangeably: deficiencies, citations, and nonconformities. They determined to standardize and assigned terms to different types of nonconformances. Deviations applied to manufacturing; out-of-specification (OOS) applied to laboratories; audit deficiencies applied to auditing activities; and citations applied to field investigations. Such standardization helps prevent misinterpretation.

26.   When Preparing Submissions, How Do We Present Our Product in the Best Light?

Writing must be both true and clear. You should always advocate the product, but not advertise. That is, use clear, unbiased language and avoid subjective terms, such as "exciting" or "interesting." Be watchful, as well, for claims that the science or technology cannot substantiate. The key is to let the science or technology speak for itself. Even if you have negative outcomes to report, you must discuss them. Do not avoid them or try to hide them. And certainly, do not omit information that a regulator will ask for at a later time and then hold up your submission.

27.   Is Our Consultant Correct in Telling Us to Write SOPs Like the Regulations?

SOPs, like all change-controlled documents, need to be easily understandable to the workforce. They should be in language common to the work environment and reflect the current operational methods. The regulations are in a form of legalese that employs "shall," which is also the future tense, so this is not the best language for working documents. SOPs are best in the present tense, since that tense is also the "habitual" tense, which reflects ongoing, repetitive action. Remember, SOPs address repetitive action with the purpose of making sure that an action yields the same results every time it occurs. "The operator measures the quantity" is better than "The operator shall measure the quantity," since the action occurs each time the total process does.

28.   Is Our Boss Correct When He Maintains that We Can Give Procedures Flexibility by Using the Word "Should" in Our Process Steps?

"Should" as a helping verb generally weakens procedures. Writers often use it to create passive directions, but in so doing they can wreak havoc with readability. Consider this procedure step in the passive voice: "When the light comes on, the tank should be filled to the 7-liter mark with purified water." The wording here is ambiguous. Does the statement mean that the operator needs to begin filling the tank or will the tank already have 7 liters of purified water in it when the light comes on? "Should" is a conditional as well, and it implies choice, so it's inappropriate in procedures, unless there is truly a condition, such as "The liquid should be pale blue." "Should" also poses mega problems in translation, since in many languages "should" means "it is preferred that," so such a statement actually gives an opinion. Your question points to another issue in procedure writing: The active voice is preferable, because it's easier to understand. Here's the same tank filling procedure step in the active voice: "When the light comes on, fill the tank to the 7-liter mark with purified water," or "When the light comes on, confirm that the tank contains 7 liters of purified water."

29.   In Procedure Writing, Is It Better to Say "Start the Packaging Run When QA Has Cleared the Line" or "When QA Has Cleared the Line, Start the Packing Run"?

The latter is preferable. Put information down in the order it occurs to prevent a premature action and possibly a deviation.

30.   How Should We Structure the Titles to Our SOPs?

Titles are important, because SOP systems need to be searchable. A title that says "Clinical Inventory Receipt, Storage, Disposition, Reconciliation, and Destruction" is much more informative than one that says "Clinical Supplies."

31.   Is It Okay to Include Peoples' Names in SOPs?

It's better to use a role or title, such as "Head of Plant Safety." If an SOP says "In case of emergency, call Joe Young," and Joe Young left the company six months ago, you have a procedure that is impossible to follow.

32.   In What Documents Can You Include Actual Peoples' Names?

Good Laboratory Practice (GLP) protocols and reports contain the name of the study director on the first page. Reports typically identify study team members as well. Documents such as validation reports identify team members by name, because the initial validation has a start and stop point, and the documentation is not subject to reversioning after completion. And documents such as chemical hygiene plans often include a safety committee, but many companies prefer to keep the safety committee in a separate short document that is cross-referenced in the plan. That way, if the new people join the committee or some move on, the main document does not have to undergo revision.

33. When We Number Sections of Documents, Do We Have to Indent for Each Level of Numbering?

No, but many companies do. The industry as a whole uses the military style of numbering: 1, 1.1, 1.1.1, and so forth. Many companies do not indent the first level and limit the number of levels since too many can make reading difficult.

34. When My Boss Tells Me I Am Too "Wordy," What Does She Mean?

She probably means that you are using more words to get your point across than you need to. A good rule of thumb for writing in this industry should be to simplify the writing, not the science or technology. So you want to provide enough detail to convince your reader of what you are saying, but you don't want to bury that detail in an avalanche or words that add nothing to the substance. William Strunk sums it up: "Vigorous writing is concise" (4). Consider this statement: "There is strong evidence shown by the current study results that hypotensive responses in normotensive rats and dogs may well have been produced by administration of Compound Z at higher doses." This sentence has 32 words. A rewrite pares 10 words with no loss of meaning: "Study results show evidence that administration of Compound Z at higher doses may have produced hypotensive responses in normotensive rats and dogs."

Other causes of wordiness are redundancies, such as "red in color" or " round in shape." Yet another contributor to wordiness is a weak sentence core. "The software package is one that is able to maintain data that are generated in the laboratory," The subject is "package" and the verb is "is," followed by "one," so the core says the "package is one." A better construction would be "The software can maintain laboratory data."

35. I Learned Never to Split Infinitives in School, But Now I See that Rule Violated Regularly, So Has It Become Acceptable?

It was never a rule, but it's not bad advice, because an infinitive has two elements: the present tense base of a verb and "to," and it's best to keep related words as close to each other as possible. That said, sometimes you may choose to split the infinitive because it is idiomatically preferable. Consider the verb "to move." If you wish to insert "really," the only place it sounds right is between the infinitive marker "to" and the verb itself: "to really move." The rule concept arose when English teachers adopted the rules of Latin and applied them to English; in Latin the infinitive is always one word, so splitting it is impossible. However, since this is a Germanic-based language, it has verbal phrases. Consider, as well, that verbs often have particles, as in "to look up." We regularly split such constructions, as in "You need to look it up in the USP."

36.   Is It Acceptable to Start Sentences with "And, "But," or "Because"?

It is perfectly acceptable to start sentences with these words. Probably the reason people think they shouldn't is that grammar school teachers often tell students not to start sentences with them. They do that because children will string sentences together with "and" or "but," or they will create sentence fragments by not completing a sentence, as in "Because I want it." That said, don't overuse these conventions either, since starting every sentence with the same words leads to a singsong rhythm in writing.

37.   Is There Any Rule About Not Ending a Sentence with a Preposition?

No. This is another false rule, but it has some merit, since prepositions in phrases take objects, and are incomplete without them. But English also has verb particles, words like "up," "down," "off," "across." It's difficult for writers to distinguish the difference. Consider "to keep down" as in "The treatment helps keep blood pressure down." Here the sentence ends in the preposition "down." If we add a phrase, you see that there are two prepositions in a row: The treatment helps keep blood pressure down in otherwise healthy adults." Even when the preposition is not working as part of the verb, managing to remove the preposition from the end of the sentence sometimes creates an awkward structure. " If we change a sentence such as this: "The doctor didn't know what he was talking about," we end up with this unwieldy rewrite: The doctor didn't know about what he was talking." Clearly such a construction is unacceptable. Nevertheless, if you don't like to end sentences with prepositions, you can simply write so that you don't as in "The doctor didn't know his topic."

38.   What's the Difference Between a Phrase and a Clause?

A clause has a subject and verb group; a phrase is a group of words that lacks either a subject or a verb. A clause that also has a complete idea is a sentence. The following sentence has a main clause, which makes it a complete sentence, a subordinate clause, and two phrases. The subject and verb groups are in bold; the phrases are in italics. "The **object** *of study IDZ-145* **was** *to validate an analytical method* **that could quantitate** a compound *in rat plasma* put in phrase italics."

39.   Why Do Our European Colleagues Use Quotation Marks Differently than We Do in the United States?

The use of quotation marks is a noted difference in punctuation. American English puts commas and periods inside quotation marks, whereas British English will place them outside or inside, depending on the context. The American style is based on a preference established by early typesetters in New England, when the British first colonized the Americas, and that preference has held. Some purists, however, advocate the British style, and it is common to see the British style in legal documents such as contracts.

40. What's the Proper Way to Punctuate Bullets?

Bullets are relatively new forms of punctuation in their own right, and were made possible with word processing. Standard punctuation was in place long before we added the ability to make bulleted lists, and there is no rule for this convention. Some writers put periods at the end of sentences, and they do not punctuate non-sentences. Other writers do not use punctuation at all with bullets. Still other writers use very formal punctuation, including commas or semicolons after all but the last bullet, which gets a period. The best advice we can give is to establish a style preference and adhere to it for consistency.

41. Which is the Correct Title: "Physicians Conference" or "Physicians' Conference"?

Both titles are equally correct. "Physicians Conference" has "physicians" as an adjective; the apostrophe simply makes it possessive.

42. Word Underlines "Which" If I Don't Put a Comma Before It; Is It Correct?

Historically, "which" and "that" have been interchange. It is really a matter of whether or not to use a comma to set off the clause that follows. "The regulation that addresses chemical hygiene is in Title 29 of the code." "That addresses chemical hygiene" is necessary to identify the topic in the regulation, so there is no comma. Consider the following sentence: "The inspection, which went well overall, produced no citations." "Which went well overall" is a subordinate element, and it adds additional, but not essential, information, hence the commas. A good thing to remember about additional information in the middle of a sentence is to set it off between two commas.

43. Our Australian Colleagues Insist on the "Oxford" Comma, But What Exactly Is It?

The Oxford comma is the same as the serial comma in the United States. The serial comma (and some in the United States would like to call it the Cambridge comma) is the last comma before "and," "or," or "but" in a list, as in the following sentence: "We tested for tablet thickness, weight range, friability, and dissolution." The journalistic style does not use it and would rather produce this sentence: "We tested for tablet thickness, weight range, friability and dissolution." Either convention is acceptable, but the serial comma helps clarify complex information and is probably preferable in scientific writing.

44. How Can We Clarify a Series Where One Unit Has Two Parts?

The following is an example of an ambiguous sentence. "The results demonstrate that the multitude of inflammatory cells, growth factors, and cytokines present in the wound bed may be modulated by application of local agents." Is it inflammatory cells and a combination of growth factors, and cytokines? Or are inflammatory cells, growth factors, and cytokines three separate entities? If it is two elements and not three, the best way to clarify is to number the elements: "The results demonstrate that (1) the multitude of inflammatory cells and (2) growth factors and cytokines present in the wound bed may be modulated by application of local agents."

45. How Do You Use Semicolons?

Semicolons are really useful marks for technical and medical writing. Unfortunately, their usage probably causes more confusion than any other mark. There are two primary uses. One is to separate sentences joined by conjunctive adverbs, which are words like "in addition," "however," "nevertheless," "thus," and "therefore." The following is a compound sentence with a semicolon: "The exact mechanism is not known; however, it is apparently not an immunosuppressant." The conjunctive adverb may be omitted in some sentences, as in the following: "The validation had two nonconformances; both were resolved and documented."

The second use is to separate items in a list when one or more items has an internal comma: "Release of a finished device for distribution requires completion of all activities in the device master record (DMR); review of testing, verification, and documentation; authorization by signature of designated individuals for acceptance and release; and dated acceptance authorization." Note that the last sentence would work equally well in bulleted form.

46. Should "In Vivo" and "In Vitro" and Other Latinate Terms Be in Italics?

There is active debate about italicizing such terms as "in vivo" and "in vitro." These words have become part of standard English usage, just like words such as "café" or "restaurant," which were borrowed from French. English constantly incorporates words from other languages, so to set them off in italics is unnecessary, although the European Medicines Agency (EMEA) advises italicizing them. Perhaps the best argument for not italicizing is this: What would we then do with "in silico," a term coined in Silicon Valley?

47. Why Do Our Eastern European and Asian Writers Leave Out Articles?

English relies on word order for meaning, unlike many other languages. Many languages, such as Chinese, Korean, and Russian, have no need for articles, so writers do not have an innate sense of their usage and, in fact, can make themselves understood without using them. Consider this sentence: "Use flat-faced wrench to tighten regulator inlet connection nut to cylinder valve outlet." The meaning is clear, but it would read better with articles: "Use a flat-faced wrench to tighten the regulator inlet connection nut to the cylinder valve outlet." The words "a," "an," and "the" are the articles in English, and they mark nouns, or naming words. They show which word in a series is the noun and which are modifiers. In the sentence "The dress code requires a clean white cotton lab coat," the article signals "coat" as the noun; the other words, some of which are nouns themselves, are acting as modifiers.

48.   In the Review Cycle, Our QA People Always Change the Article "an" to "a" Before Acronyms Like FDA and SOP, But It Doesn't Sound Right. Are They Correct?

No. Most writers have learned to use "a" before consonants and "an" before vowels, and most of the time this holds true. But the rule is that it is not the letter itself that determines the article, but the sound. So "FDA" begins with "eff," which is clearly a vowel sound, hence the article "an," as in "an FDA ruling." The word "union" provides another example, since the vowel "u" sounds like "y," hence the article "a," as in "a union contract."

49.   How Can Writers Know When to Use the Infinitive or Gerund ("to See" or "Seeing") in a Sentence?

In English, some verbs can take other verbs in the object spot. Native writers do not question whether to write "I plan to go" or "I plan going." Yet for people for whom English is not a native tongue, the two can sound exactly the same, and in fact, there is usually no issue as to meaning, just idiomatic correctness. There is only linguistic theory about how these conventions developed, and verbal phrases are idiomatic. Not all verbs can take other verbs, but many can (5). Table 10.1 shows the most common phrasal verbs that do. Note also that some verbs can take both the infinitive and the gerund, sometimes with no loss of meaning; other times with change in meaning, as in "I stopped to talk to him" and "I stopped talking to him."

50.   How Should We Comment During the Review Cycle?

Some companies still rely on paper documents during the review cycle. In such instances, people annotate their review copies and identify themselves as the reviewers (e.g., with initials). In hybrid systems, it's common to route documents for review in a PDF format; in such a system, reviewers insert comments, and the comments identify them as reviewers.

51.   Does an Author Have to Incorporate Every Suggestion that a Reviewer Makes?

No. The person who is the primary author should review all comments and incorporate those that are valid changes or additions. This process requires communication and discussion about proposed changes.

52.   How Many Review Cycles Should a Document Go Through?

If the author is diligent in getting a document as close to perfect as possible, with peer review prior to formal review, the first review cycle should pinpoint any

TABLE 10.1.   Phrasal Verbs

Verbs that Can Take Infinitive Objects

Afford	Employ	Plan
Agree	Expect	Pretend
Allow	Fail	Program
Appear	Forget[a]	Purport
Apply	Happen	Refuse
Arrange[a]	Hate[a]	Register
Ask	Have	Regret[a]
Attempt	Help	Remember[a]
Avoid	Hesitate	Seek
Be	Hope	Seem
Begin[a]	Intend	Start*
Choose	Labor	Strive
Claim	Learn	Tend
Consent	Like[a]	Think
Continue[a]	Look	Try[a]
Decide	Manage	Urge[a]
Demand	Need	Use
Deserve	Neglect	Wait
Desire	Offer	Want
Determine	Opt	Wish
	Permit	Work

We asked to go.
He attempted to dissolve the tablets.
The meeting is to start at 1 p.m.
The scientists have tried to replicate the results.
He tried to write the report.
Reviewers tend to think they have to change every document.
I want to work overtime.
I wish to speak with the director.
The scientist truly intends to find the pathway.
We always work to find the root cause.
We didn't think to look in the USP.
It happens to be the consensus.

changes. In a second review cycle, the document should be as close to perfect as possible. If the second review cycle does not yield full acceptance of the document, a concurrence meeting is in order. Nothing is gained by perpetuating the review cycle and not reaching agreement.

53.   What Should I Do When My Boss Changes Whatever I Write During the Review Cycle?

The best thing to do is step back and not take the changes as a personal criticism. The review process is part of the process of completing writing, and the regulations

TABLE 10.1. *Continued*

Verbs that Take Gerunds ("ing" Forms)

Abide	Endorse	Permit[b]
Admit	Enjoy	Promote
Admit to	Ensure	Provoke
Advise	Establish	Recommend
Allow	Finish	Regret
Anticipate	Forget	Remember[b]
Back	Forget[b]	Require
Begin[b]	Hate[b]	Risk
Continue[b]	Hear	Start[b]
Consider	Help[b]	Support
Defend	Induce	Try[b]
Demand	Like[b]	Urge[b]
Deny	Mind	Use
Do		Quit

He admits driving too fast.
We anticipate acquiring a new facility.
He continues driving too fast.
Susan remembers applying for the position.
Let's start validating the system.
I admit to doing research about our competitors.
The instructions recommend waiting a day before beginning.
Finish laying out the plan before proceeding.

[a]Can also take a gerund ("ing" form).

[b]Can also take the infinitive ("to" + base verb).

require it, so you can't escape the process. Oftentimes people think their way of saying something is better, and they edit accordingly. Sometimes changes can significantly improve a document, but they may not, and simply reflect the reviewers preference. When you disagree with a reviewer's comments, you can ask for clarification, and you may reach concurrence. Bear in mind, however, that some people have an agenda that is not related to the document itself: they are sending a message of authority. In such instances, it's best to recognize that the problem does not reside with you or the writing. The goal of review is to make the document as good as it can be, because ultimately it belongs to the company.

54. How Can I Suggest Changes to a Document that I Know Someone Has Worked Hard on Before It Goes into Review?

A good rule of thumb is to say something nice about the effort so far. Then, if you think something requires change, couch your comments in such a way that the focus is on what you think of the document. "This isn't clear to me. Do you mean A or B?" Such a comment is not an attack. A comment like "Confusing! Rewrite!" does absolutely nothing to improve a document or help the writer. Surely, such a comment can only trigger ill will. A good reviewer serves as a coach.

55. What's the Best Way to Review Documents of Non-native Writers?

Since English is a second language for a significant segment of this industry, non-native writers often bring conventions of their own languages to English, and this practice can make the writing illogical or ungrammatical. It's usually easy to insert articles and prepositions, but word order is another matter. When reviewing a document that has unclear writing, the best thing to do is sit down with the author and discuss the author's intent. Such a discussion usually works best if you acknowledge the hard work and effort the author has put into the document and then point out what you don't understand. Communication about writing, rather than just correction, is invaluable to non-native writers of English because it presents an opportunity for continual learning.

56. How Can We Speed Up the Review Process When the Regulatory Manager Moves Commas and Restructures Sentences?

Companies should establish who reviews and who approves documents by document category. Next they should establish who reviews for what purpose. For instance, patents and legal may review an abstract or journal article for correct reference to intellectual properties; medical and science groups for scientific correctness; regulatory for regulatory compliance; marketing for company image; and medical/technical writers for grammatical correctness.

57. We Hear a Lot about Metrics, But What Exactly Are They?

Metrics have to do with process and time. A process has steps, and steps can have time limits. Companies can set metrics for preparing, reviewing, revising, and completing documents by document category. For instance, a writer may have two weeks to draft an SOP; reviewers may have a week to provide comments; the writer may have a week to incorporate or reject comments; and the document reaches completion the following week with a concurrence meeting and sign off. There's a caveat, however: Don't put metrics into SOPs for document preparation, because metrics must be flexible. It's best to delegate the setting of metrics to one person, such as a project manager, working on a specific task.

58. How Can We Avoid Wasting Time Deciding Mechanical Issues Such as How to Abbreviate "Standard Deviation" as "s.d." or SD?

The best way to avoid this sort of grief is to adhere to a standard style. Many companies develop their own style guides that cover such documentation components as acronyms, abbreviations, definitions, preferred usage, mechanics, and formatting.

Creating such a guide is time-consuming, however, and once established, a guide needs to be controlled, with updates that include new documentation issues that arise. Having a style guide is beneficial, because it provides a place to park conventions that don't belong in SOPs, such as "Put one space between sentences." Such a statement in an SOP is just silly, since writers can easily violate such a convention, particularly in the revision process. If you create such a guide, make it a reference document, and don't make it binding like an SOP. Identify it as such with a qualifying statement—for example, "This style guide is a reference tool for writers to help them achieve consistency." There are also excellent published guides such as the *AMA Manual of Style* (6), *The Chicago Manual of Style* (7), and Scientific Style and Format (8).

REFERENCES

1. www.plainlanguage.gov, accessed March 5, 2009.
2. Gough, Janet. *Write It Down: Guidance for Preparing Effective and Compliant Documentation*, 2002. Taylor and Francis: Boca Raton, FL.
3. www.fda.gov
4. Strunk, William, and White, E. B. *The Elements of Style*, 3rd edition, 1979. Macmillan: New York.
5. Celce-Murcia, Marianne, and Larsen-Freeman, Diane. *The Grammar Book: An ESL/EFL Teacher's Course*, 2nd edition, 1999. Heinle & Heinle: Boston.
6. *AMA Manual of Style: A Guide for Authors and Editors*, 10th edition, 2007. Oxford University Press: New York.
7. *The Chicago Manual of Style; The Essential Guide for Writers, Editors, and Publishers*, 14th edition, 1993. Chicago and London: The University of Chicago Press.
8. *Scientific Style and Format: The CBE Manual for Authors, Editors, and Publishers*, 6th edition, 1994. Cambridge: Cambridge University Press.

CHAPTER 11

MAINTAINING THE SYSTEM

INTRODUCTION

So much emphasis is on defining and implementing a document system, but maintaining it and keeping it operating optimally is equally important. Because business models and the workforce change, the system requires careful tending so that continues to perform its critical function. Many factors can affect how a system performs: it may become burdened with files that are no longer critical; it may need to absorb new categories of documentation; or it may require allowing access to the system from remote locations. Whatever changes the company undergoes, it is likely that many will affect the system for document management. This chapter addresses managing the system over time and answers the following questions:

1. How can we continue to maintain data integrity?

2. What should we retain and archive: just paper, electronic records, or both?

3. Our new partner requires that we provide analytical test results, but can we use the public Internet to transmit them?

4. How long do validation and change control documents have to be retained?

5. How can we make sure our CROs maintain our documents compliantly?

6. When we audit our CRO's document practices, what should we look for?

7. How do you retire a computer system?

8. How do you migrate data from one system to another system?

   If a firm stops using an electronic system in favor of a different system, does it have to keep the hardware and software?

9. If a firm stops using an electronic system in favor of a different system, does it have to keep the hardware and software?

10. How do companies keep their systems compliant?

11. Our system was audited and found to be noncompliant, so does it make sense to do a gap analysis or go directly to remediation?

12. Is it QA's responsibility to make sure the computer systems remain compliant?

13. How do we ensure QA audits effectively?

14. Do we have to keep audit reports on our systems?

15. How can we make sure that our document/record-keeping systems are functioning as they should?

16. Is there an industry standard for auditing frequency of record-keeping/document management systems?

17. If we have an e-system, what happens when we don't have funds to update the version of the software and the vendor refuses to provide help-desk support?

18. What should we do if a user forgets his/her password?

19. What does scalability mean?

20. How do we best integrate the system our newly acquired company uses with our own electronic system?

21. How can we integrate our new company's SOPs with the SOPs we already have in place?

22. Can we permit consultants to use our system?

23. Can we eliminate using two passwords and use one for both logging on and approving documents?

24. If we change from using two passwords to using one, do we need to revalidate?

25. When we change our document templates, should we reformat all the documents created in the old template?

26. How can we increase user access to our system as the company grows?

27. How can we expand our EDMS system with additional software for publishing?

28. What do we do when a vice president allows his administrative assistant to access the system using his password?

29. How can a system tell if someone uses someone else's password to access a system?

30. How can we change our system to automatically lock out someone when they leave the program unattended?

31. What's the best method for backing up data?

32. Can our software provider legally refuse to give us help-desk services if we don't need or want the upgrade to the software?

33. Since we have rolling blackouts and are worried that the system may go down, are there any precautions we should take?

34. How much data can we afford to lose if the system goes down?

35. How should we approach disaster recovery?

36. If we send our paper files to a storage facility, how can we make sure the facility will keep them secure?

37. If we perform a GLP animal study and the results rule out a path for development, can we safely get rid of that documentation?

38. What should we do when people regularly print out documents from our e-system and circulate them as paper documents for review and comment?

39. How can we make our slow and difficult to access EDMS system more user friendly?

40. How can we increase the hardware capacity?

41. If we increase the hardware capacity, do we need to revalidate?

42. Now that we have added devices to our product line, how can we modify our system to house the documents and records for those products?

43. How can we change privileges so that we can add more system administrators?

44. How can we provide access to our e-system to our new partners?

45. How can we add remote access to our system for our executives who travel?

46. What can be done if a laptop containing confidential information is stolen?

47. How can we prevent reviewers from constantly rejecting documents during review cycles?

48. What should we do with people who complain about our system and make things difficult for others?

49. What is the biggest threat to our system, and what should we look for?

50. If our CRO can't produce our documents in a "timely manner," what should we do?

51. The contractor who is performing GLP studies for us is going out of business, and can't find our records, so what should we do?

52. When a software vendor issues an upgrade that it has validated, can a company install the upgrade into the production environment without additional validation?

53. If a server has dual redundant power supplies and one fails, do the users have to approve a change control document before IT makes the repair?

54. Should a company upgrade when the developer makes an upgrade available?

55. How does a company know what changes have been included in an upgrade?

1. How Can We Continue to Maintain Data Integrity?

Data integrity is part of any process, so when you validate the process, you demonstrate data integrity. But verification of data integrity must also occur on an ongoing basis as the system is used.

2. What Should We Retain and Archive: Just Paper, Electronic Records, or Both?

You should retain electronic records for sure. Once paper records are backed up to electronic files via scan, the paper can be destroyed, which saves storage costs. Archive means to move data from the original storage location to another location. You can archive anything any time because there is no net loss of data.

3.   Our New Partner Requires that We Provide Analytical Test Results, But Can We Use the Public Internet to Transmit Them?

You can use the public Internet to transmit any data provided the transmission is secure, meaning encrypted.

4.   How Long Do Validation and Change Control Documents Have to Be Retained?

Validation and change control documents have to be retained for exactly the same period as the data. Paper copies can be destroyed as long as you keep the backup electronic scan copies. That usually occurs after somewhere between 7 and 15 years.

5.   How Can We Make Sure Our CROs Maintain Our Documents Compliantly?

As with any group, internal or external, the people who own the data and use it to make decisions are responsible for controlling them. Control requires constant vigilance, a series of audits, and good procedures that transfer backup and archive copies of data at regular intervals. It is best to establish document controls with CROs after an onsite audit and include those controls in the contract.

6.   When We Audit Our CRO's Document Practices, What Should We Look For?

It's important that you affirm that a contractor producing documentation for you is handling it compliantly. The contract you have with the organization doing work for you should have spelled out the deliverables, including documentation. If the contractor is retaining original documents and you are getting copies, then you need to confirm that the original documents are maintained appropriately. A look at the documentation practices, including document preparation and review, is warranted each time you audit the organization.

7.   How Do You Retire a Computer System?

In order to retire a computer system, the users must determine how the data in the system will be accessed after retirement. The data must be kept available during the entire data retention period. If a data retention period has not been established, this is the first step. There are many options to choose from in order to access the data. The system can be retained in its entirety with the database changed to read-only access. The data can be migrated to another computer system. Data migration requires data validation of the data transfer and computer system validation of the receiving system. The data can be exported and made available in another application. Often exported data is formatted for XML (extensive markup language). Sometimes the hardware components are securely stored with the data unavailable online. In all cases, the data must be backed up and retained offsite for the entire data retention period.

Data transfer can occur by exporting the data from the old system and importing the data to the new system. If this functionality isn't available in one of the systems, custom software programs may have to be developed. These software programs must be validated to ensure that they don't omit the data entry checks that would normally occur if data were entered from the user interface screens and other data repositories. In any case, both quantity and quality checks must occur on the transferred data. Quantity checks involve counting the number of records or bytes before and after the transfer to ensure that all are accounted for. Quality checks involve manual comparison of data between fields in the old and new system. Quality checks should be performed for the first record, last record, and several complex records that represent the spectrum of data. Data migration may be documented as part of the computer system validation of the new system, as retirement of the old system, or as a separate validation project.

8. How Do You Migrate Data from One System to Another System?

Data can be migrated to another system in several ways. The first thing to consider is which data to migrate. Adding imperfect data to a new system can cause problems in the new system. Migrating or transferring old and rarely used data to a new system is both time consuming and inefficient. Data in the old system must be "clean," meaning that all requirements from the new system must be addressed. Old data may not match the requirements of the new system and may require data conversion. A common example is converting two-digit year values to four-digit year values. After the old data are cleaned and converted, they can be transferred to the new system. A simple way to transfer data is to manually enter it into the new system and perform quality checks. While this is impractical for large volumes of data, it is often overlooked as a viable and inexpensive means of data transfer. Most important, manual data entry does not bypass the data entry checks that help to ensure valid data.

9. If a Firm Stops Using an Electronic System in Favor of a Different System, Does It Have to Keep the Hardware and Software?

Part 11 says that electronic records must be retrievable throughout the record retention period established in the predicate rules. This doesn't mean, however, that the hardware and software need to be retained. Companies can convert to another technology that can read the records. What's important is that transferred records are identical to the records in the original system and that metadata, audit trails, and electronic signature record links are retained. Only if the data cannot be transferred does the original hardware and software need to be retained.

10. How Do Companies Keep Their Systems Compliant?

Computer systems evolve as their configurations, software updates, and processes change. A full validation project is followed by a series of software change controls, and major changes to the system require additional full validation projects.

11. Our System Was Audited and Found to Be Noncompliant, So Does It Make Sense to Do a Gap Analysis or Go Directly to Remediation?

Gap analysis identifies what a system should have that it doesn't have. Remediation deals with how to correct or replace a system. You can't do remediation without knowing the gaps. A gap analysis on a system usually takes less than two hours. If a gap analysis is taking much more time to accomplish, then it is the gap analysis method that is at fault.

12. Is It QA's Responsibility to Make Sure the Computer Systems Remain Compliant?

Users are responsible for computer system validation and ongoing compliance. QA is responsible for ensuring that the users are maintaining compliance but are not themselves directly responsible for the users' compliance.

13. How Do We Ensure that QA Audits Effectively?

QA audits help users to focus on deficient areas. The user groups and QA should work together, as opposed to maintaining an adversarial relationship.

14. Do We Have to Keep Audit Reports on Our Systems?

Audit reports are change-controlled documents and should be maintained in the same manner as validation documents and system SOPs.

15. How Can We Make Sure that Our Document/ Record-Keeping Systems Are Functioning as They Should?

Users ensure that computer systems are functioning as intended by performing computer system validation and then change control. All users are responsible for ensuring correct operation on a continual basis. During computer system validation, the risk assessment is used to identify hazards, resulting effects, and risk mitigations. These mitigations become part of the process and provide a check and balance so that users can determine when a system is not functioning as intended.

16. Is There an Industry Standard for Auditing Frequency of Record-Keeping/Document Management Systems?

Computer systems are included in department audits that follow the standard one- to two-year audit frequency.

17. If We Have an E-System, What Happens When We Don't Have Funds to Update the Version of the Software and the Vendor Refuses to Provide Help-Desk Support?

Based on the terms of the contract with the software vendor, most computer systems may continue in production even when maintenance fees are no longer paid. While the software is not upgraded and vendor support is no longer available, users may operate the system using internal or third-party resources for support and maintenance.

18. What Should We Do If a User Forgets His/Her Password?

Each time a user forgets his/her password, a new temporary password may be issued and upon the next login, the user changes the password. The system then automatically links the password to the user's name. The user may select any password that meets the complexity requirements and as such are easier to remember.

19. What Does Scalability Mean?

It is the ability of a system to grow with the company as the company's needs change and evolve. A system that is scalable can have additional hardware and software components installed to allow more users and more work to be performed without having to replace the software and retrain the users.

20. How Do We Best Integrate the System Our Newly Acquired Company Uses with Our Own Electronic System?

Usually when companies start working together, there is a desire to unify processes and reported information. The tendency is for the parent or bigger company to immediately demand radical changes from the other. This causes much conflict and ill will and often results in reduced productivity. Most often a slow evolutionary approach works best. First identify a minimal set of requirements that are required from both companies. Each company should explain why they need each requirement and how they obtain it. Next, develop plans as to how changes can be made. Sometimes minor changes can be made quickly, sometimes exporting data to the other company's system allows for them to report the data using their own familiar tools, while other times a system needs to be replaced.

21. How Can We Integrate Our New Company's SOPs with the SOPs We Already Have in Place?

The rule of thumb here is to go slowly and don't try to control everything with one set of SOPs. There are many things to consider, and you don't want to make radical changes that can introduce errors and a lack of productivity. The first step is to educate each company about the other company's SOPs. This process includes document number, format, content, hierarchy, review and approval process, and revision processes. Next create a cross reference that shows where commonalities exist. Doing so will make the differences easily identifiable. Finally, develop a plan that makes one small change at a time. Often cosmetic and non-content changes are a good way to start off to create harmony and familiarity. Each change will become easier to make as time goes on because over time the two companies will see themselves as one company and the natural tendency for uniformity will become the driving force.

22.   Can We Permit Consultants to Use Our System?

There are so many types of workers today that work from so many locations there is a new term that addresses all workers, the "workforce." The workforce includes employees, temporary employees, remote workers, contractors, vendors, consultants, interns, second- and third-party staff, and anyone that needs access to a company's processes and computer systems. It is extremely important to set strict security access roles appropriately.

23.   Can We Eliminate Using Two Passwords and Use One for Both Logging On and Approving Documents?

There is no need to have separate passwords within a computer system. Many systems today use single authentication; that is, all passwords are authenticated against the network operating system, such as Windows Active Directory. Doing so allows one set of password rules to be enforced for all software applications, eliminates users having to remember and change multiple passwords, and has less IT overhead.

24.   If We Change from Using Two Passwords to Using One, Do We Need to Revalidate?

Changing from two passwords to one password is usually a simple configuration option change. This change must be performed using the standard change control process and must be validated, usually in a test environment, before production roll out.

25.   When We Change Our Document Templates, Should We Reformat All the Documents Created in the Old Template?

When a document template changes, only documents created or revised after that effective date are impacted. There's no need to reformat since documents are linked by date to previous versions of templates.

26.   How Can We Increase User Access to Our System as the Company Grows?

The key to adding more users to a system is to use role-based security. With role-based security, roles are assigned access privileges, and there is only a limited set of roles. Many users can be assigned the same roles with no impact on the validation status of the security system. Be careful not to increase the number of roles to match job descriptions; instead assign multiple roles to a user.

27.   How Can We Expand Our EDMS System with Additional Software for Publishing?

Some Electronic Document Management Systems have submission publishing options that are sold from the EDMS vendor or another vendor. Some publishing applications have document input mechanisms to allow documents from the EDMS to be "copied" to the publishing application. If these options aren't available, it is common to export documents from the EDMS as PDF files and then import them into the publishing system.

28.   What Do We Do When a Vice President Allows His Administrative Assistant to Access the System Using his Password?

It is a violation of federal and state laws to allow someone to use your password or to use someone else's password. This behavior is unacceptable at any level. Most companies have Quality Assurance address corrective actions. To keep the system compliant, the audit trail record must always record the correct user for all actions.

29.   How Can a System Tell If Someone Uses Someone Else's Password to Access a System?

There are a few 21 CFR Part 11 industry standards that help users to detect that someone else has used their password. The most effective industry standard is the feature in which the user logs on and then the system tells the user the last logon and last logoff dates and times. This allows the user to identify if someone else used the system between sessions. Another industry standard requires the logon and logoff information to be captured in a log, so a user that suspects inappropriate access can review the activity log. When in doubt, users should change their password, thereby rendering the old password void.

30.   How Can We Change Our System to Automatically Lock Out Someone When They Leave the Program Unattended?

Most software applications have a configuration option that allows the system to lock out a user after a preset time of inactivity. If the application does not provide this feature, there may be an option for the web service or database components. Users should contact the software vendor to have this feature added because this is a 21 CFR Part 11 industry standard. In addition to the application locking out a user, the operating system should be configured to blank the screen and lock out a user by requiring a password for reactivation.

31.   What's the Best Method for Backing Up Data?

There are many equally effective methods for data backup. The key is to perform backup daily and store the backup in another location so that a disaster at one location doesn't cause both copies of data to be destroyed. This may mean that the backup location must be several miles or more distant from the original location. Daily backup of data has been the norm. However, large systems with many users and the increase in computerization have led companies to rethink the daily backup. The use of redundant hard drives, journaling, hourly differential backup, and automatic system failure switchover techniques are becoming commonplace for many systems.

32.   Can Our Software Provider Legally Refuse to Give Us Help-Desk Services If We Don't Need or Want the Upgrade to the Software?

The software license and maintenance agreements describe the services, payment, and conditions between two companies. This is why it is important that the software procurement process include review of contracts by legal and software purchasing specialists.

33.   Since We Have Rolling Blackouts and Are Worried that the System May Go Down, Are There Any Precautions We Should Take?

Even if you don't have inconsistent utility power, you should assume that a power failure can occur at any time. All servers and critical computer system components should be connected to uninterruptible power supplies that automatically condition the power and shut down the equipment if sufficient reserve power is unavailable.

34.   How Much Data Can We Afford to Lose If the System Goes Down?

The rule of thumb has been to perform backup daily because you can't afford to recreate more than one day's worth of data. With large systems of many users, and the tremendous increase in computer usage, companies are realizing that 24 hours of lost data can have catastrophic ramifications and that no loss of data is the best model.

35.   How Should We Approach Disaster Recovery?

The first thing to consider about disaster recovery is disaster preparation. This means that you have to figure out how to prevent and deal with the most common types of data loss and equipment damage scenarios. When a disaster strikes, your recovery is based on what is available from your preparations.

36.   If We Send Our Paper Files to a Storage Facility, How Can We Make Sure that the Facility Will Keep Them Secure?

As with any business partner, you need to have a contract that details responsibilities. You should check references and, whenever possible, perform an onsite inspection.

Most of all, periodically test the companies capabilities to ensure that they are prepared to help you when you really need assistance. As with any security system, many components work together to create a more secure environment. Don't put all your eggs in one basket.

37. If We Perform a GLP Animal Study and the Results Rule Out a Path for Development, Can We Safely Get Rid of that Documentation?

You can, but it's a better idea to retain it, since the records are part of the development of the product. They tell what you didn't do, so they indirectly support what you actually did. Often the best strategy is to retain the electronic files and scans of paper documents and to destroy the paper records.

38. What Should We Do When People Regularly Print Out Documents from Our E-System and Circulate Them as Paper Documents for Review and Comment?

It is common for people to choose paper review over electronic review. This really isn't a serious problem as long as the e-system records that documents have been reviewed and have the final approved electronic documents. You must be sure to destroy all paper documents after the review so that there is no confusion as to which document is legitimate. Many e-systems print a watermark on the printed pages to indicate that the originating data came from the e-system and that the printed document must not be relied on as the most current version of the document.

39. How Can We Make Our Slow and Difficult to Access EDMS System More User-Friendly?

The nature of an EDMS system is to be fast and easy to use. If this is not the case, then contact the software vendor and ask what is the most likely cause of the problem. Usually, outdated hardware such as processor and memory are easily replaced. A system that is difficult to use often means that the processes put in place were difficult from the start. Usually a revalidation project can be used to streamline processes and identify configuration options that improve efficiency.

40. How Can We Increase the Hardware Capacity?

There are many available options that can be used to increase data capacity and processing power. New processors are significantly faster, and tools such as virtualization software allow many virtual servers to access a collection of processors and memory that have greater performance than can be had on a single server. Data storage is both plentiful, inexpensive, and more durable than ever. Work with IT and your software vendors to determine the best solutions. And remember that technology is ever changing, so don't buy more than you need in the short term.

41. If We Increase the Hardware Capacity, Do We Need to Revalidate?

Most hardware changes do not require revalidation of the system. However, hardware changes that cause software to be reinstalled does require revalidation. If the versions of software components don't change, the revalidation effort is usually minimal and can be performed under change control. Always base the revalidation effort on risk and likelihood of failures. Many companies, when faced with revalidation, elect to upgrade the software at the same time so that they get more value for their effort.

42. Now that We Have Added Devices to Our Product Line, How Can We Modify Our System to House the Documents and Records for Those Products?

Usually all that is required is a new set of folders set up in a hierarchical structure to allow for a different collection of documents. This should be performed under change control because security roles and workflow options will most likely be affected. You can add corrective and preventive action (CAPA) software as an interface to an existing e-sytem or implement it separately as part of the overall document/records management system.

43. How Can We Change Privileges So that We Can Add More System Administrators?

Security should be role-based. The system administrator role should allow for significant access privileges but not grant all access privileges. Many user accounts can include this system administrator role. A user who performs work should only have roles needed to do that specific work. When a user needs to have multiple roles at different times—say a supervisor most of the time and a system administrator on occasion—the user should be given two user accounts, each assigned separate roles. For example, Bob Smith may log on as BSmith to work normally as a supervisor, and he may log on as BSmithAdmin to work as a system administrator. Both accounts may use the same password.

44. How Can We Provide Access to Our E-System to Our New Partners?

As with all users, access is provided through roles. New partners may mean that new roles with unique sets of privileges need to be created. These types of changes can be managed through change control.

45. How Can We Add Remote Access to Our System for Our Executives Who Travel?

Many systems today are remotely accessed in a transparent manner, meaning that the system doesn't know the user is remote. The use of virtual private networks

(VPNs) and terminal server tools allows users to work securely from anywhere on the Internet. Remote access requires encryption and greater security than when working locally.

46.   What Can Be Done If a Laptop Containing Confidential Information Is Stolen?

Once confidential information is compromised, there isn't much you can do. You must follow applicable regulatory laws for reporting potential criminal activity resulting from the lost data. This is expensive and comes with negative publicity. Prevention is the best solution. It is easy and inexpensive to encrypt a folder so that all files place in it cannot be accessed without a password. Encryption utilities come standard with almost all common operating systems.

47.   How Can We Prevent Reviewers from Constantly Rejecting Documents During Review Cycles?

The best way to prevent reviewers from rejecting someone else's work is to give them the opportunity to revise the document before the review process. An effective technique is to conduct a review meeting where the document authors and reviewers all examine a document projected at the front of the room. Any comments or changes are made in real time for all to see and discuss.

48.   What Should We Do with People Who Complain About Our System and Make Things Difficult for Others?

People who complain may have legitimate concerns or be acting out of fear or ignorance. An effective way to get these people out of complaint mode is to have them participate on a validation or revalidation project. The project allows them to express their concerns and educates them about the system. Another option is to retrain them about the system and perhaps increase their privileges so that they gain a better understanding of the system.

49.   What Is the Biggest Threat to Our System, and What Should We Look For?

The biggest threat to any system is always the users of the system. Validation by the users, SOPs written by users, training of users by other users, and change control involving users are the most effective activities to keep the people aware of system threats and gives them the tools to detect problems when they are small and manageable.

50.   If Our CRO Can't Produce Our Documents in a "Timely Manner," What Should We Do?

If your contractor fails to produce your documents within 24 hours, which is the FDA expectation for producing records stored offsite, you need to go back and revisit your contractual arrangement and reiterate your expectation that records be available on demand. If the contractor does not comply, it may be time to reconsider your choice of contractor.

51. The Contractor Who Is Performing GLP Studies for Us Is Going Out of Business and Can't Find Our Records, So What Should We Do?

The contractor is required by law to transfer your files to you. Here's what the regulations say: "If a facility conducting nonclinical testing goes out of business, all raw data, documentation, and other material specified in this section shall be transferred to the archives of the sponsor of the study. The Food and Drug Administration shall be notified in writing of such a transfer" (1).

52. When a Software Vendor Issues an Upgrade that It Has Validated, Can a Company Install the Upgrade into the Production Environment Without Additional Validation?

Users must validate all changes. An upgrade is such a change and, according to the complexity of the upgrade, may require partial-to-complete revalidation. Depending on risk, validation should occur in a test environment before making the upgrade to the production environment.

53. If a Server Has Dual Redundant Power Supplies and One Fails, Do the Users Have to Approve a Change Control Document Before IT Makes the Repair?

No, IT can make the repair in real time and fill out a maintenance event log. IT should informally notify the system administrator that a failure has occurred. Change control can occur in many ways with varying amounts of documentation commensurate with the scope of the changes made. A maintenance event deals with low risk and time critical types of changes.

54. Should a Company Upgrade When the Developer Makes an Upgrade Available?

It depends. Users should first determine if the upgrade has features that will improve the operation of their application. If there are significant changes, such as a feature that improves the audit trail, they may opt to do so. However, many companies hold off on incorporating upgrades if they don't really need them. This makes sense because changes to software often mean that there are bugs in the system that need to be worked out. Thus, waiting a while for feedback to see how the system fares in production elsewhere may be prudent.

55.   How Does a Company Know What Changes Have Been Included in an Upgrade?

The vendor should supply release notes that detail the changes made between software versions. The release notes may be posted on their website or may be included with the software upgrade package.

REFERENCE

1. 21 CFR Part 58.195 Retention of Records (h). www.accessdata.fda.gov, accessed April 9, 2009.

CHAPTER 12

MAINTAINING INSPECTION READINESS

INTRODUCTION

Who will come to inspect? Young companies know that it is unlikely that the FDA or another regulator will come to inspect. Established companies know that they will, typically every two years or before a new product approval. But inspections are not always by regulators. Companies doing contractual research or manufacturing can be certain that inspections are forthcoming. And companies under consideration for purchase or partnering can expect to see inspectors at the door. During any inspection, for whatever purpose, quality plays a key role, and this means that documentation systems will be subject to scrutiny. This chapter addresses how companies can maintain inspection readiness, and it answers the following questions.

1. What is the difference between and inspection, audit, and evaluation?
2. If we pride ourselves on being compliant, do we need to prepare for an inspection?
3. What should companies do to remain inspection ready?
4. What does "due diligence" mean?
5. How can we best prepare for an inspection?
6. We maintain inspection readiness by conducting regular self-audits, but should we keep the audit reports?
7. If we do retain our audit reports, do we have to make them available to inspectors?
8. What's the difference between corrective action and preventive action, and what records do we need to keep?
9. How does a company know when it is going to be inspected?
10. Who will inspect?
11. Can we refuse to permit an inspection?
12. Should we cancel an inspection if key personnel are not available?

Managing the Documentation Maze, By Janet Gough and David Nettleton
Copyright © 2010 John Wiley & Sons, Inc.

13. What triggers an FDA inspection related to 21 CFR Part 11?

14. If the FDA requests documents in advance of an inspection visit, should we provide copies?

15. What's the advantage of letting an inspecting group see documents in advance?

16. When undergoing any inspection (regulator/due diligence/client), what should be the protocol?

17. Who needs to participate in an actual inspection?

18. What roles should employees play during an inspection?

19. What should a person do if they don't know the answer to an inspector's question?

20. What should a person do when asked a question about something they personally did not do or observe?

21. What should people do if they know the answer to a question but that information is not desirable to be disclosed?

22. Is there any way to know what inspectors will look for during an inspection?

23. Do all inspectors look for the same things?

24. Do inspectors always look at document systems?

25. How closely will inspectors look at our document system?

26. When a company undergoes an inspection, what specific role does document management play?

27. Our company has controlled documents in multiple systems so which will inspectors look at?

28. What is considered a "high-risk system"?

29. How much evidence of validation is necessary to satisfy inspectors?

30. How should we make documents available to inspectors on site?

31. Must a company provide the FDA with electronic copies of electronic records?

32. Why, during a recent inspection, when we printed an electronic file from our hybrid document management system, did the inspector insist that we produce the original document?

33. Can a company give an inspector a password to a document system to log on and see how the system works?

34. Can inspectors request to see work instructions?

35. How can we track the documents the inspectors want to see?

36. Do inspectors ever look at training records?

37. Do you have to let inspectors take photographs?

38. Can an inspector ask us to sign an affidavit?

39. If a company complies with the text of the regulation, is it sufficient to avoid citation?

40. What are the most common citations for document systems?

41. Our processes are working well for us; but if an inspection finds something, can we just say we'll fix it?

42. What is a 483 and how does a company get one?

43. What's the difference between a warning letter and a 483?

44. Who knows who gets a 483 and a warning letter?

45. How does the FDA know to keep confidential information out of warning letters?

46. What happens if a company fails to address citations?

47. If we get a 483, does it automatically mean we'll get a warning letter?

48. If you get an observation during an inspection, what should you do?

49. How can we avoid a lot of corrective actions after the inspection?

50. Who can best respond in writing to any observations from an inspection?

51. How can we make sure that our inspection responses are sound?

52. Can we disagree with an observation?

53. Should we keep our inspection reports?

1. What Is the Difference Between and Inspection, Audit, and Evaluation?

These terms are often used interchangeably. Companies may define the subtle differences internally. Some companies define inspections to be from an external source, while audits are from an internal source. Usually an evaluation follows the same concepts as an audit without involving regulatory authority. For example, a potential software vendor is evaluated before a purchasing decision.

2. If We Pride Ourselves on Being Compliant, Do We Need to Prepare for an Inspection?

What the FDA and other organizations conducting compliance inspections want to see is that the company systematically takes inventory of its strengths and weaknesses, that it weighs what's there against the regulations in place, and that it is perpetually in a state of forward motion to stay compliant and in control. When an inspection team arrives onsite, they will respond only to what they see or don't see. And their assessment will be generated from their own expectations of what they expect to be in place. Inspectors are not privy to the culture of the company and its way of doing business. So no matter how pristine an operation is, it's the impression it makes that drives the results of an inspection. Astute companies take every safeguard to maintain and project control. And that is why your company should prepare for all inspections. Preparation involves communication and review of the plan for dealing with unforeseen issues that may arise.

3.   What Should Companies Do to Remain Inspection Ready?

The most important aspect of compliance is in following your own SOPs, keeping them current, and continually training staff as change occurs. Compliance is something that occurs gradually and continually. Trying to become compliant immediately before an inspection usually doesn't yield good results.

4.   What Does "Due Diligence" Mean?

Due diligence is a catchall term that means paying attention and doing your homework. So a company may undergo due diligence when a potential partner looks at it, or a company may say they do due diligence when they hire a consultant or contractor.

5.   How Can We Best Prepare for an Inspection?

Companies can put a preparatory team in place. Such a team has a leader who is knowledgeable in compliance, quality assurance, and regulatory affairs. This person, however, should not be expected to provide answers to every question, but should assume the role of chief spokesman during the preparation process. Team members typically come from areas that the actual inspection will focus on. These people must be able to function with a unified front, and they must be able to address issues that may be problematic and solve them.

6.   We Maintain Inspection Readiness by Conducting Regular Self-Audits, But Should We Keep the Audit Reports?

The regulations do not require that you keep the actual reports, but require that you have a record of your auditing activities. Many companies retain the last audit report until they affirm that corrective action has taken place; so at any given time, they have only one current audit report of a given area. Of course any corrective action that you undertake as a result of your audits requires documentation (1).

7.   If We Do Retain Our Audit Reports, Do We Have to Make Them Available to Inspectors?

No. The company should have a procedure in place that delineates its position on audit reports. Most companies have a policy that restricts viewing of internal audit reports to select management staff. Many companies do not retain their actual internal audit reports, but keep a record of audit activities. Other companies retain the last audit report, which they then destroy upon a new audit.

8. What's the Difference Between Corrective Action and Preventive Action, and What Records Do We Need to Keep?

CAPA discussions in the United States seem to focus more on the CA than on the PA because the CA tells how companies address their problems. Notified Bodies, who issue ISO 13485 certification, focus more on the PA component and may call for a preventive action program. In the United States, quality activities are often preventive in nature, and stem from a quality system mindset, so you can say that a preventive action program is one where staff anticipate problems and fix them before an incident occurs.

9. How Does a Company Know When It Is Going to Be Inspected?

Companies do not always know when they are going to be inspected. When an inspection is part of planned regulatory interaction, inspections are usually scheduled. Most inspections of foreign establishments are scheduled as well.

10. Who Will Inspect?

With regard to the FDA, only their inspectors may perform inspections. Any number of regulatory bodies can visit a facility to inspect. Organizations that issue certifications, such as ISO, do so regularly. Depending on the product, Drug Enforcement Agency (DEA), Environmental Protection Agency (EPA), and Occupational Safety and Health Administration (OSHA) are but some of the regulatory bodies that may inspect. Others may be regulatory bodies from countries where a company's products will be manufactured or distributed. Potential partners or clients seeking to affirm that a company's systems are worth their investment may also inspect.

11. Can We Refuse to Permit an Inspection?

Sometimes a company will choose to refuse an inspection; but if it does, the reason for doing so should be valid. For instance, if a company has recently completed the construction of a new facility, but is not yet actively using it, it may indicate that an inspection would be more efficacious at a later date. Refusing an inspection is always noted, however, and if it happens frequently, it can be a red flag for the inspecting body.

12. Should We Cancel an Inspection If Key Personnel Are Not Available?

When an inspection is initially scheduled, you should avoid times when key personnel are unavailable. Within a short time after scheduling an inspection, you may want to reschedule if key personnel are to be unavailable. However, you should not cancel or attempt to reschedule an inspection within a couple of weeks of the target date or after inspectors have made travel arrangements. Your quality system must be such that you can manage an inspection at any time. Key personnel must have backup personnel that can perform duties in their place. Remember, too, that many inspections take place with no advanced notification at all.

13.   What Triggers an FDA Inspection Related to 21 CFR Part 11?

During the course of FDA audits, inspectors review both paper and electronic records. The use of electronic records, and especially electronic signatures, can result in the inspection delving into documentation related to 21 CFR Part 11. Specifically, inspectors are interested in computer system validation, risk mitigations that depend on software controls, and software functionality related to industry standards.

14.   If the FDA Requests Documents in Advance of an Inspection Visit, Should We Provide Copies?

If your company has an SOP that states that copies are not to be removed from the premises, then inspectors will usually abide by that requirement. But most companies will send a limited set of documents to the FDA in advance. A standard is to print "controlled copy" on each document, maintain a list of documents to be issued, send by return receipt mail with the signature of recipient, and account for each document as it is returned. Companies should not send electronic copies of documents to the FDA. Original documents should never leave the site.

15.   What's the Advantage of Letting an Inspecting Group See Documents in Advance?

A preview of documents can facilitate the inspection. If a company does make them available, it may request a signed receipt for them. This practice shows control, not only of original documents, but of all copies too. Document copies are usually returned at the time of the inspection.

16.   When Undergoing Any Inspection (Regulator/Due Diligence/Client), What Should Be the Protocol?

Every company should have an SOP that explains how to conduct an inspection. The idea is to limit the inspectors' access to the facilities and documentation, answer only questions asked, avoid volunteering information, be respectful and courteous, and assist in making the inspection as short as possible. Such an SOP should address preparation before the inspection, how to verify credentials of inspectors, how to notify all staff of the inspection, how to keep track of documentation given to the inspectors, how to respond to 483, how a summary of the events is issued to management, and how follow-up actions are to be determined.

17. Who Needs to Participate in an Actual Inspection?

It's important for companies to have a host or host group ready to receive inspectors and stay with them throughout the process or be available as necessary. Companies typically assign an official host, a person who knows the company's systems and procedures as well as the binding regulations. This is the person who greets inspectors at the door and acts as the official spokesperson for the company through the inspection. Other key participants include a scribe—a person who documents all discussions, questions, and answers. The scribe should keep of log of documents provided in the sequence in which they are made available to which inspector. The scribe needs to be able to write succinctly, quickly, and accurately. The scribe's role is to record, but not interpret. It's also smart to assign a person to be the data facilitator. This person doesn't need to be visible to the inspection team, but assumes the duty of making requested documents available during the inspection.

18. What Roles Should Employees Play During an Inspection?

Staff should be available to answer questions from inspectors, demonstrate processes, and provide supporting documentation (2).

19. What Should a Person Do If They Don't Know the Answer to an Inspector's Question?

Any time you don't know the answer to a question, simply state that you don't know the answer. Do not try to explain why you don't know. Immediately inform the inspection team of this incident so they can prepare an answer if the question is raised again.

20. What Should a Person Do When Asked a Question About Something They Personally Did Not Do or Observe?

Answer any question asked of you about what you did or observed truthfully and succinctly. Do not volunteer information and elaborate about the event. If you did not participate in the events related to the question, simply state that you did not participate and request that the question be directed to the inspection team so an appropriate person can be sought.

21. What Should People Do If They know the Answer to a Question But that Information Is Not Desirable to be Disclosed?

Whenever you are asked a question that discloses information that is undesirable, it is best to tell the inspector that you need a few minutes to research the answer. Leave the room and notify the inspection team of the issue. Together you will present supporting documentation and truthfully disclose the answer. Remember that presentation counts, so be sure to present the facts in a way that minimizes the undesirable disclosure of information.

22. Is There Any Way to Know What Inspectors Will Look For During an Inspection?

Inspections always address quality systems, known failures, and issues from previous inspections. Other than that, it is difficult to know which areas will be inspected. This is intentional so that companies will be diligent in their compliance in all areas.

23. Do All Inspectors Look For the Same Things?

Inspectors in general consider all the processes in place at your facility and examine those that are of interest to them. This means that the individual inspector makes a big difference in the focus of an inspection. Inspectors don't have time to look at everything, so they pick what is important to them at the time. They almost always will look at your quality awareness. They always address observations from previous inspections. Sometimes they select a system or process that has not be inspected before. Often they pick systems for which they have considerable knowledge.

24. Do Inspectors Always Look at Document Systems?

Document systems are usually included in every inspection. Because the document system is the repository for the processes and provides the controls for their maintenance, it is the foundation for all other systems. If a company has been previously inspected and there were no observations related to the document system, an inspector may choose to focus elsewhere, but surely it will be a focus of a subsequent inspection.

25. How Closely Will Inspectors Look At Our Document System?

Document systems are part of a company's quality system, and as such they will almost certainly receive focus during an inspection. The "proof" of compliant operations resides in the documentation a company keeps, and companies are required to produce it. Inspections usually include review of document approval, version control, and observation of document distribution.

26. When a Company Undergoes an Inspection, What Specific Role Does Document Management Play?

The document management role must ensure that documents are known, controlled, and retrievable and provide requested documents to inspectors. The assumption is that companies will have document controls in place. Failure to produce documents in a "timely manner" during an inspection signals that the company does not have good controls in place. Failure to produce documents during an inspection is a frequent 483 citation.

27. Our Company Has Controlled Documents in Multiple Systems, So Which Ones Will Inspectors Look at?

Let's say that your company has three laboratory information management systems (LIMS) plus an electronic document management system where you keep SOPs, methods, work instructions, protocols, and reports. Typically, inspectors will first look at the document management systems to ensure that a foundation is present for other systems. Often the inspector will select one of the LIMS systems to focus on next. Due to time constraints, the remaining LIMS systems will most likely be addressed in subsequent inspections. Since you don't know which LIMS will be inspected, all must be in a constant state of control.

28. What Is Considered a "High-Risk System"?

How systems are determined to be high risk has not been clearly defined by the FDA, and the regulations require companies to make this determination for themselves. The FDA uses the company's determination and their experience to focus on high-risk systems, which are systems that in comparison to other systems have a greater impact on the safety and efficacy of products or services. Companies perform a risk assessment by inventorying systems, assessing the criticality of the data the systems handle, performing gap analysis to identify deficiencies from industry standards, and assigning relative risk levels. Risk is based on the outcome of a failure on product and patient safety as well as accuracy of electronic data.

29. How Much Evidence of Validation is Necessary to Satisfy Inspectors?

Software systems require validation, so be prepared to produce your validation records that show that your system is suitable for its intended performance. Validation must address all the functions of the system that are use and configuration selected.

Validation documents must include requirements, installation, specification, risk assessment, testing, SOPs, training, and formal system release.

30. How Should We Make Documents Available to Inspectors on Site?

A company should designate a specific area for the inspectors to review documents such as a conference room. Inspectors should not be left unaccompanied. And all documents issued to inspectors should be logged.

31. Must a Company Provide the FDA with Electronic Copies of Electronic Records?

The company must provide electronic copies of electronic records if the FDA demands them. However, if access to electronic records requires training, it may be too cumbersome for the company and the inspector to go through the training process, so controlled, printed copies may suffice, provided that they are printed from the electronic records on demand, and not previously printed.

32. Why, During a Recent Inspection, When We Printed an Electronic File from Our Hybrid Document Management System, Did the Inspector Insist that We Produce the Original Document?

The FDA is clear that it accepts electronic copies in place of paper, provided that your document system is compliant with Part 11. It may be that the inspector was gauging how long it would take you to retrieve an original document and whether it was in a "timely manner," which industry has interpreted to mean within 24 hours.

33. Can a Company Give an Inspector a Password to a Document System to Log On and See How the System Works?

Absolutely not. If an inspector says "I know how to use your system," this doesn't mean that he or she can have access. A secure system requires passwords known only to the user to log on, and the inspector can't possibly have one for your system. Furthermore, any system, whether fully electronic or hybrid, requires training, and the inspector has not been trained in your system.

34. Can Inspectors Request to See Work Instructions?

Yes. Work instructions function like SOPs and must have the same controls. It doesn't matter what you call your procedural documents; they are all binding. Work instructions, workstation instructions, guidelines, procedures, and training documents are examples of documents that must be controlled as SOPs are controlled. The most common difference is the number and level of approvals required.

35. How Can We Track the Documents the Inspectors Want to See?

It's good practice to compile an inspection book to contain the batch records, procedures, and other documents that the investigators have requested or are mostly likely going to request. If the firm has made copies of documents available to the inspection team prior to the actual event, it should log that data in the inspection book as well. Having documents ready shortens the time necessary to produce the copies during the inspection, and it also shortens the time the inspectors are in the facility. The key is to log all documents issued and returned from an inspector.

36. Do Inspectors Ever Look at Training Records?

Yes. Training records are often a focus during inspections and have a high rate of noncompliance. It is one thing to have controlled processes and documents; but if you fail to train the people that implement those processes, you don't have a quality system. Training records must address training of all SOPs, work instructions, and other process documents. Often companies do not maintain a system for ensuring that each person is trained on all the documents that apply to their work. Also, many companies fail to retrain people when new versions of controlled documents are reissued.

37. Do You Have to Let Inspectors Take Photographs?

If your company has an SOP that states that photographs are not to be taken, then inspectors will usually abide by that requirement. However, the FDA has almost unlimited rights to search and record information if they suspect illegal activity. The FDA frequently employs the Department of Justice and Federal Bureau of Investigation to assist in cases where criminal activity is suspected.

38. Can an Inspector Ask Us to Sign an Affidavit?

Yes, they can ask, but under no circumstances should a host sign an affidavit that compromises the company's position. Every company should have an SOP that explains how to conduct an inspection. When subject to such a request, the host should politely refer to the company's policy on signing affidavits.

39. If a Company Complies with the Text of the Regulation, Is It Sufficient to Avoid Citation?

No. The FDA issues the regulation, but it takes time for industry to respond and establish standards for compliance. These standards evolve from discussions between companies and the agency. The FDA can therefore cite a company for failure to comply with industry standards. Note that the FDA regulations are so high level that they encompass virtually every industry standard that can arise.

40. What Are the Most Common Citations for Document Systems?

The most common citations for document systems include lack of computer system validation by representatives of the users of the system, lack of training records, noncompliance with SOPs for document controls and approvals, and inadequate security related to 21 CFR Part 11 industry standards. Since many document systems use a hybrid approach where documents are retained electronically and are also printed and approved with a handwritten signature, there are many failures related to keeping both copies of the document in synchronization.

41. Our Processes Are Working Well for Us; But If an Inspection Finds Something, Can We Just Say We'll Fix It?

Such a position shows a willingness to comply with what the inspection team wants to see, but it's not the wisest course of action. To wait for an assessment by a regulatory group or proposed partner or customer can be costly. It is a reactive approach, and it can force excessive man-hour expenditure, and possibly even production downtime if the company has to aggressively respond to inspection observations. It's much better to assess internal systems and processes before an outside body has the chance to do so. A proactive company anticipates what a compliance inspection may entail and pinpoints any potential observations and addresses them before inspectors come on site.

42. What Is a 483 and How Does a Company Get One?

FDA form 483 is a document that an inspector fills out to make an observation. The FDA issues 483s at the end of an inspection.

43. What's the Difference Between a Warning Letter and a 483?

While a 483 records an observation and does not require a corrective action, a warning letter is a formal notice that the company must make a corrective action. Most companies make corrective actions in response to 483s to avoid escalating the issue to a warning letter.

44. Who Knows Who Gets a 483 and a Warning Letter?

Both 483s and warning letters are public information and can be obtained by anyone through the Freedom of Information Act.

45. How Does the FDA Know to Keep Confidential Information Out of Warning Letters?

The FDA attempts to redact personal information, trade secrets, and other types of confidential information from documents made public. Often a thick black pen is used to cross out text in a document. There is no assurance that everything you consider important will be redacted. It is in your best interest to limit disclosure of confidential information and immediately verify any information made publicly available so that you can request additional redaction if necessary.

46. What Happens If a Company Fails to Address Citations?

If a company fails to respond to a warning letter or does not make required corrective actions, the FDA may take many courses of actions including product recalls, loss of right to market products, fines, and product seizure. Failure to address 483 observations from a previous inspection may result in addition 483 observations and warning letters in subsequent inspections.

47. If We Get a 483, Does It Automatically Mean We'll Get a Warning Letter?

No. Most 483s are observations that are not systemic. Companies should try to make corrections actions for the issue noted and make procedural changes to avoid the issue from recurring. Often corrective actions can be made during the course of the inspection; but care should be taken to not act too quickly, thereby making the issue bigger.

48. If You Get an Observation During an Inspection, What Should You Do?

All observations, 483s and others, should be recorded and reviewed with management. A plan for making corrective actions should be written and approved in advance of making the corrective actions. Communication and correspondence with the FDA should be managed by a single channel, most often with the Regulatory Affairs department.

49. How Can We Avoid a Lot of Corrective Actions After the Inspection?

It's always a good idea to look at the company's past interaction with an inspection team to highlight what the company needs to do to prepare itself. What was the outcome, and why? What were the expectations, and how did the company fail to meet them? Since the inspectors are sure to focus on past issues, it is best to be proactive in those areas.

50.   Who Can Best Respond in Writing to Any Observations from an Inspection?

This task usually falls to the official host or a member of the host group. Sometimes preparation of the response becomes a team effort, particularly if the inspection has been lengthy and the findings many. Writing a response to observations is never easy, because so much hangs on the result. The primary writer has the task of promising action on behalf of the company, and doing it in such a way that the proposed activities are clearly defined. But the good news is this: This type of writing is "just the facts." Regulatory Affairs and senior management should approve all correspondence.

51.   How Can We Make Sure that Our Inspection Responses Are Sound?

There are several pitfalls in this type of writing:

- **Elevating the Tone.** Some writers try to sound officious and elevate the tone by using long and obscure words when they are not necessary. It's better to be direct and say, "We agree with this observation and will do the following...."

- **Placing Blame.** It's tempting to point to individuals and say that they are culpable for any deficiencies. Saying "The former manager did not require a signature on the batch record" does nothing good. What readers of such a response want to know is that the company assumes responsibility and will fix a problem.

- **Explaining Too Much.** Nothing is gained by adding information the inspectors already know.

- **Making Unrealistic Promises.** This pitfall is usually related to the third one, but it goes beyond being redundant. If a response offers more than is necessary, the company may find itself committed to accomplishing what it has laid out on paper, perhaps in an untenable timeframe, even though the observation does not call for it.

52.   Can We Disagree with an Observation?

You may disagree with an inspection finding and politely try to explain why the observation is not correct. However, it is not in your best interest to be argumentative or defensive. It is best to formulate a written response, collect supporting documents, have it approved by management, and send it through the proper communication channel.

53.   Should We Keep Our Inspection Reports?

Every company should have an SOP that explains how to conduct an inspection. This SOP should require all inspections to be recorded in a log. It should be part of the company's policy to retain each inspection report, whether it has multiple observations or none, because each report is part of the history of the company. In addition, the company needs to retain copies of inspection responses and other official correspondence.

REFERENCES

1. Title 21 CRF Part s211.180 and 211.188, www.fda.gov, accessed April 9, 2009.
2. Gough, Janet. *Hosting a Compliance Inspection*, 2001. Bethesda, MD: Parenteral Drug Association and Godalming, Surrey, UK: Davis Horwood International.

CHAPTER 13

RESOURCES

INTRODUCTION

The process of managing documents is not static. New technologies and better pro-cesses continually evolve, and companies struggle to stay abreast of innovative possibilities and implement changes to benefit systems in place. Changing business models as well may dictate changes to systems. Mergers, acquisitions, company spinoffs, and new or changing product lines will surely drive adjustments to docu-mentation and documentation practices. How then can people involved in managing document systems keep current? Attending conferences and participating in training outside the company offer opportunities for learning about industry practices and problem resolution. Journals and newsletters also provide information about the industry. And, of course, staying abreast of regulatory agency initiatives and rulings helps document managers in keeping systems both current and compliant. This chapter provides resources for staying current with the regulations and industry standards. They were current when this book went to press. This chapter answers the following questions:

1. How can we keep our systems current with the regulatory requirements?
2. How can we make sure our staff can manage our systems over time?
3. Are their any online information sources that we can access through our e-mail system?
4. What do you suggest we subscribe to in order to keep up to date with industry issues and practices?
5. Is training available from local colleges and universities?
6. Are there local organizations that provide education and training?
7. Is there any funding available for training to help companies meet their train-ing needs?

1. How Can We Keep Our Systems Current with the Regulatory Requirements?

The regulations are the foundation for data management. To maintain compliant documents and records, it's important to know which regulations apply to a

Managing the Documentation Maze, By Janet Gough and David Nettleton
Copyright © 2010 John Wiley & Sons, Inc.

company's operations. The regulations for individual countries provide a high-level overview of requirements, and companies must stay abreast of regulations that apply to their business model. Guidance documents are also a valuable resource for interpreting the regulations. (See Table 13.1.)

TABLE 13.1.   Global Regulatory Agency Resources

www.fda.gov/oia/agencies.htm
www.rainfo.com

TABLE 13.2.   Resources for Industry Training/Education

Organization	Address
American Association of Medical Writers	www.amwa.org
American Association of Pharmaceutical Scientists	www.AAPSPharmaceutica.com
American Conference Institute	www.americanconference.com
Association of Clinical Research Professional	www.acrpnet.org
Barnett International	www.barnettinternational.com
CDER Learn	www.fda.gov/cder/learn/CDERLearn/default.htm
Center for Business Intelligence	www.cbinet.com
Center for Professional Innovation and Education	www.cfpie.com
Computer System Validation, author David Nettleton	www.ComputerSystemValidation.com
Drug Information Association	www.DIAhome.org
Food and Drug Law Institute	www.fdli.org
GxP Documentation, author Janet Gough	www.gxpdocumentation.com
Institute for International Research	www.iirusa.com
Institute of Validation Technology	www.ivthome.com/
International Quality and Productivity Center	www.iqpc.com
Management Forum	www.management-forum.co.uk
Parenteral Drug Association	www.pda@pda.org
Pharma Conference	www.pharmaconference.com
Pharmaceutical Education and Research Institute	www.peri.org
Pharmaceutical Education Associates	www.pharmedassociates.com
Pharmaceutical Training Institute	www.pti-international.com
Regulatory Affairs Professional Society	www.raps.org
Secure Software Certificate	www.SecureSoftwareCertificate.com
Society of Clinical Research Associates	www.socra.org

2. How Can We Make Sure Our Staff Can Manage Our Systems Over Time?

Training your staff is the key to maintaining your systems. People who manage systems do it best if they understand the activities a company conducts and the documentation those activities require. Training can be in-house, via an onsite trainer or webinar. There are many consultants and organizations that provide in-house tailored training. In addition, many organizations conduct seminars and conferences that address a wide range of industry topics. (See Table 13.2.)

3. Are Their Any Online Information Sources that We Can Access Through Our E-Mail System?

Yes. Many organizations provide industry information via free e-mails. Some are provided daily, weekly, and monthly. Other industry providers may require a fee. (See Tables 13.2 and 13.3.)

TABLE 13.3.   Resources for Newsletters, Updates, and Subscriptions

Source	Address
American Medical Writers Association Journal	www.amwa.org
Applied Clinical Trials	www.appliedclinicaltrialsonline.com
Bio IT World	www.bioitworld.com
Bio World Today	www.bioworld.org
BioPharm International	www.biopharminternational.com
DIA Daily	www.diahome.org
DIA Dispatch	
Drug Information Association Journal	
Drug Delivery Technology	www.drugdeliverytech.com
FDA Library	www.fdainfo.com
FDA News	www.FDAnews.com
FDA Update	www.fdaupdate.com
FDA Weekly Digest Bulletin	www.fda.gov
FDAWeek/Inside Health Policy	www.insidehealthpolicy.com
Fierce Vaccines	www.fiercevaccines.com
FOI Services	www.foiservices.com
Health News Daily	www.healthnewsdaily.com
In-Pharma Technologist	www.inhyphen;pharmatechnologist.com
List Servers	list.nih.gov
	www.fda.gov/emaillist.html
Pharmaceutical Technology	www.pharmtech.com
PharmaEd Resources	www.pharmaedresources.com
Regulatory Focus	www.raps.org
Scrip Worldwide Pharmaceutical News	www.pjbpubs.com
Therapeutics Daily	www.therapeuticsdaily.com

4. What Do You Suggest We Subscribe to in Order to Keep Up to Date with Industry Issues and Practices?

There are many excellent publications available to the industry. Many are free within the United States; others require subscription or membership fees. (See Tables 13.2 and 13.3.)

5. Is Training Available from Local Colleges and Universities?

In many cases, state and private educational facilities offer courses related to the pharmaceutical, medical device, and biotech industries. Often these courses are provided via continuing education services. For example, the University of California Santa Cruz Extension at Cupertino provides comprehensive training and certification programs for the San Francisco bay area medical technology industry (www.ucsc-extension.edu).

6. Are There Local Organizations that Provide Education and Training?

Most large industry organizations have local chapters that meet periodically in addition to the annual national meetings. In addition, there are many small organizations that focus on a specific geographic area. For example, the Organization of Regulatory and Clinical Associates is active in the Seattle, WA area (www.orcanw.org).

7. Is There Any Funding Available for Training to Help Companies Meet Their Training Needs?

Yes, many states, such as Pennsylvania and New Jersey, grant money for training. In such states, human resource personnel from several companies have asked for and received funding to conduct collective training.

APPENDIX

FEDERAL REGISTER

PART II   Thursday, March 20, 1997

Department of Health and Human Services

Food and Drug Administration

21 CFR Part 11
Electronic Records; Electronic Signatures; Final Rule
Electronic Submissions; Establishment of Public Docket; Notice

DEPARTMENT OF HEALTH AND HUMAN SERVICES

Food and Drug Administration

21 CFR Part 11

[Docket No. 92N–0251]

RIN 0910–AA29

Electronic Records; Electronic Signatures

AGENCY: Food and Drug Administration, HHS.

ACTION: Final rule.

SUMMARY: The Food and Drug Administration (FDA) is issuing regulations that provide criteria for acceptance by FDA, under certain circumstances, of electronic records, electronic signatures, and handwritten signatures executed to electronic records as equivalent to paper records and handwritten signatures executed on paper. These regulations, which apply to all FDA program areas, are intended to permit the widest possible use of electronic technology, compatible with FDA's responsibility to promote and protect public health. The use of electronic records as well as their submission to FDA is voluntary. Elsewhere in this issue of the Federal Register, FDA is publishing a document providing information concerning submissions that the agency is prepared to accept electronically.

DATES: Effective August 20, 1997. Submit written comments on the information collection provisions of this final rule by May 19, 1997.

ADDRESSES: Submit written comments on the information collection provisions of this final rule to the Dockets Management Branch (HFA–305), Food and Drug Administration, 12420 Parklawn Dr., rm. 1–23, Rockville, MD 20857.

The final rule is also available electronically via Internet: http://www.fda.gov.

FOR FURTHER INFORMATION CONTACT:

Paul J. Motise, Center for Drug Evaluation and Research (HFD–325), Food and Drug Administration, 7520 Standish Pl., Rockville, MD 20855, 301–594–1089. E-mail address via Internet: Motise@CDER.FDA.GOV, or

Tom M. Chin, Division of Compliance Policy (HFC–230), Food and Drug Administration, 5600 Fishers Lane, Rockville, MD 20857, 301–827–0410. E-mail address via Internet: TChin@FDAEM.SSW.DHHS.GOV

SUPPLEMENTARY INFORMATION:

I. BACKGROUND

In 1991, members of the pharmaceutical industry met with the agency to determine how they could accommodate paperless record systems under the current good manufacturing practice (CGMP) regulations in parts 210 and 211 (21 CFR parts 210 and 211). FDA created a Task Force on Electronic Identification/Signatures to develop a uniform approach by which the agency could accept electronic signatures and records in all program areas. In a February 24, 1992, report, a task force subgroup, the Electronic Identification/Signature Working Group, recommended publication of an advance notice of proposed rulemaking (ANPRM) to obtain public comment on the issues involved.

In the Federal Register of July 21, 1992 (57 FR 32185), FDA published the ANPRM, which stated that the agency was considering the use of electronic identification/signatures, and requested comments on a number of related topics and concerns. FDA received 53 comments on the ANPRM. In the Federal Register of August 31, 1994 (59 FR 45160), the agency published a proposed rule that incorporated many of the comments to the ANPRM, and requested that comments on the proposed regulation be submitted by November 29, 1994. A complete discussion of the options considered by FDA and other background information on the agency's policy on electronic records and electronic signatures can be found in the ANPRM and the proposed rule.

FDA received 49 comments on the proposed rule. The commenters represented a broad spectrum of interested parties: Human and veterinary pharmaceutical companies as well as biological products, medical device, and food interest groups, including 11 trade associations, 25 manufacturers, and 1 Federal agency.

II. HIGHLIGHTS OF THE FINAL RULE

The final rule provides criteria under which FDA will consider electronic records to be equivalent to paper records, and electronic signatures equivalent to traditional handwritten signatures. Part 11 (21 CFR part 11) applies to any paper records required by statute or agency regulations and supersedes any existing paper record requirements by providing that electronic records may be used in lieu of paper records. Electronic signatures which meet the requirements of the rule will be considered to be equivalent to full handwritten signatures, initials, and other general signings required by agency regulations.

Section 11.2 provides that records may be maintained in electronic form and electronic signatures may be used in lieu of traditional signatures. Records and signatures submitted to the agency may be presented in an electronic form provided the requirements of part 11 are met and the records have been identified in a public docket as the type of submission the agency accepts in an electronic form. Unless records are identified in this docket as appropriate for electronic submission, only paper records will be regarded as official submissions.

Section 11.3 defines terms used in part 11, including the terms: Biometrics, closed system, open system, digital signature, electronic record, electronic signature, and handwritten signature.

Section 11.10 describes controls for closed systems, systems to which access is controlled by persons responsible for the content of electronic records on that system. These controls include measures designed to ensure the integrity of system operations and information stored in the system. Such measures include: (1) Validation; (2) the ability to generate accurate and complete copies of records; (3) archival protection of records; (4) use of computer-generated, time-stamped audit trails; (5) use of appropriate controls over systems documentation; and (6) a determination that persons who develop, maintain, or use electronic records and signature systems have the education, training, and experience to perform their assigned tasks.

Section 11.10 also addresses the security of closed systems and requires that: (1) System access be limited to authorized individuals; (2) operational system checks be used to enforce permitted sequencing of steps and events as appropriate; (3) authority checks be used to ensure that only authorized individuals can use the system, electronically sign a record, access the operation or computer system input or output device, alter a record, or perform operations; (4) device (e.g., terminal) checks be used to determine the validity of the source of data input or operation instruction; and (5) written policies be established and adhered to holding individuals accountable and responsible for actions initiated under their electronic signatures, so as to deter record and signature falsification.

Section 11.30 sets forth controls for open systems, including the controls required for closed systems in § 11.10 and additional measures such as document encryption and use of appropriate digital signature standards to ensure record authenticity, integrity, and confidentiality.

Section 11.50 requires signature manifestations to contain information associated with the signing of electronic records. This information must include the printed name of the signer, the date and time when the signature was executed, and the meaning (such as review, approval, responsibility, and authorship) associated with the signature. In addition, this information is subject to the same controls as for electronic records and must be included in any human readable forms of the electronic record (such as electronic display or printout).

Under § 11.70, electronic signatures and handwritten signatures executed to electronic records must be linked to their respective records so that signatures cannot be excised, copied, or otherwise transferred to falsify an electronic record by ordinary means.

Under the general requirements for electronic signatures, at § 11.100, each electronic signature must be unique to one individual and must not be reused by, or reassigned to, anyone else. Before an organization establishes, assigns, certifies, or otherwise sanctions an individual's electronic signature, the organization shall verify the identity of the individual.

Section 11.200 provides that electronic signatures not based on biometrics must employ at least two distinct identification components such as an identification code and password. In addition, when an individual executes a series of signings during a single period of controlled system access, the first signing must be executed using all electronic signature components and the subsequent signings must be executed using at least one component designed to be used only by that individual. When an individual executes one or more signings not performed during a single period of controlled system access, each signing must be executed using all of the electronic signature components.

Electronic signatures not based on biometrics are also required to be used only by their genuine owners and administered and executed to ensure that attempted use of an individual's electronic signature by anyone else requires the collaboration of two or more individuals. This would make it more difficult for anyone to forge an electronic signature. Electronic signatures based upon biometrics must be designed to ensure that such signatures cannot be used by anyone other than the genuine owners.

Under § 11.300, electronic signatures based upon use of identification codes in combination with passwords must employ controls to ensure security and integrity. The controls must include the following provisions: (1) The uniqueness of each combined identification code and password must be maintained in such a way that no two individuals have the same combination of identification code and password; (2) persons using identification codes and/or passwords must ensure that they are periodically recalled or revised; (3) loss management procedures must be followed to deauthorize lost, stolen, missing, or otherwise potentially compromised tokens, cards, and other devices that bear or generate identification codes or password information; (4) transaction safeguards must be used to prevent unauthorized use of passwords and/or identification codes, and to detect and report any attempt to misuse such codes; (5) devices that bear or generate identification codes or password information, such as tokens or cards, must be tested initially and periodically to ensure that they function properly and have not been altered in an unauthorized manner.

III. COMMENTS ON THE PROPOSED RULE

A. General Comments

1. Many comments expressed general support for the proposed rule. Noting that the proposal's regulatory approach incorporated several suggestions submitted by industry in comments on

the ANPRM, a number of comments stated that the proposal is a good example of agency and industry cooperation in resolving technical issues.

Several comments also noted that both industry and the agency can realize significant benefits by using electronic records and electronic signatures, such as increasing the speed of information exchange, cost savings from the reduced need for storage space, reduced errors, data integration/trending, product improvement, manufacturing process streamlining, improved process control, reduced vulnerability of electronic signatures to fraud and abuse, and job creation in industries involved in electronic record and electronic signature technologies.

One comment noted that, when part 11 controls are satisfied, electronic signatures and electronic records have advantages over paper systems, advantages that include: (1) Having automated databases that enable more advanced searches of information, thus obviating the need for manual searches of paper records; (2) permitting information to be viewed from multiple perspectives; (3) permitting determination of trends, patterns, and behaviors; and (4) avoiding initial and subsequent document misfiling that may result from human error.

There were several comments on the general scope and effect of proposed part 11. These comments noted that the final regulations will be viewed as a standard by other Government agencies, and may strongly influence the direction of electronic record and electronic signature technologies. One comment said that FDA's position on electronic signatures/electronic records is one of the most pressing issues for the pharmaceutical industry and has a significant impact on the industry's future competitiveness. Another comment said that the rule constitutes an important milestone along the Nation's information superhighway.

FDA believes that the extensive industry input and collaboration that went into formulating the final rule is representative of a productive partnership that will facilitate the use of advanced technologies. The agency acknowledges the potential benefits to be gained by electronic record/electronic signature systems. The agency expects that the magnitude of these benefits should significantly outweigh the costs of making these systems, through compliance with part 11, reliable, trustworthy, and compatible with FDA's responsibility to promote and protect public health. The agency is aware of the potential impact of the rule, especially regarding the need to accommodate and encourage new technologies while maintaining the agency's ability to carry out its mandate to protect public health. The agency is also aware that other Federal agencies share the same concerns and are addressing the same issues as FDA; the agency has held informal discussions with other Federal agencies and participated in several interagency groups on electronic records/electronic signatures and information technology issues. FDA looks forward to exchanging information and experience with other agencies for mutual benefit and to promote a consistent Federal policy on electronic records and signatures. The agency also notes that benefits, such as the ones listed by the comments, will help to offset any system modification costs that persons may incur to achieve compliance with part 11.

B.   Regulations Versus Guidelines

2. Several comments addressed whether the agency's policy on electronic signatures and electronic records should be issued as a regulation or recommended in a guideline. Most comments supported a regulation, citing the need for a practical and workable approach for criteria to ensure that records can be stored in electronic form and are reliable, trustworthy, secure, accurate, confidential, and authentic. One comment specifically supported a single regulation covering all FDA-regulated products to ensure consistent requirements across all product lines. Two comments asserted that the agency should only issue guidelines or "make the regulations voluntary." One of these comments said that by issuing regulations, the agency is shifting from creating tools to enhance communication (technological quality) to creating tools for enforcement (compliance quality).

The agency remains convinced, as expressed in the preamble to the proposed rule (59 FR 45160 at 45165), that a policy statement, inspection guide, or other guidance would be an inappropriate means for enunciating a comprehensive policy on electronic signatures and records. FDA has concluded that regulations are necessary to establish uniform, enforceable, baseline standards for accepting electronic signatures and records. The agency believes, however, that supplemental guidance documents would be useful to address controls in greater detail than would be appropriate for regulations. Accordingly, the agency anticipates issuing supplemental guidance as needed and will afford all interested parties the opportunity to comment on the guidance documents.

The need for regulations is underscored by several opinions expressed in the comments. For example, one comment asserted that it should be acceptable for supervisors to remove the signatures of their subordinates from signed records and replace them with their own signatures. Although the agency does not object to the use of a supervisor's signature to endorse or confirm a subordinate's actions, removal of an original signature is an action the agency views as falsification. Several comments also argued that an electronic signature should consist of only a password, that passwords need not be unique, that it is acceptable for people to use passwords associated with their personal lives (like the names of their children or their pets), and that passwords need only be changed every 2 years. FDA believes that such procedures would greatly increase the possibility that a password could be compromised and the chance that any resulting impersonation and/or falsification would continue for a long time. Therefore, an enforceable regulation describing the acceptable characteristics of an electronic signature appears necessary.

C. Flexibility and Specificity

3. Several comments addressed the flexibility and specificity of the proposed rule. The comments contended that agency acceptance of electronic records systems should not be based on any particular technology, but rather on the adequacy of the system controls under which they are created and managed. Some comments claimed that the proposed rule was overly prescriptive and that it should not specify the mechanisms to be used, but rather only require owners/users to design appropriate safeguards and validate them to reasonably ensure electronic signature integrity and authenticity. One comment commended the agency for giving industry the freedom to choose from a variety of electronic signature technologies, while another urged that the final rule be more specific in detailing software requirements for electronic records and electronic notebooks in research and testing laboratories.

The agency believes that the provisions of the final rule afford firms considerable flexibility while providing a baseline level of confidence that records maintained in accordance with the rule will be of high integrity. For example, the regulation permits a wide variety of existing and emerging electronic signature technologies, from use of identification codes in conjunction with manually entered passwords to more sophisticated biometric systems that may necessitate additional hardware and software. While requiring electronic signatures to be linked to their respective electronic records, the final rule affords flexibility in achieving that link through use of any appropriate means, including use of digital signatures and secure relational database references. The final rule accepts a wide variety of electronic record technologies, including those based on optical storage devices. In addition, as discussed in comment 40 of this document, the final rule does not establish numerical standards for levels of security or validation, thus offering firms flexibility in determining what levels are appropriate for their situations. Furthermore, while requiring operational checks, authority checks, and periodic testing of identifying devices, persons have the flexibility of conducting those controls by any suitable method. When the final rule calls for a certain control, such as periodic testing of identification tokens, persons have the option of determining the frequency.

D. Controls for Electronic Systems Compared with Paper Systems

4. Two comments stated that any controls that do not apply to paper-based document systems and handwritten signatures should not apply to electronic record and signature systems unless those controls are needed to address an identified unique risk associated with electronic record systems. One comment expressed concern that FDA was establishing a much higher standard for electronic signatures than necessary.

In attempting to establish minimum criteria to make electronic signatures and electronic records trustworthy and reliable and compatible with FDA's responsibility to promote and protect public health (e.g., by hastening the availability of new safe and effective medical products and ensuring the safety of foods), the agency has attempted to draw analogies to handwritten signatures and paper records wherever possible. In doing so, FDA has found that the analogy does not always hold because of the differences between paper and electronic systems. The agency believes some of those differences necessitate controls that will be unique to electronic technology and that must be addressed on their own merits and not evaluated on the basis of their equivalence to controls governing paper documents.

The agency found that some of the comments served to illustrate the differences between paper and electronic record technologies and the need to address controls that may not generally be found in paper record systems. For example, several comments pointed out that electronic records built upon information databases, unlike paper records, are actually transient views or representations of information that is dispersed in various parts of the database. (The agency notes that the databases themselves may be geographically dispersed but linked by networks.) The same software that generates representations of database information on a screen can also misrepresent that information, depending upon how the software is written (e.g., how a query is prepared). In addition, database elements can easily be changed at any time to misrepresent information, without evidence that a change was made, and in a manner that destroys the original information. Finally, more people have potential access to electronic record systems than may have access to paper records.

Therefore, controls are needed to ensure that representations of database information have been generated in a manner that does not distort data or hide noncompliant or otherwise bad information, and that database elements themselves have not been altered so as to distort truth or falsify a record. Such controls include: (1) Using time-stamped audit trails of information written to the database, where such audit trails are executed objectively and automatically rather than by the person entering the information, and (2) limiting access to the database search software. Absent effective controls, it is very easy to falsify electronic records to render them indistinguishable from original, true records.

The traditional paper record, in comparison, is generally a durable unitized representation that is fixed in time and space. Information is recorded directly in a manner that does not require an intermediate means of interpretation. When an incorrect entry is made, the customary method of correcting FDA-related records is to cross out the original entry in a manner that does not obscure the prior data. Although paper records may be falsified, it is relatively difficult (in comparison to falsification of electronic records) to do so in a nondetectable manner. In the case of paper records that have been falsified, a body of evidence exists that can help prove that the records had been changed; comparable methods to detect falsification of electronic records have yet to be fully developed.

In addition, there are significant technological differences between traditional handwritten signatures (recorded on paper) and electronic signatures that also require controls unique to electronic technologies. For example, the traditional handwritten signature cannot be readily compromised by being "loaned" or "lost," whereas an electronic signature based on a password in combination with an identification code can be compromised by being "loaned" or "lost." By contrast, if one person attempts to write the handwritten signature of another person, the falsification would be difficult to execute and a long-standing body of investigational techniques would be available to detect the falsification. On the other hand, many electronic signatures are relatively easy to falsify and methods of falsification almost impossible to detect.

Accordingly, although the agency has attempted to keep controls for electronic record and electronic signatures analogous to traditional paper systems, it finds it necessary to establish certain controls specifically for electronic systems.

E. FDA Certification of Electronic Signature Systems

5. One comment requested FDA certification of what it described as a low-cost, biometric-based electronic signature system, one which uses dynamic signature verification with a parameter code recorded on magnetic stripe cards.

The agency does not anticipate the need to certify individual electronic signature products. Use of any electronic signature system that complies with the provisions of part 11 would form the basis for agency acceptance of the system regardless of what particular technology or brand is used. This approach is consistent with FDA's policy in a variety of program areas. The agency, for example, does not certify manufacturing equipment used to make drugs, medical devices, or food.

F. Biometric Electronic Signatures

6. One comment addressed the agency's statement in the proposed rule (59 FR 45160 at 45168) that the owner of a biometric/behavioral link could not lose or give it away. The comment stated that it was possible for an owner to "lend" the link for a file to be opened, as a collaborative fraudulent gesture, or to unwittingly assist a fraudulent colleague in an "emergency," a situation, the comment said, that was not unknown in the computer industry.

The agency acknowledges that such fraudulent activity is possible and that people determined to falsify records may find a means to do so despite whatever technology or preventive measures are in place. The controls in part 11 are intended to deter such actions, make it difficult to execute falsification by mishap or casual misdeed, and to help detect such alterations when they occur (see § 11.10 (introductory paragraph and especially §§ 11.10(j) and 11.200(b)).

G. Personnel Integrity

7. A few comments addressed the role of individual honesty and trust in ensuring that electronic records are reliable, trustworthy, and authentic. One comment noted that firms must rely in large measure upon the integrity of their employees. Another said that subpart C of part 11, Electronic Signatures, appears to have been written with the belief that pharmaceutical manufacturers have an incentive to falsify electronic signatures. One comment expressed concern about possible signature falsification when an employee leaves a company to work elsewhere and the employee uses the electronic signature illegally.

The agency agrees that the integrity of any electronic signature/electronic record system depends heavily upon the honesty of employees and that most persons are not motivated to falsify records. However, the agency's experience with various types of records and signature falsification demonstrates that some people do falsify information under certain circumstances. Among those circumstances are situations in which falsifications can be executed with ease and have little likelihood of detection. Part 11 is intended to minimize the opportunities for readily executing falsifications and to maximize the chances of detecting falsifications.

Concerning signature falsification by former employees, the agency would expect that upon the departure of an employee, the assigned electronic signature would be "retired" to prevent the former employee from falsely using the signature.

H. Security of Industry Electronic Records Submitted to FDA

8. Several comments expressed concern about the security and confidentiality of electronic records submitted to FDA. One suggested that submissions be limited to such read-only formats as CD–ROM with raw data for statistical manipulation provided separately on floppy diskette. One comment suggested that in light of the proposed rule, the agency should review its own internal security procedures. Another addressed electronic records that may be disclosed under the Freedom of Information Act and expressed concern regarding agency deletion of trade secrets. One comment anticipated FDA's use of open systems to access industry records (such as medical device production and control records) and suggested that such access should be restricted to closed systems.

The agency is well aware of its legal obligation to maintain the confidentiality of trade secret information in its possession, and is committed to meet that obligation regardless of the form (paper or electronic) a record takes. The procedures used to ensure confidentiality are consistent with the provisions of part 11. FDA is also examining other controls, such as use of digital signatures, to ensure submission integrity. To permit legitimate changes to be made, the agency does not believe that it is necessary to restrict submissions to those maintained in read-only formats in all cases; each agency receiving unit retains the flexibility to determine whatever format is most suitable. Those intending to submit material are expected to consult with the appropriate agency receiving unit to determine the acceptable formats.

Although FDA access to electronic records on open systems maintained by firms is not anticipated in the near future, the agency believes it would be inappropriate to rule out such a procedure. Such access can be a valuable inspection tool and can enhance efficiencies by reducing the time investigators may need to be on site. The agency believes it is important to develop appropriate procedures and security measures in cooperation with industry to ensure that such access does not jeopardize data confidentiality or integrity.

I. Effective Date/Grandfathering

9. Several comments addressed the proposed effective date of the final rule, 90 days after publication in the Federal Register, and suggested potential exemptions (grandfathering) for systems now in use. Two comments requested an expedited effective date for the final rule. One comment requested an effective date at least 18 months after publication of the final rule to permit firms to modify and validate their systems. One comment expressed concern about how the rule, in general, will affect current systems, and suggested that the agency permit firms to continue to use existing electronic record systems that otherwise conform to good manufacturing or laboratory practices until these firms make major modifications to those systems or until 5 years have elapsed, whichever comes first. Several other comments requested grandfathering for specific sections of the proposed rule.

The agency has carefully considered the comments and suggestions regarding the final rule's effective date and has concluded that the effective date should be 5 months after date of publication in the Federal Register. The agency wishes to accommodate firms that are prepared now to comply with part 11 or will be prepared soon, so as to encourage and foster new technologies in a manner that ensures that electronic record and electronic signature systems are reliable, trustworthy, and compatible with FDA's responsibility to promote and protect public health. The agency believes that firms that have consulted with FDA before adopting new electronic record and electronic signature technologies (especially technologies that may impact on the ability of the agency to conduct its work effectively) will need to make few, if any, changes to systems used to maintain records required by FDA.

The agency believes that the provisions of part 11 represent minimal standards and that a general exemption for existing systems that do not meet these provisions would be inappropriate and not in the public interest because such systems are likely to generate electronic records and electronic signatures that are unreliable, untrustworthy, and not compatible with FDA's responsibility to promote and protect public health. Such an exemption might, for example, mean that a firm could: (1) Deny FDA inspectional access to electronic record systems, (2) permit unauthorized access to those systems, (3) permit individuals to share identification codes and passwords, (4) permit systems to go unvalidated, and (5) permit records to be falsified in many ways and in a manner that goes undetected.

The agency emphasizes that these regulations do not require, but rather permit, the use of electronic records and signatures. Firms not confident that their electronic systems meet the minimal requirements of these regulations are free to continue to use traditional signatures and paper documents to meet recordkeeping requirements.

J.   Comments by Electronic Mail (e-mail) and Electronic Distribution of FDA Documents

10. One comment specifically noted that the agency has accepted comments by e-mail and that this provides an additional avenue for public participation in the rulemaking process. Another comment encouraged FDA to expand the use of electronic media to provide information by such open systems as bulletin boards.

The agency intends to explore further the possibility of continuing to accept public comments by e-mail and other electronic means. For this current experiment, the agency received only one comment by e-mail. The comment that addressed this issue was, itself, transmitted in a letter. The agency recognizes the benefits of distributing information electronically, has expanded that activity, and intends to continue that expansion. Although only one e-mail comment was received, the agency does not attribute that low number to a lack of ability to send e-mail because the agency received e-mail from 198 persons who requested the text of the proposed rule, including requests from people outside the United States.

K.   Submissions by Facsimile (Fax)

11. One comment said that part 11 should include a provision for FDA acceptance of submissions by fax, such as import form FDA 2877. The comment noted that the U.S. Customs Service accepts fax signatures on its documents, and claimed that FDA's insistence on hard copies of form FDA 2877 is an impediment to imports.

The agency advises that part 11 permits the unit that handles import form FDA 2877 to accept that record in electronic form when it is prepared logistically to do so. As noted in the discussion on § 11.1(b) in comment 21 of this document, the agency recognizes that faxes can be in paper or electronic form, based on the capabilities of the sender and recipient.

L.   Blood Bank Issues

12. Two comments addressed blood bank issues in the context of electronic records and electronic signatures and said the agency should clarify that part 11 would permit electronic crossmatching by a central blood center for individual hospitals. One comment stated that remote blood center and transfusion facilities should be permitted to rely on electronically communicated information, such as authorization for labeling/issuing units of blood, and that the electronic signature of the supervisor in the central testing facility releasing the product for labeling and issuance should be sufficient because the proposed rule guards against security and integrity problems.

One comment questioned whether, under part 11, electronic signatures would meet the signature requirements for the release of units of blood, and if there would be instances where a full signature would be required instead of a technician's identification. Another comment asserted that it is important to clarify how the term "batch" will be interpreted under part 11, and suggested that the term used in relation to blood products refers to a series of units of blood having undergone common manufacturing processes and recorded on the same computerized document. The comment contrasted this to FDA's current view that each unit of blood be considered a batch.

The agency advises that part 11 permits release records now in paper form to be in electronic form and traditional handwritten signatures to be electronic signatures. Under part 11, the name of the technician must appear in the record display or printout to clearly identify the technician. The appearance of the technician's identification code alone would not be sufficient. The agency also advises that the definition of a "batch" for blood or other products is not affected by part 11, which addresses the trustworthiness and reliability of electronic records and electronic signatures, regardless of how a batch, which is the subject of those records and signatures, is defined.

M.   Regulatory Flexibility Analysis

13. One comment said that, because part 11 will significantly impact a substantial number of small businesses, even though the impact would be beneficial, FDA is required to perform a regulatory flexibility analysis and should publish such an analysis in the Federal Register before a final rule is issued.

The comment states that the legislative history of the Regulatory Flexibility Act is clear that, "significant economic impact," as it appears at 5 U.S.C. 605(b) is neutral with respect to whether such impact is beneficial or adverse.

Contrary to the comment's assertion, the legislative history is not dispositive of this matter. It is well established that the task of statutory construction must begin with the actual language of the statute. (See *Bailey* v. *United States*, 116 S. Ct. 595, 597 (1996).) A statutory term must not be construed in isolation; a provision that may seem ambiguous in isolation is often clarified by the remainder of the statute. (See *Dept. Of Revenue of Oregon* v. *ACF Industries*, 114 S. Ct. 843, 850 (1994).) Moreover, it is a fundamental canon of statutory construction that identical terms within the same statute must bear the same meaning. (See *Reno* v. *Koray*, 115 S. Ct. 2021, 2026 (1995).)

In addition to appearing in 5 U.S.C. 605(b), the term "significant economic impact" appears elsewhere in the statute. The legislation is premised upon the congressional finding that alternative regulatory approaches may be available which "minimize the significant economic impact" of rules (5 U.S.C. 601 note). In addition, an initial regulatory flexibility analysis must describe significant regulatory alternatives that "minimize any significant economic impact" (5 U.S.C. 603(c)). Similarly, a final regulatory flexibility analysis must include a description of the steps the agency has taken to "minimize any significant economic impact" (5 U.S.C. 604(a)(5)). The term appeared as one of the elements of a final regulatory flexibility analysis, as originally enacted in 1980. (See Pub. L. No. 96–354, 3(a), 94 Stat. 1164, 1167 (1980) (formerly codified at 5 U.S.C. 604(a)(3)).) In addition, when Congress amended the elements of a final regulatory flexibility analysis in 1996, it re-enacted the term, as set forth above. (See Pub. L. 104–121, 241(b), 110 Stat. 857, 865 (1996) (codified at 5 U.S.C.604(a)(5)).)

Unless the purpose of the statute was intended to increase the economic burden of regulations by minimizing positive or beneficial effects, "significant economic impact" cannot include such effects. Because it is beyond dispute that the purpose of the statute is not increasing economic burdens, the plain meaning of "significant economic impact" is clear and necessarily excludes beneficial or positive effects of regulations. Even where there are some limited contrary indications in the statute's legislative history, it is inappropriate to resort to legislative history to cloud a statutory text that is clear on its face. (See *Ratzlaff* v. *United States*, 114 S. Ct. 655, 662 (1994).) Therefore, the agency concludes that a final regulatory flexibility analysis is not required for this regulation or any regulation for which there is no significant adverse economic impact on small entities. Notwithstanding these conclusions, FDA has nonetheless considered the impact of the rule on small entities. (See section XVI. of this document.)

N.   Terminology

14. One comment addressed the agency's use of the word "ensure" throughout the rule and argued that the agency should use the word "assure" rather than "ensure" because "ensure" means "to guarantee or make certain" whereas "assure" means "to make confident." The comment added that "assure" is also more consistent with terminology in other regulations.

The agency wishes to emphasize that it does not intend the word "ensure" to represent a guarantee. The agency prefers to use the word "ensure" because it means to make certain.

O.   General Comments Regarding the Prescription Drug Marketing Act of 1987 (PDMA)

15. Three comments addressed the use of handwritten signatures that are recorded electronically (SRE's) under part 11 and PDMA. One firm described its delivery information acquisition device and noted its use of time stamps to record when signatures are executed. The comments requested clarification that SRE's would be acceptable under the PDMA regulations. One comment assumed that subpart C of part 11 (Electronic Signatures) would not apply to SRE's, noting that it was not practical under PDMA (given the large number of physicians who may be eligible to receive drug product samples) to use such alternatives as identification codes combined with passwords.

The agency advises that part 11 applies to handwritten signatures recorded electronically and that such signatures and their corresponding electronic records will be acceptable for purposes of meeting PDMA's requirements when the provisions of part 11 are met. Although subpart C of part 11 does not apply to handwritten signatures recorded electronically, the agency advises that controls related to electronic records (subpart B), and the general provisions of subpart A, do apply to electronic records in the context of PDMA. The agency emphasizes, however, that part 11 does not restrict PDMA signings to SRE's, and that organizations retain the option of using electronic signatures in conformance with part 11. Furthermore, the agency believes that the number of people in a given population or organization should not be viewed as an insurmountable obstacle to use of electronic signatures. The agency is aware, for example, of efforts by the American Society of Testing and Materials to develop standards for electronic medical records in which digital signatures could theoretically be used on a large scale.

P.   Comments on the Unique Nature of Passwords

16. Several comments noted, both generally and with regard to §§ 11.100(a), 11.200(a), and 11.300, that the password in an electronic signature that is composed of a combination of password and identification code is not, and need not be, unique. Two comments added that passwords may be known to system security administrators who assist people who forget passwords and requested that the rule acknowledge that passwords need not be unique. One comment said that the rule should describe how uniqueness is to be determined.

The agency acknowledges that when an electronic signature consists of a combined identification code and password, the password need not be unique. It is possible that two persons in the same organization may have the same password. However, the agency believes that where good password practices are implemented, such coincidence would be highly unlikely. As discussed in section XIII. of this document in the context of comments on proposed § 11.300, records are less trustworthy and reliable if it is relatively easy for someone to deduce or execute, by chance, a person's electronic signature where the identification code of the signature is not confidential and the password is easily guessed.

The agency does not believe that revising proposed § 11.100(a) is necessary because what must remain unique is the electronic signature, which, in the case addressed by the comments, consists not of the password alone, but rather the password in combination with an identification code. If the combination is unique, then the electronic signature is unique.

The agency does not believe that it is necessary to describe in the regulations the various ways of determining uniqueness or achieving compliance with the requirement. Organizations thereby maintain implementation flexibility.

The agency believes that most system administrators or security managers would not need to know passwords to help people who have forgotten their own. This is because most administrators or managers have global computer account privileges to resolve such problems.

IV.   SCOPE (§ 11.1)

17. One comment suggested adding a new paragraph to proposed § 11.1 that would exempt computer record maintenance software installed before the effective date of the final rule, and that would exempt electronic records maintained before that date. The comment argued that such exemptions were needed for economic and constitutional reasons because making changes to existing systems would be costly and because the imposition of additional requirements after the fact could be regarded as an ex post facto rule. The comment said firms have been using electronic systems that have demonstrated reliability and security for many years before the agency's publication of the ANPRM, and that the absence of FDA's objections in inspectional form FDA 483 was evidence of the agency's acceptance of the system.

As discussed in section III.I. of this document, the agency is opposed to "grandfathering" existing systems because such exemptions may perpetuate environments that provide opportunities for record falsification and impair FDA's ability to protect and promote public health. However, the agency wishes to avoid any confusion regarding the application of the provisions of part 11 to systems and electronic records in place before the rule's effective date. Important distinctions need to be made relative to an electronic record's creation, modification, and maintenance because various portions of part 11 address matters relating to these actions. Those provisions apply depending upon when a given electronic record is created, modified, or maintained.

Electronic records created before the effective date of this rule are not covered by part 11 provisions that relate to aspects of the record's creation, such as the signing of the electronic record. Those records would not, therefore, need to be altered retroactively. Regarding records that were first created before the effective date, part 11 provisions relating to modification of records, such as audit trails for record changes and the requirement that original entries not be obscured, would apply only to those modifications made on or after the rule's effective date, not to modifications made earlier. Likewise, maintenance provisions of part 11, such as measures to ensure that electronic records can be retrieved throughout their retention periods, apply to electronic records that are being maintained on or after the rule's effective date. The hardware and software, as well as operational procedures used on or after the rule's effective date, to create, modify, or maintain electronic records must comply with the provisions of part 11.

The agency does not agree with any suggestion that FDA endorsement or acceptance of an electronic record system can be inferred from the absence of objections in an inspection report. Before this rulemaking, FDA did not have established criteria by which it could determine the reliability and trustworthiness of electronic records and electronic signatures and could not sanction electronic alternatives when regulations called for signatures. A primary reason for issuing part 11 is to develop and codify such criteria. FDA will assess the acceptability of electronic records and electronic signatures created prior to the effective date of part 11 on a case-by-case basis.

18. One comment suggested that proposed § 11.1 exempt production of medical devices and in vitro diagnostic products on the grounds that the subject was already adequately addressed in the medical device CGMP regulations currently in effect in § 820.195 (21 CFR 820.195), and that additional regulations would be confusing and would limit compliance.

The agency believes that part 11 complements, and is supportive of, the medical device CGMP regulations and the new medical device quality system regulation, as well as other regulations, and that compliance with one does not confound compliance with others. Before publication of the ANPRM, the agency determined that existing regulations, including the medical device CGMP regulations, did not adequately address electronic records and electronic signatures. That determination was reinforced in the comments to the ANPRM, which focused on the need to identify what makes electronic records reliable, trustworthy, and compatible with FDA's responsibility to promote and protect public health. For example, the provision cited by the comment, § 820.195, states "When automated data processing is used for manufacturing or quality assurance purposes, adequate checks shall be designed and implemented to prevent inaccurate data output, input, and programming errors." This section does not address the many issues addressed by part 11, such as electronic signatures, record falsification, or FDA access to electronic records. The relationship between the quality system regulation and part 11 is discussed at various points in the preamble to the quality system regulation.

19. One comment asserted that for purposes of PDMA, the scope of proposed part 11 should be limited to require only those controls for assessing signatures in paper-based systems because physicians' handwritten signatures are executed to electronic records. The comment further asserted that, because drug manufacturers' representatives carry computers

into physicians' offices (where the physicians then sign sample requests and receipts), only closed system controls should be needed.

The agency believes that, for purposes of PDMA, controls needed for electronic records bearing handwritten signatures are no different from controls needed for the same kinds of records and signatures used elsewhere, and that proposed § 11.1 need not make any such distinction.

In addition, the agency disagrees with the implication that all PDMA electronic records are, in fact, handled within closed systems. The classification of a system as open or closed in a particular situation depends on what is done in that situation. For example, the agency agrees that a closed system exists where a drug producer's representative (the person responsible for the content of the electronic record) has control over access to the electronic record system by virtue of possessing the portable computer and controlling who may use the computer to sign electronic records. However, should the firm's representative transfer copies of those records to a public online service that stores them for the drug firm's subsequent retrieval, the agency considers such transfer and storage to be within an open system because access to the system holding the records is controlled by the online service, which is not responsible for the record's content. Activities in the first example would be subject to closed system controls and activities in the second example would be subject to open system controls.

20. One comment urged that proposed § 11.1 contain a clear statement of what precedence certain provisions of part 11 have over other regulations.

The agency believes that such statements are found in § 11.1(c):

> Where electronic signatures and their associated records meet the requirements of this part, the agency will consider the electronic signatures to be equivalent to full handwritten signatures, initials, and other general signings as required under agency regulations unless specifically excepted by regulations * * *.

and §11.1(d) ("Electronic records that meet the requirements of this part may be used in lieu of paper records, in accordance with § 11.2, unless paper records are specifically required."). These provisions clearly address the precedence of part 11 and the equivalence of electronic records and electronic signatures.

To further clarify the scope of the rule, FDA has revised § 11.1 to apply to electronic records submitted to the agency under requirements of the Federal Food, Drug, and Cosmetic Act (the act) and the Public Health Service Act (the PHS Act). This clarifies the point that submissions required by these statutes, but not specifically mentioned in the Code of Federal Regulations (CFR), are subject to part 11.

21. Proposed § 11.1(b) stated that the regulations would apply to records in electronic form that are created, modified, maintained, or transmitted, under any records requirements set forth in Chapter I of Title 21. One comment suggested that the word "transmitted" be deleted from proposed § 11.1(b) because the wording would inappropriately apply to paper documents that are transmitted by fax. The comment noted that if the records are in machine readable form before or after transmission, they would still be covered by the revised wording.

The agency does not intend part 11 to apply to paper records even if such records are transmitted or received by fax. The agency notes that the records transmitted by fax may be in electronic form at the sender, the recipient, or both. Part 11 would apply whenever the record is in electronic form. To remedy the problem noted by the comment, the agency has added a sentence to § 11.1(b) stating that part 11 does not apply to paper records that are, or have been, transmitted by electronic means.

22. One comment asked whether paper records created by computer would be subject to proposed part 11. The comment cited, as an example, the situation in which a computer system collects toxicology data that are printed out and maintained as "raw data."

Part 11 is intended to apply to systems that create and maintain electronic records under FDA's requirements in Chapter I of Title 21, even though some of those electronic records may be printed on paper at certain times. The key to determining part 11 applicability, under § 11.1(b), is the nature of the system used to create, modify, and maintain records, as well as the nature of the records themselves.

Part 11 is not intended to apply to computer systems that are merely incidental to the creation of paper records that are subsequently maintained in traditional paper-based systems. In such cases, the computer systems would function essentially like manual typewriters or pens and any signatures would be traditional handwritten signatures. Record storage and retrieval would be of the traditional "file cabinet" variety. More importantly, overall reliability, trustworthiness, and FDA's ability to access the records would derive primarily from well-established and generally accepted procedures and controls for paper records. For example, if a person were to use word processing software to generate a paper submission to FDA, part 11 would not apply to the computer system used to generate the submission, even though, technically speaking, an electronic record was initially created and then printed on paper.

When records intended to meet regulatory requirements are in electronic form, part 11 would apply to all the relevant aspects of managing those records (including their creation, signing, modification, storage, access, and retrieval). Thus, the software and hardware used to create records that are retained in electronic form for purposes of meeting the regulations would be subject to part 11.

Regarding the comment about "raw data," the agency notes that specific requirements in existing regulations may affect the particular records at issue, regardless of the form such records take. For example, "raw data," in the context of the good laboratory practices regulations (21 CFR part 58), include computer printouts from automated instruments as well as the same data recorded on magnetic media. In addition, regulations that cover data acquisition systems generally include requirements intended to ensure the trustworthiness and reliability of the collected data.

23. Several comments on proposed § 11.1(b) suggested that the phrase "or archived and retrieved" be added to paragraph (b) to reflect more accurately a record's lifecycle.

The agency intended that record archiving and retrieval would be part of record maintenance, and therefore already covered by § 11.1(b). However, for added clarity, the agency has revised § 11.1(b) to add "archived and retrieved."

24. One comment suggested that, in describing what electronic records are within the scope of part 11, proposed § 11.1(b) should be revised by substituting "processed" for "modified" and "communicated" for "transmitted" because "communicated" reflects the fact that the information was dispatched and also received. The comment also suggested substituting "retained" for "maintained," or adding the word "retained," because "maintain" does not necessarily convey the retention requirement.

The agency disagrees. The word "modified" better describes the agency's intent regarding changes to a record; the word "processed" does not necessarily infer a change to a record. FDA believes "transmitted" is preferable to "communicated" because "communicated" might infer that controls to ensure integrity and authenticity hinge on whether the intended recipient actually received the record. Also, as discussed in comment 22 of this document, the agency intends for the term "maintain" to include records retention.

25. Two comments suggested that proposed § 11.1(b) explicitly state that part 11 supersedes all references to handwritten signatures in 21 CFR parts 211 through 226 that pertain to a drug, and in 21 CFR parts 600 through 680 that pertain to biological products for human use. The comments stated that the revision should clarify coverage and permit blood centers and transfusion services to take full advantage of electronic systems that provide process controls.

The agency does not agree that the revision is necessary because, under § 11.1(b) and (c), part 11 permits electronic records or submissions under all FDA regulations in Chapter I of Title 21 unless specifically excepted by future regulations.

26. Several comments expressed concern that the proposed rule had inappropriately been expanded in scope from the ANPRM to address electronic records as well as electronic signatures. One comment argued that the scope of part 11 should be restricted only to those records that are currently required to be signed, witnessed, or initialed, and that the agency should not require electronic records to contain electronic signatures where the corresponding paper records are not required to be signed.

The agency disagrees with the assertion that part 11 should address only electronic signatures and not electronic records for several reasons. First, based on comments on the ANPRM, the agency is convinced that the reliability and trustworthiness of electronic signatures depend in large measure on the reliability and trustworthiness of the underlying electronic records. Second, the agency has concluded that electronic records, like paper records, need to be trustworthy, reliable, and compatible with FDA's responsibility to promote and protect public health regardless of whether they are signed. In addition, records falsification is an issue with respect to both signed and unsigned records. Therefore, the agency concludes that although the ANPRM focused primarily on electronic signatures, expansion of the subject to electronic records in the proposed rule was fully justified.

The agency stresses that part 11 does not require that any given electronic record be signed at all. The requirement that any record bear a signature is contained in the regulation that mandates the basic record itself. Where records are signed, however, by virtue of meeting a signature requirement or otherwise, part 11 addresses controls and procedures intended to help ensure the reliability and trustworthiness of those signatures.

27. Three comments asked if there were any regulations, including CGMP regulations, that might be excepted from part 11 and requested that the agency identify such regulations.

FDA, at this time, has not identified any current regulations that are specifically excepted from part 11. However, the agency believes it is prudent to provide for such exceptions should they become necessary in the future. It is possible that, as the agency's experience with part 11 increases, certain records may need to be limited to paper if there are problems with the electronic versions of such records.

28. One comment requested clarification of the meaning of the term "general signings" in proposed § 11.1(c), and said that the distinction between "full handwritten" signatures and "initials" is unnecessary because handwritten includes initials in all common definitions of handwritten signature. The comment also suggested changing the term "equivalent" to "at least equivalent" because electronic signatures are not precise equivalents of handwritten signatures and computer-based signatures have the potential of being more secure.

The agency advises that current regulations that require records to be signed express those requirements in different ways depending upon the agency's intent and expectations. Some regulations expressly state that records must be signed using "full handwritten" signatures, whereas other regulations state that records must be "signed or initialed;" still other regulations implicitly call for some kind of signing by virtue of requiring record approvals or endorsements. This last broad category is addressed by the term "general signings" in § 11.1(c).

Where the language is explicit in the regulations, the means of meeting the requirement are correspondingly precise. Therefore, where a regulation states that a signature must be recorded as "full handwritten," the use of initials is not an acceptable substitute. Furthermore, under part 11, for an electronic signature to be acceptable in place of any of these signings, the agency only needs to consider them as equivalent; electronic signatures need not be superior to those other signings to be acceptable.

29. Several comments requested clarification of which FDA records are required to be in paper form, and urged the agency to allow and promote the use of electronic records in all cases. One comment suggested that proposed § 11.1(d) be revised to read, in part, "* * * unless the use of electronic records is specifically prohibited."

The agency intends to permit the use of electronic records required to be maintained but not submitted to the agency (as noted in § 11.2(a)) provided that the requirements of part 11 are met and paper records are not specifically required. The agency also wishes to encourage electronic submissions, but is limited by logistic and resource constraints. The agency is unaware of "maintenance records" that are currently explicitly required to be in paper form (explicit mention of paper is generally unnecessary because, at the time most regulations were prepared, only paper-based technologies were in use) but is providing for that possibility in the future. For purposes of part 11, the agency will not consider that a regulation requires "maintenance" records to be in paper form where the regulation is silent on the form the record must take. FDA believes that the comments' suggested wording does not offer sufficient advantages to adopt the change.

However, to enable FDA to accept as many electronic submissions as possible, the agency is amending § 11.1(b) to include those submissions that the act and the PHS Act specifically require, even though such submissions may not be identified in agency regulations. An example of such records is premarket submissions for Class I and Class II medical devices, required by section 510(k) of the act (21 U.S.C. 360(k)).

30. Several comments addressed various aspects of the proposed requirement under § 11.1(e) regarding FDA inspection of electronic record systems. Several comments objected to the proposal as being too broad and going beyond the agency's legal inspectional authority. One comment stated that access inferred by such inspection may include proprietary financial and sales data to which FDA is not entitled. Another comment suggested adding the word "authorized" before "inspection." Some comments suggested revising proposed § 11.1(e) to limit FDA inspection only to the electronic records and electronic signatures themselves, thus excluding inspection of hardware and software used to manage those records and signatures. Other comments interpreted proposed § 11.1(e) as requiring them to keep supplanted or retired hardware and software to enable FDA inspection of those outdated systems.

The agency advises that FDA inspections under part 11 are subject to the same legal limitations as FDA inspections under other regulations. The agency does not believe it is necessary to restate that limitation by use of the suggested wording. However, within those limitations, it may be necessary to inspect hardware and software used to generate and maintain electronic records to determine if the provisions of part 11 are being met. Inspection of resulting records alone would be insufficient. For example, the agency may need to observe the use and maintenance of tokens or devices that contain or generate identification information. Likewise, to assess the adequacy of systems validation, it is generally necessary to inspect hardware that is being used to determine, among other things, if it matches the system documentation description of such hardware. The agency has concluded that hardware and software used to generate and maintain electronic records and signatures are "pertinent equipment" within the meaning of section 704 of the act (21 U.S.C. 374).

The agency does not expect persons to maintain obsolete and supplanted computer systems for the sole purpose of enabling FDA inspection. However, the agency does expect firms to maintain and have available for inspection documentation relevant to those systems, in terms of compliance with part 11, for as long as the electronic records are required by other relevant regulations. Persons should also be mindful of the need to keep appropriate computer systems that are capable of reading electronic records for as long as those records must be retained. In some instances, this may mean retention of otherwise outdated and supplanted systems, especially where the old records cannot be converted to a form readable by the newer systems. In most cases, however, FDA believes that where electronic records are accurately and completely transcribed from one system to another, it would not be necessary to maintain older systems.

31. One comment requested that proposed part 11 be revised to give examples of electronic records subject to FDA inspection, including pharmaceutical and medical device production records, in order to reduce the need for questions.

The agency does not believe that it is necessary to include examples of records it might inspect because the addition of such examples might raise questions about the agency's intent to inspect other records that were not identified.

32. One comment said that the regulation should state that certain security related information, such as private keys attendant to cryptographic implementation, is not intended to be subject to inspection, although procedures related to keeping such keys confidential can be subject to inspection.

The agency would not routinely seek to inspect especially sensitive information, such as passwords or private keys, attendant to security systems. However, the agency reserves the right to conduct such inspections, consistent with statutory limitations, to enforce the provisions of the act and related statutes. It may be necessary, for example, in investigating cases of suspected fraud, to access and determine passwords and private keys, in the same manner as the agency may obtain specimens of handwritten signatures ("exemplars"). Should there be any reservations about such inspections, persons may, of course, change their passwords and private keys after FDA inspection.

33. One comment asked how persons were expected to meet the proposed requirement, under § 11.1(e), that computer systems be readily available for inspection when such systems include geographically dispersed networks. Another comment said FDA investigators should not be permitted to access industry computer systems as part of inspections because investigators would be untrained users.

The agency intends to inspect those parts of electronic record or signature systems that have a bearing on the trustworthiness and reliability of electronic records and electronic signatures under part 11. For geographically dispersed systems, inspection at a given location would extend to operations, procedures, and controls at that location, along with interaction of that local system with the wider network. The agency would inspect other locations of the network in a separate but coordinated manner, much the same way the agency currently conducts inspections of firms that have multiple facilities in different parts of the country and outside of the United States.

FDA does not believe it is reasonable to rule out computer system access as part of an inspection of electronic record or signature systems. Historically, FDA investigators observe the actions of establishment employees, and (with the cooperation of establishment management) sometimes request that those employees perform some of their assigned tasks to determine the degree of compliance with established requirements. However, there may be times when FDA investigators need to access a system directly. The agency is aware that such access will generally require the cooperation of and, to some degree, instruction by the firms being inspected. As new, complex technologies emerge, FDA will need to develop and implement new inspectional methods in the context of those technologies.

V. IMPLEMENTATION (§ 11.2)

34. Proposed § 11.2(a) stated that for "records required by chapter I of this title to be maintained, but not submitted to the agency, persons may use electronic records/signatures in lieu of paper records/conventional signatures, in whole or in part, * * *."

Two comments requested clarification of the term "conventional signatures." One comment suggested that the term "traditional signatures" be used instead. Another suggested rewording in order to clarify the slash in the phrase "records/signatures."

The agency advises that the term "conventional signature" means handwritten signature. The agency agrees that the term "traditional signature" is preferable, and has revised § 11.2(a) and (b) accordingly. The agency has also clarified proposed § 11.2(a) by replacing the slash with the word "or."

35. One comment asked if the term "persons" in proposed § 11.2(b) would include devices because computer systems frequently apply digital time stamps on records automatically, without direct human intervention.

The agency advises that the term "persons" excludes devices. The agency does not consider the application of a time stamp to be the application of a signature.

36. Proposed § 11.2(b)(2) provides conditions under which electronic records or signatures could be submitted to the agency in lieu of paper. One condition is that a document, or part of a document, must be identified in a public docket as being the type of submission the agency will accept in electronic form. Two comments addressed the nature of the submissions to the public docket. One comment asked that the agency provide specifics, such as the mechanism for updating the docket and the frequency of such updates. One comment suggested making the docket available to the public by electronic means. Another comment suggested that acceptance procedures be uniform among agency units and that electronic mail be used to hold consultations with the agency. One comment encouraged the agency units receiving the submissions to work closely with regulated industry to ensure that no segment of industry is unduly burdened and that agency guidance is widely accepted.

The agency intends to develop efficient electronic records acceptance procedures that afford receiving units sufficient flexibility to deal with submissions according to their capabilities. Although agencywide uniformity is a laudable objective, to attain such flexibility it may be necessary to accommodate some differences among receiving units. The agency considers of primary importance, however, that all part 11 submissions be trustworthy, reliable, and in keeping with FDA regulatory activity. The agency expects to work closely with industry to help ensure that the mechanics and logistics of accepting electronic submissions do not pose any undue burdens. However, the agency expects persons to consult with the intended receiving units on the technical aspects of the submission, such as media, method of transmission, file format, archiving needs, and technical protocols. Such consultations will ensure that submissions are compatible with the receiving units' capabilities. The agency has revised proposed § 11.2(b)(2) to clarify this expectation.

Regarding the public docket, the agency is not at this time establishing a fixed schedule for updating what types of documents are acceptable for submission because the agency expects the docket to change and grow at a rate that cannot be predicted. The agency may, however, establish a schedule for updating the docket in the future. The agency agrees that making the docket available electronically is advisable and will explore this option. Elsewhere in this issue of the Federal Register, FDA is providing further information on this docket.

VI. DEFINITIONS (§ 11.3)

37. One comment questioned the incorporation in proposed § 11.3(a) of definitions under section 201 of the act (21 U.S.C. 321), noting that other FDA regulations (such as 21 CFR parts 807 and 820) lack such incorporation, and suggested that it be deleted.

The agency has retained the incorporation by reference to definitions under section 201 of the act because those definitions are applicable to part 11.

38. One comment suggested adding the following definition for the term "digital signature:" "data appended to, or a cryptographic transformation of, a data unit that allows a recipient of the data unit to prove the source and integrity of the data unit and protect against forgery, e.g., by the recipient."

The agency agrees that the term digital signature should be defined and has added new § 11.3(b)(5) to provide a definition for digital signature that is consistent with the Federal Information Processing Standard 186, issued May 19, 1995, and effective December 1, 1995, by the U.S. Department of Commerce, National Institute of Standards and Technology (NIST). Generally, a digital signature is "an electronic signature based upon cryptographic methods of originator authentication, computed by using a set of rules and a set of parameters such that the identity of the signer and the integrity of the data can be verified." FDA advises that the set of rules and parameters is established in each digital signature standard.

39. Several comments suggested various modifications of the proposed definition of biometric/behavioral links, and suggested revisions that would exclude typing a password or identification code which, the comments noted, is a repeatable action. The comments suggested that actions be unique and measurable to meet the intent of a biometric method.

The agency agrees that the proposed definition of biometric/behavioral links should be revised to clarify the agency's intent that repetitive actions alone, such as typing an identification code and password, are not considered to be biometric in nature. Because comments also indicated that it would be preferable to simplify the term, the agency is changing the term "biometric/behavioral link" to "biometrics." Accordingly, § 11.3(b)(3) defines the term "biometrics" to mean "a method of verifying an individual's identity based on measurement of the individual's physical feature(s) or repeatable action(s) where those features and/or actions are both unique to that individual and measurable."

40. One comment said that the agency should identify what biometric methods are acceptable to verify a person's identity and what validation acceptance criteria the agency has used to determine that biometric technologies are superior to other methods, such as use of identification codes and passwords.

The agency believes that there is a wide variety of acceptable technologies, regardless of whether they are based on biometrics, and regardless of the particular type of biometric mechanism that may be used. Under part 11, electronic signatures that employ at least two distinct identification components such as identification codes and passwords, and electronic signatures based on biometrics are equally acceptable substitutes for traditional handwritten signatures. Furthermore, all electronic record systems are subject to the same requirements of subpart B of part 11 regardless of the electronic signature technology being used. These provisions include requirements for validation.

Regarding the comment's suggestion that FDA apply quantitative acceptance criteria, the agency is not seeking to set specific numerical standards or statistical performance criteria in determining the threshold of acceptability for any type of technology. If such standards were to be set for biometrics-based electronic signatures, similar numerical performance and reliability requirements would have to be applied to other technologies as well. The agency advises, however, that the differences between system controls for biometrics-based electronic signatures and other electronic signatures are a result of the premise that biometrics-based electronic signatures, by their nature, are less prone to be compromised than other methods such as identification codes and passwords. Should it become evident that additional controls are warranted for biometrics-based electronic signatures, the agency will propose to revise part 11 accordingly.

41. Proposed § 11.3(b)(4) defined a closed system as an environment in which there is communication among multiple persons, and where system access is restricted to people who are part of the organization that operates the system.

Many comments requested clarification of the term "organization" and stated that the rule should account for persons who, though not strictly employees of the operating organization, are nonetheless obligated to it in some manner, or who would otherwise be granted system access by the operating organization. As examples of such persons, the comments cited outside contractors, suppliers, temporary employees, and consultants. The comments suggested a variety of alternative wording, including a change of emphasis from organizational membership to organizational control over system access. One comment requested clarification of whether the rule intends to address specific disciplines within a company.

Based on the comments, the agency has revised the proposed definition of closed system to state "an environment in which system access is controlled by persons who are responsible for the content of electronic records that are on the system." The agency agrees that the most important factor in classifying a system as closed or open is whether the persons responsible for the content of the electronic records control access to the system containing those records. A system is closed if access is controlled by persons responsible for the content of the records. If those persons do not control such access, then the system is open because the records may be read, modified, or compromised by others to the possible detriment of the persons responsible for record content. Hence, those responsible for the records would need to take appropriate additional measures in an open system to protect those records from being read, modified, destroyed, or otherwise compromised by unauthorized and potentially unknown parties. The agency does not believe it is necessary to codify the basis or criteria for authorizing system access, such as existence of a fiduciary responsibility or contractual relationship. By being silent on such criteria, the rule affords maximum flexibility to organizations by permitting them to determine those criteria for themselves.

42. Concerning the proposed definition of closed system, one comment suggested adding the words "or devices" after "persons" because communications may involve nonhuman entities.

The agency does not believe it is necessary to adopt the suggested revision because the primary intent of the regulation is to address communication among humans, not devices.

43. One comment suggested defining a closed system in terms of functional characteristics that include physical access control, having professionally written and approved procedures with employees and supervisors trained to follow them, conducting investigations when abnormalities may have occurred, and being under legal obligation to the organization responsible for operating the system.

The agency agrees that the functional characteristics cited by the comment are appropriate for a closed system, but has decided that it is unnecessary to include them in the definition. The functional characteristics themselves, however, such as physical access controls, are expressed as requirements elsewhere in part 11.

44. Two comments said that the agency should regard as closed a system in which dial-in access via public phone lines is permitted, but where access is authorized by, and under the control of, the organization that operates the system.

The agency advises that dial-in access over public phone lines could be considered part of a closed system where access to the system that holds the electronic records is under the control of the persons responsible for the content of those records. The agency cautions, however, that, where an organization's electronic records are stored on systems operated by third parties, such as commercial online services, access would be under control of the third parties and the agency would regard such a system as being open. The agency also cautions that, by permitting access to its systems by public phone lines, organizations lose the added security that results from restricting physical access to computer terminal and other input devices. In such cases, the agency believes firms would be prudent to implement additional security measures above and beyond those controls that the organization would use if the access device was within its facility and commensurate with the potential consequences of such unauthorized access. Such additional controls might include, for example, use of input device checks, caller identification checks (phone caller identification), call backs, and security cards.

45. Proposed § 11.3(b)(5) defined electronic record as a document or writing comprised of any combination of text, graphic representation, data, audio information, or video information, that is created, modified, maintained, or transmitted in digital form by a computer or related system. Many comments suggested revising the proposed definition to reflect more accurately the nature of electronic records and how they differ from paper records. Some comments suggested distinguishing between machine readable records and paper records created by machine. Some comments noted that the term "document or writing" is inappropriate for electronic records because electronic records could be any combination of pieces of information assembled (sometimes on a transient basis) from many noncontiguous places, and because the term does not accurately describe such electronic information as raw data or voice mail. Two comments suggested that the agency adopt definitions of electronic record that were established, respectively, by the United Nations Commission on International Trade Law (UNCITRAL) Working Group on Electronic Data Interchange, and the American National Standards Institute/Institute of Electrical and Electronic Engineers Software Engineering (ANSI/IEEE) Standard (729–1983).

The agency agrees with the suggested revisions and has revised the definition of "electronic record" to emphasize this unique nature and to clarify that the agency does not regard a paper record to be an electronic record simply because it was created by a computer system. The agency has removed "document or writing" from this definition and elsewhere in part 11 for the sake of clarity, simplicity, and consistency.

However, the agency believes it is preferable to adapt or modify the words "document" and "writing" to electronic technologies rather than discard them entirely from the lexicon of computer technology. The agency is aware that the terms "document" and "electronic document" are used in contexts that clearly do not intend to describe paper. Therefore, the agency considers the terms "electronic record" and "electronic document" to be generally synonymous and may use the terms "writing," "electronic document," or "document" in other publications to describe records in electronic form. The agency believes that such usage is a prudent conservation of language and is consistent with the use of other terms and expressions that have roots in older technologies, but have nonetheless been adapted to newer technologies. Such terms include telephone "dialing," internal combustion engine "horse power," electric light luminance expressed as "foot candles," and (more relevant to computer technology) execution of a "carriage return."

Accordingly, the agency has revised the definition of electronic record to mean "any combination of text, graphics, data, audio, pictorial, or other information representation in digital form that is created, modified, maintained, archived, retrieved, or distributed by a computer system."

46. Proposed § 11.3(b)(6) defined an electronic signature as the entry in the form of a magnetic impulse or other form of computer data compilation of any symbol or series of symbols, executed, adopted or authorized by a person to be the legally binding equivalent of the person's handwritten signature. One comment supported the definition as proposed, noting its consistency with dictionary definitions (*Random House Dictionary of the English Language*, Unabridged Ed. 1983, and *American Heritage Dictionary*, 1982). Several other comments, however, suggested revisions. One comment suggested replacing "electronic signature" with "computer based signature," "authentication," or "computer based authentication" because "electronic signature" is imprecise and lacks clear and recognized meaning in the information

security and legal professions. The comment suggested a definition closer to the UNCITRAL draft definition:

(1) [a] method used to identify the originator of the data message and to indicate the originator's approval of the information contained therein; and (2) that method is as reliable as was appropriate for the purpose for which the data message was generated or communicated, in the light of all circumstances, including any agreement between the originator and the addressee of the data message.

One comment suggested replacing "electronic signature" with "electronic identification" or "electronic authorization" because the terms include many types of technologies that are not easily distinguishable and because the preamble to the proposed rule gave a rationale for using "electronic signature" that was too "esoteric for practical consideration."

The agency disagrees that "electronic signature" as proposed should be replaced with other terms and definitions. As noted in the preamble to the proposed rule, the agency believes that it is vital to retain the word "signature" to maintain the equivalence and significance of various electronic technologies with the traditional handwritten signature. By not using the word "signature," people may treat the electronic alternatives as less important, less binding, and less in need of controls to prevent falsification. The agency also believes that use of the word signature provides a logical bridge between paper and electronic technologies that facilitates the general transition from paper to electronic environments. The term helps people comply with current FDA regulations that specifically call for signatures. Nor does the agency agree that this reasoning is beyond the reach of practical consideration.

The agency declines to accept the suggested UNCITRAL definition because it is too narrow in context in that there is not always a specified message addressee for electronic records required by FDA regulations (e.g., a batch production record does not have a specific "addressee").

47. Concerning the proposed definition of "electronic signature," other comments suggested deletion of the term "magnetic impulse" to render the term media neutral and thus allow for such alternatives as an optical disk. Comments also suggested that the term "entry" was unclear and recommended its deletion. Two comments suggested revisions that would classify symbols as an electronic signature only when they are committed to permanent storage because not every computer entry is a signature and processing to permanent storage must occur to indicate completion of processing.

The agency advises that the proposal did not limit electronic signature recordings to "magnetic impulse" because the proposed definition added, "or other form of computer data * * *." However, in keeping with the agency's intent to accept a broad range of technologies, the terms "magnetic impulse" and "entry" have been removed from the proposed definition. The agency believes that recording of computer data to "permanent" storage is not a necessary or warranted qualifier because it is not relevant to the concept of equivalence to a handwritten signature. In addition, use of the qualifier regarding permanent storage could impede detection of falsified records if, for example, the signed falsified record was deleted after a predetermined period (thus, technically not recorded to "permanent" storage). An individual could disavow a signature because the record had ceased to exist.

For consistency with the proposed definition of handwritten signature, and to clarify that electronic signatures are those of individual human beings, and not those of organizations (as included in the act's definition of "person"), FDA is changing "person" to "individual" in the final rule.

Accordingly, § 11.3(b)(7) defines electronic signature as a computer data compilation of any symbol or series of symbols executed, adopted, or authorized by an individual to be the legally binding equivalent of the individual's handwritten signature.

48. Proposed § 11.3(b)(7) (redesignated § 11.3(b)(8) in the final rule) defined "handwritten signature" as the name of an individual, handwritten in script by that individual, executed or adopted with the present intention to authenticate a writing in a permanent form. The act of signing with a writing or marking instrument such as a pen or stylus is preserved. The proposed definition also stated that the scripted name, while conventionally applied to paper, may also be applied to other devices which capture the written name.

Many comments addressed this proposed definition. Two comments suggested that it be deleted on the grounds it is redundant and that, when handwritten signatures are recorded electronically, the result fits the definition of electronic signature.

The agency disagrees that the definition of handwritten signature should be deleted. In stating the criteria under which electronic signatures may be used in place of traditional handwritten signatures, the agency believes it is necessary to define handwritten signature. In addition, the agency believes that it is necessary to distinguish handwritten signatures from electronic signatures because, with handwritten signatures, the traditional act of signing one's name is preserved. Although the handwritten signature recorded electronically and electronic signatures, as defined in part 11, may both ultimately result in magnetic impulses or other forms of computerized symbol representations, the means of achieving those recordings and, more importantly, the controls needed to ensure their reliability and trustworthiness are quite different. In addition, the agency believes that a definition for handwritten signature is warranted to accommodate persons who wish to implement record systems that are combinations of paper and electronic technologies.

49. Several comments suggested replacing the reference to "scripted name" in the proposed definition of handwritten signature with "legal mark" so as to accommodate individuals who are physically unable to write their names in script. The comments asserted that the term "legal mark" would bring the definition to closer agreement with generally recognized legal interpretations of signature.

The agency agrees and has added the term "legal mark" to the definition of handwritten signature.

50. One comment recommended that the regulation state that, when the handwritten signature is not the result of the act of signing with a writing or marking instrument, but is applied to another device that captures the written name, a system should verify that the owner of the signature has authorized the use of the handwritten signature.

The agency declines to accept this comment because, if the act of signing or marking is not preserved, the type of signature would not be considered a handwritten signature. The comment appears to be referring to instances in which one person authorizes someone else to use his or her stamp or device. The agency views this as inappropriate when the signed record does not clearly show that the stamp owner did not actually execute the signature. As discussed elsewhere in this preamble, the agency believes that where one person authorizes another to sign a document on his or her behalf, the second person must sign his or her own name (not the name of the first person) along with some notation that, in doing so, he or she is acting in the capacity, or on behalf, of the first person.

51. One comment suggested that where handwritten signatures are captured by devices, there should be a register of manually written signatures to enable comparison for authenticity and the register also include the typed names of individuals.

The agency agrees that the practice of establishing a signature register has merit, but does not believe that it is necessary, in light of other part 11 controls. As noted elsewhere in this preamble (in the discussion of proposed § 11.50), the agency agrees that human readable displays of electronic records must display the name of the signer.

52. Several comments suggested various editorial changes to the proposed definition of handwritten signature including: (1) Changing the word "also" in the last sentence to

"alternatively," (2) clarifying the difference between the words "individual" and "person," (3) deleting the words "in a permanent form," and (4) changing "preserved" to "permitted." One comment asserted that the last sentence of the proposed definition was unnecessary.

The agency has revised the definition of handwritten signature to clarify its intent and to keep the regulation as flexible as possible. The agency believes that the last sentence of the proposed definition is needed to address devices that capture handwritten signatures. The agency is not adopting the suggestion that the word "preserved" be changed to "permitted" because "preserved" more accurately states the agency's intent and is a qualifier to help distinguish handwritten signatures from others. The agency advises that the word "individual" is used, rather than "person," because the act's definition of person extends beyond individual human beings to companies and partnerships. The agency has retained the term "permanent" to discourage the use of pencils, but recognizes that "permanent" does not mean eternal.

53. One comment asked whether a signature that is first handwritten and then captured electronically (e.g., by scanning) is an electronic signature or a handwritten signature, and asked how a handwritten signature captured electronically (e.g., by using a stylus-sensing pad device) that is affixed to a paper copy of an electronic record would be classified.

FDA advises that when the act of signing with a stylus, for example, is preserved, even when applied to an electronic device, the result is a handwritten signature. The subsequent printout of the signature on paper would not change the classification of the original method used to execute the signature.

54. One comment asserted that a handwritten signature recorded electronically should be considered to be an electronic signature, based on the medium used to capture the signature. The comment argued that the word signature should be limited to paper technology.

The agency disagrees and believes it is important to classify a signature as handwritten based upon the preserved action of signing with a stylus or other writing instrument.

55. One comment asked if the definition of handwritten signature encompasses hand-written initials.

The agency advises that, as revised, the definition of handwritten signature includes handwritten initials if the initials constitute the legal mark executed or adopted with the present intention to authenticate a writing in a permanent form, and where the method of recording such initials involves the act of writing with a pen or stylus.

56. Proposed § 11.3(b)(8) (redesignated as § 11.3(b)(9) in the final rule) defined an open system as an environment in which there is electronic communication among multiple persons, where system access extends to people who are not part of the organization that operates the system.

Several comments suggested that, for simplicity, the agency define "open system" as any system that does not meet the definition of a closed system. One comment suggested that the definition be deleted on the grounds it is redundant, and that it is the responsibility of individual firms to take appropriate steps to ensure the validity and security of applications and information, regardless of whether systems are open or closed. Other comments suggested definitions of "open system" that were opposite to what they suggested for a closed system.

The agency has revised the definition of open system to mean "an environment in which system access is not controlled by persons who are responsible for the content of electronic records that are on the system." The agency believes that, for clarity, the definition should stand on its own rather than as any system that is not closed. The agency rejects the suggestion that the term need not be defined at all because FDA believes that controls for open systems merit distinct provisions in part 11 and defining the term is basic to understanding which requirements apply to a given system. The agency agrees that companies have the responsibility to take steps to ensure the validity and security of their applications and information. However, FDA finds it necessary to establish part 11 as minimal requirements to help ensure that those steps are, in fact, acceptable.

VII. ELECTRONIC RECORDS—CONTROLS FOR CLOSED SYSTEMS (§ 11.10)

The introductory paragraph of proposed § 11.10 states that:

> Closed systems used to create, modify, maintain, or transmit electronic records shall employ procedures and controls designed to ensure the authenticity, integrity, and confidentiality of electronic records, and to ensure that the signer cannot readily repudiate the signed record as not genuine. * * *

The rest of the section lists specific procedures and controls.

57. One comment expressed full support for the list of proposed controls, calling them generally appropriate and stated that the agency is correctly accommodating the fluid nature of various electronic record and electronic signature technologies. Another comment, however, suggested that controls should not be implemented at the time electronic records are first created, but rather only after a document is accepted by a company.

The agency disagrees with this suggestion. To ignore such controls at a stage before official acceptance risks compromising the record. For example, if "preacceptance" records are signed by technical personnel, it is vital to ensure the integrity of their electronic signatures to prevent record alteration. The need for such integrity is no less important at preacceptance stages than at later stages when managers officially accept the records. The possibility exists that some might seek to disavow, or avoid FDA examination of, pertinent records by declaring they had not been formally "accepted." In addition, FDA routinely can and does inspect evolving paper documents (e.g., standard operating procedures and validation protocols) even though they have yet to receive a firm's final acceptance.

58. One comment said proposed § 11.10 contained insufficient requirements for firms to conduct periodic inspection and monitoring of their own systems and procedures to ensure compliance with the regulations. The comment also called for a clear identification of the personnel in a firm who would be responsible for system implementation, operation, change control, and monitoring.

The agency does not believe it is necessary at this time to codify a self-auditing requirement, as suggested by the comment. Rather, the agency intends to afford organizations flexibility in establishing their own internal mechanisms to ensure compliance with part 11. Self-audits, however, may be considered as a general control, within the context of the introductory paragraph of § 11.10. The agency encourages firms to conduct such audits periodically as part of an overall approach to ensure compliance with FDA regulations generally. Likewise, the agency does not believe it is necessary or practical to codify which individuals in an organization should be responsible for compliance with various provisions of part 11. However, ultimate responsibility for part 11 will generally rest with persons responsible for electronic record content, just as responsibility for compliance with paper record requirements generally lies with those responsible for the record's content.

59. Several comments interpreted proposed § 11.10 as applying all procedures and controls to closed systems and suggested revising it to permit firms to apply only those procedures and controls they deem necessary for their own operations, because some requirements are excessive in some cases.

The agency advises that, where a given procedure or control is not intended to apply in all cases, the language of the rule so indicates. Specifically, use of operational checks (§ 11.10(f)) and device checks (§ 11.10(h)) is not required in all cases. The remaining requirements do apply in all cases and are, in the agency's opinion, the minimum needed to ensure the trustworthiness and reliability of electronic record systems. In addition, certain controls that firms deem adequate for their routine internal operations might nonetheless leave records vulnerable to manipulation and, thus, may be incompatible with FDA's responsibility to protect public health. The suggested revision would effectively permit firms to implement various controls selectively and possibly shield records from FDA, employ unqualified personnel, or permit employees to evade responsibility for fraudulent use of their electronic signatures.

The agency believes that the controls in § 11.10 are vital, and notes that almost all of them were suggested by comments on the ANPRM. The agency believes the wording of the regulation nonetheless permits firms maximum flexibility in how to meet those requirements.

60. Two comments suggested that the word "confidentiality" in the introductory paragraph of proposed § 11.10 be deleted because it is unnecessary and inappropriate. The comments stated that firms should determine if certain records need to be confidential, and that as long as records could not be altered or deleted without appropriate authority, it would not matter whether they could read the records.

The agency agrees that not all records required by FDA need to be kept confidential within a closed system and has revised the reference in the introductory paragraph of § 11.10 to state "* * * and, when appropriate, the confidentiality of electronic records." The agency believes, however that the need for retaining the confidentiality of certain records is not diminished because viewers cannot change them. It may be prudent for persons to carefully assess the need for record confidentiality. (See, e.g., 21 CFR 1002.42, Confidentiality of records furnished by dealers and distributors, with respect to certain radiological health products.) In addition, FDA's obligation to retain the confidentiality of information it receives in some submissions hinges on the degree to which the submitter maintains confidentiality, even within its own organization. (See, e.g., 21 CFR 720.8(b) with respect to cosmetic ingredient information in voluntary filings of cosmetic product ingredient and cosmetic raw material composition statements.)

61. One comment asked if the procedures and controls required by proposed § 11.10 were to be built into software or if they could exist in written form.

The agency expects that, by their nature, some procedures and controls, such as use of time-stamped audit trails and operational checks, will be built into hardware and software. Others, such as validation and determination of personnel qualifications, may be implemented in any appropriate manner regardless of whether the mechanisms are driven by, or are external to, software or hardware. To clarify this intent, the agency has revised the introductory paragraph of proposed § 11.10 to read, in part, "Persons who use closed systems to create, modify * * *." Likewise, for clarity and consistency, the agency is introducing the same phrase, "persons who use * * *" in §§ 11.30 and 11.300.

62. One comment contended that the distinction between open and closed systems should not be predominant because a $100,000 transaction in a closed system should not have fewer controls than a $1 transaction in an open system.

The agency believes that, within part 11, firms have the flexibility they need to adjust the extent and stringency of controls based on any factors they choose, including the economic value of the transaction. The agency does not believe it is necessary to modify part 11 at this time so as to add economic criteria.

63. One comment suggested that the reference to repudiation in the introductory paragraph of § 11.10 should be deleted because repudiation can occur at any time in legal proceedings. Another comment, noting that the proposed rule appeared to address only nonrepudiation of a signer, said the rule should address nonrepudiation of record "genuineness" or extend to nonrepudiation of submission, delivery, and receipt. The comment stated that some firms provide nonrepudiation services that can prevent someone from successfully claiming that a record has been altered.

In response to the first comment, the agency does not agree that the reference to repudiation should be deleted because reducing the likelihood that someone can readily repudiate an electronic signature as not his or her own, or that the signed record had been altered, is vital to the agency's basic acceptance of electronic signatures. The agency is aware that the need to deter such repudiation has been addressed in many forums and publications that discuss electronic signatures. Absent adequate controls, FDA believes some people would be more likely to repudiate an electronically-signed record because of the relative ease with which electronic records may be altered and the ease with which one individual could impersonate another. The agency notes, however, that the rule does not call for nonrepudiation as an absolute guarantee, but requires that the signer cannot "readily" repudiate the signature.

In response to the second comment, the agency agrees that it is also important to establish nonrepudiation of submission, delivery, and receipt of electronic records, but advises that, for purposes of § 11.10, the agency's intent is to limit nonrepudiation to the genuineness of the signer's record. In other words, an individual should not be able to readily say that: (1) He or she did not, in fact, sign the record; (2) a given electronic record containing the individual's signature was not, in fact, the record that the person signed; or (3) the originally signed electronic record had been altered after having been signed.

64. Proposed § 11.10(a) states that controls for closed systems are to include the validation of systems to ensure accuracy, reliability, consistent intended performance, and the ability to conclusively discern invalid or altered records.

Many comments objected to this proposed requirement because the word "conclusively" inferred an unreasonably high and unattainable standard, one which is not applied to paper records.

The agency intends to apply the same validation concepts and standards to electronic record and electronic signature systems as it does to paper systems. As such, FDA does not intend the word "conclusively" to suggest an unattainable absolute and has, therefore, deleted the word from the final rule.

65. One comment suggested qualifying the proposed validation requirement in § 11.10(a) to state that validation be performed "where necessary" and argued that validation of commercially available software is not necessary because such software has already been thoroughly validated. The comment acknowledged that validation may be required for application programs written by manufacturers and others for special needs.

The agency disagrees with the comment's claim that all commercial software has been validated. The agency believes that commercial availability is no guarantee that software has undergone "thorough validation" and is unaware of any regulatory entity that has jurisdiction over general purpose software producers. The agency notes that, in general, commercial software packages are accompanied not by statements of suitability or compliance with established standards, but rather by disclaimers as to their fitness for use. The agency is aware of the complex and sometimes controversial issues in validating commercial software. However, the need to validate such software is not diminished by the fact that it was not written by those who will use the software.

In the future, the agency may provide guidance on validation of commercial software used in electronic record systems. FDA has addressed the matter of software validation in general in such documents as the "Draft Guideline for the Validation of Blood Establishment Computer Systems," which is available from the Manufacturers Assistance and Communications Staff, Center for Biologics Evaluation and Research (HFM–42), Food and Drug Administration, 1401 Rockville Pike, Rockville, MD 20852–1448, 301–594–2000. This guideline is also available by sending e-mail to the following Internet address: CBER_INFO@A1.CBER.FDA. GOV). For the purposes of part 11, however, the agency believes it is vital to retain the validation requirement.

66. One comment requested an explanation of what was meant by the phrase "consistent intended" in proposed § 11.10(a) and why "consistent performance" was not used instead. The comment suggested that the rule should distinguish consistent intended performance from well-recognized service "availability."

The agency advises that the phrase "consistent intended performance" relates to the general principle of validation that planned and expected performance is based upon predetermined design specifications (hence, "intended"). This concept is in accord with the agency's 1987 "Guideline on General Principles of Process Validation," which is available from the Division of Manufacturing and Product Quality, Center for Drug Evaluation and Research (HFD–320), Food and Drug Administration, 7520 Standish Pl., Rockville, MD 20855, 301–594–0093). This guideline defines validation as establishing documented evidence that provides a high degree of assurance that a specific process will consistently produce a product meeting its predetermined specifications and quality attributes. The agency believes that the comment's concepts are accommodated by this definition to the extent that system "availability" may be one of the predetermined specifications or quality attributes.

67. One comment said the rule should indicate whether validation of systems does, or should, require any certification or accreditation.

The agency believes that although certification or accreditation may be a part of validation of some systems, such certification or accreditation is not necessary in all cases, outside of the context of any such approvals within an organization itself. Therefore, part 11 is silent on the matter.

68. One comment said the rule should clarify whether system validation should be capable of discerning the absence of electronic records, in light of agency concerns about falsification. The comment added that the agency's concerns regarding invalid or altered records can be mitigated by use of cryptographically enhanced methods, including secure time and date stamping.

The agency does not believe that it is necessary at this time to include an explicit requirement that systems be capable of detecting the absence of records. The agency advises that the requirement in § 11.10(e) for audit trails of operator actions would cover those actions intended to delete records. Thus, the agency would expect firms to document such deletions, and would expect the audit trail mechanisms to be included in the validation of the electronic records system.

69. Proposed § 11.10(b) states that controls for closed systems must include the ability to generate true copies of records in both human readable and electronic form suitable for inspection, review, and copying by the agency, and that if there were any questions regarding the ability of the agency to perform such review and copying, persons should contact the agency.

Several comments objected to the requirement for "true" copies of electronic records. The comments asserted that information in an original record (as may be contained in a database) may be presented in a copy in a different format that may be more usable. The comments concluded that, to generate precise "true" copies of electronic records, firms may have to retain the hardware and software that had been used to create those records in the first place (even when such hardware and software had been replaced by newer systems). The comments pointed out that firms may have to provide FDA with the application logic for "true" copies, and that this may violate copyright provisions. One comment illustrated the difference between "true" copies and other equally reliable, but not exact, copies of electronic records by noting that pages from FDA's paper publications (such as the CFR and the Compliance Policy Guidance Manual) look quite different from electronic copies posted to FDA's bulletin board. The comments suggested different wording that would effectively require accurate and complete copies, but not necessarily "true" copies.

The agency agrees that providing exact copies of electronic records in the strictest meaning of the word "true" may not always be feasible. The agency nonetheless believes it is vital that copies of electronic records provided to FDA be accurate and complete. Accordingly, in § 11.10(b), "true" has been replaced with "accurate and complete." The agency expects that this revision should obviate the potential problems noted in the comments. The revision should also reduce the costs of providing copies by making clear that firms need not maintain obsolete equipment in order to make copies that are "true" with respect to format and computer system.

70. Many comments objected to the proposed requirement that systems be capable of generating electronic copies of electronic records for FDA inspection and copying, although they generally agreed that it was appropriate to provide FDA with readable paper copies. Alternative wording was suggested that would make providing electronic copies optional, such that persons could provide FDA with nothing but paper copies if they so wished. The comments argued that providing FDA with electronic copies was unnecessary, unjustified, not practical considering the different types of computer systems that may be in use, and would unfairly limit firms in their selection of hardware and software if they could only use systems that matched FDA's capabilities (capabilities which, it was argued, would not be uniform throughout the United States). One comment suggested that the rule specify a particular format, such as ASCII, for electronic copies to FDA.

The agency disagrees with the assertion that FDA need only be provided with paper copies of electronic records. To operate effectively, the agency must function on the same technological plane as the industries it regulates. Just as firms realize efficiencies and benefits in the use of electronic records, FDA should be able to conduct audits efficiently and thoroughly using the same technology. For example, where firms perform computerized trend analyses of electronic records to improve their processes, FDA should be able to use computerized methods to audit electronic records (on site and off, as necessary) to detect trends, inconsistencies, and potential problem areas. If FDA is restricted to reviewing only paper copies of those records, the results would severely impede its operations. Inspections would take longer to complete, resulting in delays in approvals of new medical products, and expenditure of additional resources both by FDA (in performing the inspections and transcribing paper records to electronic format) and by the inspected firms, which would generate the paper copies and respond to questions during the resulting lengthened inspections.

The agency believes that it also may be necessary to require that persons furnish certain electronic copies of electronic records to FDA because paper copies may not be accurate and complete if they lack certain audit trail (metadata) information. Such information may have a direct bearing on record trustworthiness and reliability. These data could include information, for example, on when certain items of electronic mail were sent and received.

The agency notes that people who use different computer systems routinely provide each other with electronic copies of electronic records, and there are many current and developing tools to enable such sharing. For example, at a basic level, records may be created in, or transferred to, the ASCII format. Many different commercial programs have the capability to import from, and export to, electronic records having different formats. Firms use electronic data interchange (commonly known as EDI) and agreed upon transaction set formats to enable them to exchange copies of electronic records effectively. Third parties are also developing portable document formats to enable conversion among several diverse formats.

Concerning the ability of FDA to handle different formats of electronic records, based upon the emergence of format conversion tools such as those mentioned above, the agency's experience with electronic submissions such as computer assisted new drug applications (commonly known as CANDA's), and the agency's planned Submissions Management and Review Tracking System (commonly known as SMART), FDA is confident that it can work with firms to minimize any formatting difficulties. In addition, substitution of the words "accurate and complete" for "true," as discussed in comment 69, should make it easier for firms to provide FDA with electronic copies of their electronic records. FDA does not believe it is necessary to specify any particular format in part 11 because it prefers, at this time, to afford industry and the agency more flexibility in deciding which formats meet the capabilities of all parties. Accordingly, the agency has revised proposed § 11.10(b) to read:

> The ability to generate accurate and complete copies of records in both human readable and electronic form suitable for inspection, review, and copying by the agency. Persons should contact the agency if there are any questions regarding the ability of the agency to perform such review and copying of the electronic records.

71. Proposed § 11.10(c) states that procedures and controls for closed systems must include the protection of records to enable their accurate and ready retrieval throughout the records retention period.

One firm commented that, because it replaces systems often (about every 3 years), it may have to retain supplanted systems to meet these requirements. Another comment suggested that the rule be modified to require records retention only for as long as "legally mandated."

The agency notes that, as discussed in comment 70 of this document, persons would not necessarily have to retain supplanted hardware and software systems provided they implemented conversion capabilities when switching to replacement technologies. The agency does not believe it is necessary to add the qualifier "legally mandated" because the retention period for a given record will generally be established by the regulation that requires the record. Where the regulations do not specify a given time, the agency would expect firms to establish their own retention periods. Regardless of the basis for the retention period, FDA believes that the requirement that a given electronic record be protected to permit it to be accurately and readily retrieved for as long as it is kept is reasonable and necessary.

72. Proposed § 11.10(e) would require the use of time-stamped audit trails to document record changes, all write-to-file operations, and to independently record the date and time of operator entries and actions. Record changes must not obscure previously recorded information and such audit trail documentation must be retained for a period at least as long as required for the subject electronic documents and must be available for agency review and copying.

Many comments objected to the proposed requirement that all write-to-file operations be documented in the audit trail because it is unnecessary to document all such operations. The comments said that this would require audit trails for such automated recordings as those made to internal buffers, data swap files, or temporary files created by word processing programs. The comments suggested revising § 11.10(e) to require audit trails only for operator entries and actions.

Other comments suggested that audit trails should cover: (1) Operator data inputs but not actions, (2) only operator changes to records, (3) only critical write-to-file information, (4) operator changes as well as all actions, (5) only new entries, (6) only systems where data can be altered, (7) only information recorded by humans, (8) information recorded by both humans and devices, and (9) only entries made upon adoption of the records as official. One comment said audit trails should not be required for data acquisition systems, while another comment said audit trails are critical for data acquisition systems.

It is the agency's intent that the audit trail provide a record of essentially who did what, wrote what, and when. The write-to-file operations referenced in the proposed rule were not intended to cover the kind of "background" nonhuman recordings the comments identified.

The agency considers such operator actions as activating a manufacturing sequence or turning off an alarm to warrant the same audit trail coverage as operator data entries in order to document a thorough history of events and those responsible for such events. Although FDA acknowledges that not every operator "action," such as switching among screen displays, need be covered by audit trails, the agency is concerned that revising the rule to cover only "critical" operations would result in excluding much information and actions that are necessary to document events thoroughly.

The agency believes that, in general, the kinds of operator actions that need to be covered by an audit trail are those important enough to memorialize in the electronic record itself. These are actions which, for the most part, would be recorded in corresponding paper records according to existing recordkeeping requirements.

The agency intends that the audit trail capture operator actions (e.g., a command to open a valve) at the time they occur, and operator information (e.g., data entry) at the time the information is saved to the recording media (such as disk or tape), in much the same manner as such actions and information are memorialized on paper. The audit trail need not capture every keystroke and mistake that is held in a temporary buffer before those commitments. For example, where an operator records the lot number of an ingredient by typing the lot number, followed by the "return key" (where pressing the return key would cause the information to be saved to a disk file), the audit trail need not record every "backspace delete" key the operator may have previously pressed to correct a typing error. Subsequent "saved" corrections made after such a commitment, however, must be part of the audit trail.

At this time, the agency's primary concern relates to the integrity of human actions. Should the agency's experience with part 11 demonstrate a need to require audit trails of device operations and entries, the agency will propose appropriate revisions to these regulations. Accordingly, the agency has revised proposed § 11.10(e) by removing reference to all write-to-file operations and clarifying that the audit trail is to cover operator entries and actions that create, modify, or delete electronic records.

73. A number of comments questioned whether proposed § 11.10(e) mandated that the audit trail be part of the electronic record itself or be kept as a separate record. Some comments interpreted the word "independently" as requiring a separate record. Several comments focused on the question of whether audit trails should be generated manually under operator control or automatically without operator control. One comment suggested a revision that would require audit trails to be generated by computer, because the system, not the operator, should record the audit trail. Other comments said the rule should facilitate date and time recording by software, not operators, and that the qualifier "securely" be added to the language describing the audit trail. One comment, noting that audit trails require validation and qualification to ensure that time stamps are accurate and independent, suggested that audit trails be required only when operator actions are witnessed.

The agency advises that audit trail information may be contained as part of the electronic record itself or as a separate record. FDA does not intend to require one method over the other. The word "independently" is intended to require that the audit trail not be under the control of the operator and, to prevent ready alteration, that it be created independently of the operator.

To maintain audit trail integrity, the agency believes it is vital that the audit trail be created by the computer system independently of operators. The agency believes it would defeat the purpose of audit trails to permit operators to write or change them. The agency believes that, at this time, the source of such independent audit trails may effectively be within the organization that creates the electronic record. However, the agency is aware of a situation under which time and date stamps are provided by trusted third parties outside of the creating organization. These third parties provide, in effect, a public electronic notary service. FDA will monitor development of such services in light of part 11 to determine if a requirement for such third party services should be included in these regulations. For now, the agency considers the advent of such services as recognition of the need for strict objectivity in recording time and date stamps.

The agency disagrees with the premise that only witnessed operator actions need be covered by audit trails because the opportunities for record falsification are not limited to cases where operator actions are witnessed. Also, the need for validating audit trails does not diminish the need for their implementation.

FDA agrees with the suggestion that the proposed rule be revised to require a secure audit trail—a concept inherent in having such a control at all. Accordingly, proposed § 11.10(e) has been revised to require use of "secure, computer-generated" audit trails.

74. A few comments objected to the requirement that time be recorded, in addition to dates, and suggested that time be recorded only when necessary and feasible. Other comments specifically supported the requirement for recording time, noting that time stamps make electronic signatures less vulnerable to fraud and abuse. The comments noted that, in any setting, there is a need to identify the date, time, and person responsible for adding to or changing a value. One of the comments suggested that the rule require recording the reason for making changes to electronic records. Other comments implicitly supported recording time.

FDA believes that recording time is a critical element in documenting a sequence of events. Within a given day a number of events and operator actions may take place, and without recording time, documentation of those events would be incomplete. For example, without time stamps, it may be nearly impossible to determine such important sequencing as document approvals and revisions and the addition of ingredients in drug production. Thus, the element of time becomes vital to establishing an electronic record's trustworthiness and reliability.

The agency notes that comments on the ANPRM frequently identified use of date/time stamps as an important system control. Time recording, in the agency's view, can also be an effective deterrent to records falsification. For example, event sequence codes alone would not necessarily document true time in a series of events, making falsification of that sequence easier if time stamps are not used. The agency believes it should be very easy for firms to implement time stamps because there is a clock in every computer and document management software, electronic mail systems and other electronic record/electronic applications, such as digital signature programs, commonly apply date and time stamps. The agency does not intend that new technologies, such as cryptographic technologies, will be needed to comply with this requirement. The agency believes that implementation of time stamps should be feasible in virtually all computer systems because effective computer operations depend upon internal clock or timing mechanisms and, in the agency's experience, most computer systems are capable of precisely recording such time entries as when records are saved.

The agency is implementing the time stamp requirement based on the understanding that all current computers, electronic document software, electronic mail, and related electronic record systems include such technologies. The agency also understands that time stamps are applied automatically by these systems, meaning firms would not have to install additional hardware, software, or incur additional burden to implement this control. In recognition of this, the agency wishes to clarify that a primary intent of this provision is to ensure that people take reasonable measures to ensure that those built in time stamps are accurate and that people do not alter them casually so as to readily mask unauthorized record changes.

The agency advises that, although part 11 does not specify the time units (e.g., tenth of a second, or even the second) to be used, the agency expects the unit of time to be meaningful in terms of documenting human actions.

The agency does not believe part 11 needs to require recording the reason for record changes because such a requirement, when needed, is already in place in existing regulations that pertain to the records themselves.

75. One comment stated that proposed § 11.10(e) should not require an electronic signature for each write-to-file operation.

The agency advises that § 11.10(e) does not require an electronic signature as the means of authenticating each write-to-file operation. The agency expects the audit trail to document who did what and when, documentation that can be recorded without electronic signatures themselves.

76. Several comments, addressing the proposed requirement that record changes not obscure previously recorded information, suggested revising proposed § 11.10(e) to apply only to those entries intended to update previous information.

The agency disagrees with the suggested revision because the rewording is too narrow. The agency believes that some record changes may not be "updates" but significant modifications or falsifications disguised as updates. All changes to existing records need to be documented, regardless of the reason, to maintain a complete and accurate history, to document individual responsibility, and to enable detection of record falsifications.

77. Several comments suggested replacing the word "document" with "record" in the phrase "Such audit trails shall be retained for a period at least as long as required for the subject electronic documents * * *" because not all electronic documents are electronic records and because the word document connotes paper.

As discussed in section III.D. of this document, the agency equates electronic documents with electronic records, but for consistency, has changed the phrase to read "Such audit trail documentation shall be retained for a period at least as long as that required for the subject electronic records * * *."

78. Proposed § 11.10(k)(ii) (§ 11.10(k)(2) in this regulation) addresses electronic audit trails as a systems documentation control. One comment noted that this provision appears to be the same as the audit trail provision of proposed § 11.10(e) and requested clarification.

The agency wishes to clarify that the kinds of records subject to audit trails in the two provisions cited by the comment are different. Section 11.10(e) pertains to those records that are required by existing regulations whereas § 11.10(k)(2) covers the system documentation records regarding overall controls (such as access privilege logs, or system operational specification diagrams). Accordingly, the first sentence of § 11.10(e) has been revised to read "Use of secure, computer-generated, time-stamped audit trails to independently record and date the time of operator entries and actions that create, modify, or delete electronic records."

79. Proposed § 11.10(f) states that procedures and controls for closed systems must include the use of operational checks to enforce permitted sequencing of events, as appropriate.

Two comments requested clarification of the agency's intent regarding operational checks.

The agency advises that the purpose of performing operational checks is to ensure that operations (such as manufacturing production steps and signings to indicate initiation or completion of those steps) are not executed outside of the predefined order established by the operating organization.

80. Several comments suggested that, for clarity, the phrase "operational checks" be modified to "operational system checks."

The agency agrees that the added modifier "system" more accurately reflects the agency's intent that operational checks be performed by the computer systems and has revised proposed § 11.10(f) accordingly.

81. Several comments suggested revising proposed § 11.10(f) to clarify what is to be checked. The comments suggested that "steps" in addition to "events" be checked, only critical steps be checked, and that "records" also be checked.

The agency intends the word "event" to include "steps" such as production steps. For clarity, however, the agency has revised proposed § 11.10(f) by adding the word "steps." The agency does not, however, agree that only critical steps need be subject to operational checks because a given specific step or event may not be critical, yet it may be very important that the step be executed at the proper time relative to other steps or events. The agency does not believe it necessary to add the modifier "records" to proposed § 11.10(f) because creation, deletion, or modification of a record is an event. Should it be necessary to create, delete, or modify records in a particular sequence, operational system checks would ensure that the proper sequence is followed.

82. Proposed § 11.10(g) states that procedures and controls for closed systems must include the use of authority checks to ensure that only authorized individuals use the system, electronically sign a record, access the operation or device, alter a record, or perform the operation at hand.

One comment suggested that the requirement for authority checks be qualified with the phrase "as appropriate," on the basis that it would not be necessary for certain parts of a system, such as those not affecting an electronic record. The comment cited pushing an emergency stop button as an example of an event that would not require an authority check. Another comment suggested deleting the requirement on the basis that some records can be read by all employees in an organization.

The agency advises that authority checks, and other controls under § 11.10, are intended to ensure the authenticity, integrity, and confidentiality of electronic records, and to ensure that signers cannot readily repudiate a signed record as not genuine. Functions outside of this context, such as pressing an emergency stop button, would not be covered. However, even in this example, the agency finds it doubtful that a firm would permit anyone, such as a stranger from outside the organization, to enter a facility and press the stop button at will regardless of the existence of an emergency. Thus, there would likely be some generalized authority checks built into the firm's operations.

The agency believes that few organizations freely permit anyone from within or without the operation to use their computer system, electronically sign a record, access workstations, alter records, or perform operations. It is likely that authority checks shape the activities of almost every organization. The nature, scope, and mechanism of performing such checks is up to the operating organization. FDA believes, however, that performing such checks is one of the most fundamental measures to ensure the integrity and trustworthiness of electronic records.

Proposed § 11.10(g) does not preclude all employees from being permitted to read certain electronic records. However, the fact that some records may be read by all employees would not justify deleting the requirement for authority checks entirely. The agency believes it is highly unlikely that all of a firm's employees would have authority to read, write, and sign all of its electronic records.

83. One comment said authority checks are appropriate for document access but not system access, and suggested that the phrase "access the operation or device" be deleted. The comment added, with respect to authority checks on signing records, that in many organizations, more than one individual has the authority to sign documents required under FDA regulations and that such authority should be vested with the individual as designated by the operating organization. Another comment said proposed § 11.10(g) should explicitly require access authority checks and suggested that the phrase "use the system" be changed to "access and use the system." The comment also asked for clarification of the term "device."

The agency disagrees that authority checks should not be required for system access because, as discussed in comment 82 of this document, it is unlikely that a firm would permit any unauthorized individuals to access its computer systems. System access control is a basic security function because system integrity may be impeached even if the electronic records themselves are not directly accessed. For example, someone could access a system and change password requirements or otherwise override important security measures, enabling individuals to alter electronic records or read information that they were not authorized to see. The agency does not believe it necessary to add the qualifier "access and" because § 11.10(d) already requires that system access be limited to authorized individuals. The agency intends the word "device" to mean a computer system input or output device and has revised proposed § 11.10(g) to clarify this point.

Concerning signature authority, FDA advises that the requirement for authority checks in no way limits organizations in authorizing individuals to sign multiple records. Firms may use any appropriate mechanism to implement such checks. Organizations do not have to embed a list of authorized signers in every record to perform authority checks. For example, a record may be linked to an authority code that identifies the title or organizational unit of people who may sign the record. Thus, employees who have that corresponding code, or belong to that unit, would be able to sign the record. Another way to implement controls would be to link a list of authorized records to a given individual, so that the system would permit the individual to sign only records in that list.

84. Two comments addressed authority checks within the context of PDMA and suggested that such checks not be required for drug sample receipt records. The comments said that different individuals may be authorized to accept drug samples at a physician's office, and that the large number of physicians who would potentially qualify to receive samples would be too great to institute authority checks.

The agency advises that authority checks need not be automated and that in the context of PDMA such checks would be as valid for electronic records as they are for paper sample requests because only licensed practitioners or their designees may accept delivery of drug samples. The agency, therefore, acknowledges that many individuals may legally accept samples and, thus, have the authority to sign electronic receipts. However, authority checks for electronic receipts could nonetheless be performed by sample manufacturer representatives by using the same procedures as the representatives use for paper receipts. Accordingly, the agency disagrees with the comment that proposed § 11.10(g) should not apply to PDMA sample receipts.

The agency also advises that under PDMA, authority checks would be particularly important in the case of drug sample request records because only licensed practitioners may request drug samples.

Accordingly, proposed § 11.10(g) has been revised to read: "Use of authority checks to ensure that only authorized individuals can use the system, electronically sign a record, access the operation or computer system input or output device, alter a record, or perform the operation at hand."

85. Proposed § 11.10(h) states that procedures and controls for closed systems must include the use of device (e.g., terminal) location checks to determine, as appropriate, the validity of the source of data input or operational instruction. Several comments objected to this proposed requirement and suggested its deletion because it is: (1) Unnecessary (because the data source is always known by virtue of system design and validation); (2) problematic with respect to mobile devices, such as those connected by modem; (3) too much of a "how to;" (4) not explicit enough to tell firms what to do; (5) unnecessary in the case of PDMA; and (6) technically challenging. One comment stated that a device's identification, in addition to location, may be important and suggested that the proposed rule be revised to require device identification as well.

FDA advises that, by use of the term "as appropriate," it does not intend to require device checks in all cases. The agency believes that these checks are warranted where only certain devices have been selected as legitimate sources of data input or commands. In such cases, the device checks would be used to determine if the data or command source was authorized. In a network, for example, it may be necessary for security reasons to limit issuance of critical commands to only one authorized workstation. The device check would typically interrogate the source of the command to ensure that only the authorized workstation, and not some other device, was, in fact, issuing the command.

The same approach applies for remote sources connected by modem, to the extent that device identity interrogations could be made automatically regardless of where the portable devices were located. To clarify this concept, the agency has removed the word "location" from proposed § 11.10(h). Device checks would be necessary under PDMA when the source of commands or data is relevant to establishing authenticity, such as when licensed practitioners order drug samples directly from the manufacturer or authorized distributor without the intermediary of a sales representative. Device checks may also be useful to firms in documenting and identifying which sales representatives are transmitting drug sample requests from licensed practitioners.

FDA believes that, although validation may demonstrate that a given terminal or workstation is technically capable of sending information from one point to another, validation alone would not be expected to address whether or not such device is authorized to do so.

86. Proposed § 11.10(i) states that procedures and controls for closed systems must include confirmation that persons who develop, maintain, or use electronic record or signature systems have the education, training, and experience to perform their assigned tasks.

Several comments objected to the word "confirmation" because it is redundant with, or more restrictive than, existing regulations, and suggested alternate wording, such as "evidence." Two comments interpreted the proposed wording as requiring that checks of personnel qualifications be performed automatically by computer systems that perform database type matches between functions and personnel training records.

The agency advises that, although there may be some overlap in proposed § 11.10(i) and other regulations regarding the need for personnel to be properly qualified for their duties, part 11 is specific to functions regarding electronic records, an issue that other regulations may or may not adequately address. Therefore, the agency is retaining the requirement.

The agency does not intend to require that the check of personnel qualifications be performed automatically by a computer system itself (although such automation is desirable). The agency has revised the introductory paragraph of § 11.10, as discussed in section VII. of this document, to clarify this point. The agency agrees that another word should be used in place of "confirmation," and for clarity has selected "determination."

87. One comment suggested that the word "training" be deleted because it has the same meaning as "education" and "experience," and objected to the implied requirement for records of employee training. Another comment argued that applying this provision to system developers was irrelevant so long as systems perform as required and have been appropriately validated. The comment suggested revising proposed § 11.10(i) to require employees to be trained only "as necessary." One comment, noting that training and experience are very important, suggested expanding proposed § 11.10(i) to require appropriate examination and certification of persons who perform certain high-risk, high-trust functions and tasks.

The agency regards this requirement as fundamental to the proper operation of a facility. Personnel entrusted with important functions must have sufficient training to do their jobs. In FDA's view, formal education (e.g., academic studies) and general industry experience would not necessarily prepare someone to begin specific, highly technical tasks at a given firm. Some degree of on-the-job training would be customary and expected. The agency believes that documentation of such training is also customary and not unreasonable.

The agency also disagrees with the assertion that personnel qualifications of system developers are irrelevant. The qualifications of personnel who develop systems are relevant to the expected performance of the systems they build and their ability to explain and support these systems. Validation does not lessen the need for personnel to have the education, training, and experience to do their jobs properly. Indeed, it is highly unlikely that poorly qualified developers would be capable of producing a system that could be validated. The agency advises that, although the intent of proposed § 11.10(i) is to address qualifications of those personnel who develop systems within an organization, rather than external "vendors" per se, it is nonetheless vital that vendor personnel are likewise qualified to do their work. The agency agrees that periodic examination or certification of personnel who perform certain critical tasks is desirable. However, the agency does not believe that at this time a specific requirement for such examination and certification is necessary.

88. Proposed § 11.10(j) states that procedures and controls for closed systems must include the establishment of, and adherence to, written policies that hold individuals accountable and liable for actions initiated under their electronic signatures, so as to deter record and signature falsification.

Several comments suggested changing the word "liable" to "responsible" because the word "responsible" is broader, more widely understood by employees, more positive and inclusive of elements of honesty and trust, and more supportive of a broad range of disciplinary measures. One comment argued that the requirement would not deter record or signature falsification because employee honesty and integrity cannot be regulated.

The agency agrees because, although the words "responsible" and "liable" are generally synonymous, "responsible" is preferable because it is more positive and supportive of a broad range of disciplinary measures. There may be a general perception that electronic records and electronic signatures (particularly identification codes and passwords) are less significant and formal than traditional paper records and handwritten signatures. Individuals may therefore not fully equate the seriousness of electronic record falsification with paper record falsification. Employees need to understand the gravity and consequences of signature or record falsification. Although FDA agrees that employee honesty cannot be ensured by requiring it in a regulation, the presence of strong accountability and responsibility policies is necessary to ensure that employees understand the importance of maintaining the integrity of electronic records and signatures.

89. Several comments expressed concern regarding employee liability for actions taken under their electronic signatures in the event that such signatures are compromised, and requested "reasonable exceptions." The comments suggested revising proposed § 11.10(j) to hold people accountable only where there has been intentional falsification or corruption of electronic data.

The agency considers the compromise of electronic signatures to be a very serious matter, one that should precipitate an appropriate investigation into any causative weaknesses in an organization's security controls. The agency nonetheless recognizes that where such compromises occur through no fault or knowledge of individual employees, there would be reasonable limits on the extent to which disciplinary action would be taken. However, to maintain emphasis on the seriousness of such security breeches and deter the deliberate fabrication of "mistakes," the agency believes § 11.10 should not provide for exceptions that may lessen the import of such a fabrication.

90. One comment said the agency should consider the need for criminal law reform because current computer crime laws do not address signatures when unauthorized access or

computer use is not an issue. Another comment argued that proposed § 11.10(j) should be expanded beyond "individual" accountability to include business entities.

The agency will consider the need for recommending legislative initiatives to address electronic signature falsification in light of the experience it gains with this regulation. The agency does not believe it necessary to address business entity accountability specifically in § 11.10 because the emphasis is on actions and accountability of individuals, and because individuals, rather than business entities, apply signatures.

91. One comment suggested that proposed § 11.10(j) should be deleted because it is unnecessary because individuals are presumably held accountable for actions taken under their authority, and because, in some organizations, individuals frequently delegate authority to sign their names.

As discussed in comments 88 to 90 of this document, the agency has concluded that this section is necessary. Furthermore it does not limit delegation of authority as described in the comment. However, where one individual signs his or her name on behalf of someone else, the signature applied should be that of the delegatee, with some notation of that fact, and not the name of the delegator. This is the same procedure commonly used on paper documents, noted as "X for Y."

92. Proposed § 11.10(k) states that procedures and controls for closed systems must include the use of appropriate systems documentation controls, including: (1) Adequate controls over the distribution, access to, and use of documentation for system operation and maintenance; and (2) records revision and change control procedures to maintain an electronic audit trail that documents time-sequenced development and modification of records. Several comments requested clarification of the type of documents covered by proposed § 11.10(k). One comment noted that this section failed to address controls for record retention. Some comments suggested limiting the scope of systems documentation to application and configurable software, or only to software that could compromise system security or integrity. Other comments suggested that this section should be deleted because some documentation needs wide distribution within an organization, and that it is an onerous burden to control user manuals.

The agency advises that § 11.10(k) is intended to apply to systems documentation, namely, records describing how a system operates and is maintained, including standard operating procedures. The agency believes that adequate controls over such documentation are necessary for various reasons. For example, it is important for employees to have correct and updated versions of standard operating and maintenance procedures. If this documentation is not current, errors in procedures and/or maintenance are more likely to occur. Part 11 does not limit an organization's discretion as to how widely or narrowly any document is to be distributed, and FDA expects that certain documents will, in fact, be widely disseminated. However, some highly sensitive documentation, such as instructions on how to modify system security features, would not routinely be widely distributed. Hence, it is important to control distribution of, access to, and use of such documentation.

Although the agency agrees that the most critical types of system documents would be those directly affecting system security and integrity, FDA does not agree that control over system documentation should only extend to security related software or to application or configurable software. Documentation that relates to operating systems, for example, may also have an impact on security and day-to-day operations. The agency does not agree that it is an onerous burden to control documentation that relates to effective operation and security of electronic records systems. Failure to control such documentation, as discussed above, could permit and foster records falsification by making the enabling instructions for these acts readily available to any individual.

93. Concerning the proposed requirement for adequate controls over documentation for system operation and maintenance, one comment suggested that it be deleted because it is under the control of system vendors, rather than operating organizations. Several comments suggested that the proposed provision be deleted because it duplicates § 11.10(e) with respect to audit trails. Some comments also objected to maintaining the change control procedures in electronic form and suggested deleting the word "electronic" from "electronic audit trails."

The agency advises that this section is intended to apply to systems documentation that can be changed by individuals within an organization. If systems documentation can only be changed by a vendor, this provision does not apply to the vendor's customers. The agency acknowledges that systems documentation may be in paper or electronic form. Where the documentation is in paper form, an audit trail of revisions need not be in electronic form. Where systems documentation is in electronic form, however, the agency intends to require the audit trail also be in electronic form, in accordance with § 11.10(e). The agency acknowledges that, in light of the comments, the proposed rule may not have been clear enough regarding audit trails addressed in § 11.10(k) compared to audit trails addressed in § 11.10(e) and has revised the final rule to clarify this matter.

The agency does not agree, however, that the audit trail provisions of § 11.10(e) and (k), as revised, are entirely duplicative. Section 11.10(e) applies to electronic records in general (including systems documentation); § 11.10(k) applies exclusively to systems documentation, regardless of whether such documentation is in paper or electronic form.

As revised, § 11.10(k) now reads as follows:

(k) Use of appropriate controls over systems documentation including:

    (1) Adequate controls over the distribution of, access to, and use of documentation for system operation and maintenance.

    (2) Revision and change control procedures to maintain an audit trail that documents time-sequenced development and modification of systems documentation.

VIII. ELECTRONIC RECORDS — CONTROLS FOR OPEN SYSTEMS (§ 11.30)

Proposed § 11.30 states that: "Open systems used to create, modify, maintain, or transmit electronic records shall employ procedures and controls designed to ensure the authenticity, integrity and confidentiality of electronic records from the point of their creation to the point of their receipt." In addition, § 11.30 states:

> * * * Such procedures and controls shall include those identified in § 11.10, as appropriate, and such additional measures as document encryption and use of established digital signature standards acceptable to the agency, to ensure, as necessary under the circumstances, record authenticity, integrity, and confidentiality.

94. One comment suggested that the reference to digital signature standards be deleted because the agency should not be setting standards and should not dictate how to ensure record authenticity, integrity, and confidentiality. Other comments requested clarification of the agency's expectations with regard to digital signatures: (1) The kinds that would be acceptable, (2) the mechanism for announcing which standards were acceptable (and whether that meant FDA would be certifying particular software), and (3) a definition of digital signature. One comment asserted that FDA should accept international standards for digital signatures. Some comments also requested a definition of encryption. One comment encouraged the agency to further define open systems.

The agency advises that § 11.30 requires additional controls, beyond those identified in § 11.10, as needed under the circumstances, to ensure record authenticity, integrity, and confidentiality for open systems. Use of digital signatures is one measure that may be used, but is not specifically required. The agency wants to ensure that the digital signature standard used is, in fact, appropriate. Development of digital signature standards is a complex undertaking, one FDA does not expect to be performed by individual firms on an ad hoc basis, and one FDA does not now seek to perform.

The agency is nonetheless concerned that such standards be robust and secure. Currently, the agency is aware of two such standards, the RSA (Rivest-Shamir-Adleman), and NIST's Digital Signature Standard (DSS). The DSS became Federal Information Processing Standard (FIPS) 186 on December 1, 1994. These standards are incorporated in different software programs. The agency does not seek to certify or otherwise approve of such programs, but expects people who use such programs to ensure that they are suitable for their intended use. FDA is aware that NIST provides certifications regarding mathematical conformance to the DSS core algorithms, but does not formally evaluate the broader programs that contain those algorithms. The agency has revised the final rule to clarify its intent that firms retain the flexibility to use any appropriate digital signature as an additional system control for open systems. FDA is also including a definition of digital signature under § 11.3(b)(5).

The agency does not believe it necessary to codify the term "encryption" because, unlike the term digital signature, it has been in general use for many years and is generally understood to mean the transforming of a writing into a secret code or cipher. The agency is aware that there are several commercially available software programs that implement both digital signatures and encryption.

95. Two comments noted that use of digital signatures and encryption is not necessary in the context of PDMA, where access to an electronic record is limited once it is signed and stored. One of the comments suggested that proposed § 11.30 be revised to clarify this point.

As discussed in comment 94 of this document, use of digital signatures and encryption would be an option when extra measures are necessary under the circumstances. In the case of PDMA records, such measures may be warranted in certain circumstances, and unnecessary in others. For example, if electronic records were to be transmitted by a firm's representative by way of a public online service to a central location, additional measures would be necessary. On the other hand, where the representative's records are hand delivered to that location, or transferred by direct connection between the representative and the central location, such additional measures to ensure record authenticity, confidentiality, and integrity may not be necessary. The agency does not believe that it is practical to revise § 11.30 to elaborate on every possible situation in which additional measures would or would not be needed.

96. One comment addressed encryption of submissions to FDA and asked if people making those submissions would have to give the agency the appropriate "keys" and, if so, how the agency would protect the security of such information.

The agency intends to develop appropriate procedures regarding the exchange of "keys" attendant to use of encryption and digital signatures, and will protect those keys that must remain confidential, in the same manner as the agency currently protects trade secrets. Where the agency and a submitter agree to use a system that calls for the exchange of secret keys, FDA will work with submitters to achieve mutually agreeable procedures. The agency notes, however, that not all encryption and digital signature systems require that enabling keys be secret.

97. One comment noted that proposed § 11.30 does not mention availability and non-repudiation and requested clarification of the term "point of receipt." The comment noted that, where an electronic record is received at a person's electronic mailbox (which resides

on an open system), additional measures may be needed when the record is transferred to the person's own local computer because such additional transfer entails additional security risks. The comment suggested wording that would extend open system controls to the point where records are ultimately retained.

The agency agrees that, in the situation described by the comment, movement of the electronic record from an electronic mailbox to a person's local computer may necessitate open system controls. However, situations may vary considerably as to the ultimate point of receipt, and FDA believes proposed § 11.30 offers greater flexibility in determining open system controls than revisions suggested by the comment. The agency advises that the concept of nonrepudiation is part of record authenticity and integrity, as already covered by § 11.10(c). Therefore, FDA is not revising § 11.30 as suggested.

IX. ELECTRONIC RECORDS—SIGNATURE MANIFESTATIONS (§ 11.50)

Proposed § 11.50 requires that electronic records that are electronically signed must display in clear text the printed name of the signer, and the date and time when the electronic signature was executed. This section also requires that electronic records clearly indicate the meaning (such as review, approval, responsibility, and authorship) associated with their attendant signatures.

98. Several comments suggested that the information required under proposed § 11.50 need not be contained in the electronic records themselves, but only in the human readable format (screen displays and printouts) of such records. The comments explained that the records themselves need only contain links, such as signature attribute codes, to such information to produce the displays of information required.

The comments noted, for example, that, where electronic signatures consist of an identification code in combination with a password, the combined code and password itself would not be part of the display. Some comments suggested that proposed § 11.50 be revised to clarify what items are to be displayed.

The agency agrees and has revised proposed § 11.50 accordingly. The intent of this section is to require that human readable forms of signed electronic records, such as computer screen displays and printouts bear: (1) The printed name of the signer (at the time the record is signed as well as whenever the record is read by humans); (2) the date and time of signing; and (3) the meaning of the signature. The agency believes that revised § 11.50 will afford persons the flexibility they need to implement the display of information appropriate for their own electronic records systems, consistent with other system controls in part 11, to ensure record integrity and prevent falsification.

99. One comment stated that the controls in proposed § 11.50 would not protect against inaccurate entries.

FDA advises that the purpose of this section is not to protect against inaccurate entries, but to provide unambiguous documentation of the signer, when the signature was executed, and the signature's meaning. The agency believes that such a record is necessary to document individual responsibility and actions.

In a paper environment, the printed name of the individual is generally present in the signed record, frequently part of a traditional "signature block." In an electronic environment, the person's name may not be apparent, especially where the signature is based on identification codes combined with passwords. In addition, the meaning of a signature is generally apparent in a paper record by virtue of the context of the record or, more often, explicit phrases such as "approved by," "reviewed by," and "performed by." Thus, the agency believes that for clear documentation purposes it is necessary to carry such meanings into the electronic record environment.

100. One comment suggested that proposed § 11.50 should apply only to those records that are required to be signed, and that the display of the date and time should be performed in a secure manner.

The agency intends that this section apply to all signed electronic records regardless of whether other regulations require them to be signed. The agency believes that if it is important enough that a record be signed, human readable displays of such records must include the printed name of the signer, the date and time of signing, and the meaning of the signature. Such information is crucial to the agency's ability to protect public health. For example, a message from a firm's management to employees instructing them on a particular course of action may be critical in litigation. This requirement will help ensure clear documentation and deter falsification regardless of whether the signature is electronic or handwritten.

The agency agrees that the display of information should be carried out in a secure manner that preserves the integrity of that information. The agency, however, does not believe it is necessary at this time to revise § 11.50 to add specific security measures because other requirements of part 11 have the effect of ensuring appropriate security.

Because signing information is important regardless of the type of signature used, the agency has revised § 11.50 to cover all types of signings.

101. Several comments objected to the requirement in proposed § 11.50(a) that the time of signing be displayed in addition to the date on the grounds that such information is: (1) Unnecessary, (2) costly to implement, (3) needed in the electronic record for auditing purposes, but not needed in the display of the record, and (4) only needed in critical applications. Some comments asserted that recording time should be optional. One comment asked whether the time should be local to the signer or to a central network when electronic record systems cross different time zones.

The agency believes that it is vital to record the time when a signature is applied. Documenting the time when a signature was applied can be critical to demonstrating that a given record was, or was not, falsified. Regarding systems that may span different time zones, the agency advises that the signer's local time is the one to be recorded.

102. One comment assumed that a person's user identification code could be displayed instead of the user's printed name, along with the date and time of signing.

This assumption is incorrect. The agency intends that the printed name of the signer be displayed for purposes of unambiguous documentation and to emphasize the importance of the act of signing to the signer. The agency believes that because an identification code is not an actual name, it would not be a satisfactory substitute.

103. One comment suggested that the word "printed" in the phrase "printed name" be deleted because the word was superfluous. The comment also stated that the rule should state when the clear text must be created or displayed because some computer systems, in the context of electronic data interchange transactions, append digital signatures to records before, or in connection with, communication of the record.

The agency disagrees that the word "printed" is superfluous because the intent of this section is to show the name of the person in an unambiguous manner that can be read by anyone. The agency believes that requiring the printed name of the signer instead of codes or other manifestations, more effectively provides clarity.

The agency has revised this section to clarify the point at which the signer's information must be displayed, namely, as part of any human readable form of the electronic record. The revision, in the agency's view, addresses the comment's concern regarding the application of digital signatures. The agency advises that under § 11.50, any time after an electronic record has been signed, individuals who see the human readable form of the record will be able to immediately tell who signed the record, when it was signed, and what the signature meant. This includes the signer who, as with a traditional signature to paper, will be able to review the signature instantly.

104. One comment asked if the operator would have to see the meaning of the signature, or if the information had to be stored on the physical electronic record.

As discussed in comment 100 of this document, the information required by § 11.50(b) must be displayed in the human readable format of the electronic record. Persons may elect to store that information directly within the electronic record itself, or in logically associated records, as long as such information is displayed any time a person reads the record.

105. One comment noted that proposed § 11.50(b) could be interpreted to require lengthy explanations of the signatures and the credentials of the signers. The comment also stated that this information would more naturally be contained in standard operating procedures, manuals, or accompanying literature than in the electronic records themselves.

The agency believes that the comment misinterprets the intent of this provision. Recording the meaning of the signature does not infer that the signer's credentials or other lengthy explanations be part of that meaning. The statement must merely show what is meant by the act of signing (e.g., review, approval, responsibility, authorship).

106. One comment noted that the meaning of a signature may be included in a (digital signature) public key certificate and asked if this would be acceptable. The comment also noted that the certificate might be easily accessible by a record recipient from either a recognized database or one that might be part of, or associated with, the electronic record itself. The comment further suggested that FDA would benefit from participating in developing rules of practice regarding certificate-based public key cryptography and infrastructure with the Information Security Committee, Section of Science and Technology, of the American Bar Association (ABA).

The intent of this provision is to clearly discern the meaning of the signature when the electronic record is displayed in human readable form. The agency does not expect such meaning to be contained in or displayed by a public key certificate because the public key is generally a fixed value associated with an individual. The certificate is used by the recipient to authenticate a digital signature that may have different meanings, depending upon the record being signed. FDA acknowledges that it is possible for someone to establish different public keys, each of which may indicate a different signature meaning. Part 11 would not prohibit multiple "meaning" keys provided the meaning of the signature itself was still clear in the display of the record, a feature that could conceivably be implemented by software.

Regarding work of the ABA and other standard-setting organizations, the agency welcomes an open dialog with such organizations, for the mutual benefit of all parties, to establish and facilitate the use of electronic record/ electronic signature technologies. FDA's participation in any such activities would be in accordance with the agency's policy on standards stated in the Federal Register of October 11, 1995 (60 FR 53078).

Revised § 11.50, signature manifestations, reads as follows:

(a) Signed electronic records shall contain information associated with the signing that clearly indicates all of the following:

(1) The printed name of the signer;

(2) The date and time when the signature was executed; and

(3) The meaning (such as review, approval, responsibility, or authorship) associated with the signature.

(b) The items identified in paragraphs (a)(1), (a)(2), and (a)(3) of this section shall be subject to the same controls as for electronic records and shall be included as part of any human readable form of the electronic record (such as electronic display or printout).

X. ELECTRONIC RECORDS — SIGNATURE/RECORD LINKING (§ 11.70)

107. Proposed § 11.70 states that electronic signatures and handwritten signatures executed to electronic records must be verifiably bound to their respective records to ensure that signatures could not be excised, copied, or otherwise transferred to falsify another electronic record.

Many comments objected to this provision as too prescriptive, unnecessary, unattainable, and excessive in comparison to paper-based records. Some comments asserted that the objectives of the section could be attained through appropriate procedural and administrative controls. The comments also suggested that objectives of the provision could be met by appropriate software (i.e., logical) links between the electronic signatures and electronic records, and that such links are common in systems that use identification codes in combination with passwords. One firm expressed full support for the provision, and noted that its system implements such a feature and that signature-to-record binding is similar to the record-locking provision of the proposed PDMA regulations.

The agency did not intend to mandate use of any particular technology by use of the word "binding." FDA recognizes that, because it is relatively easy to copy an electronic signature to another electronic record and thus compromise or falsify that record, a technology based link is necessary. The agency does not believe that procedural or administrative controls alone are sufficient to ensure that objective because such controls could be more easily circumvented than a straightforward technology based approach. In addition, when electronic records are transferred from one party to another, the procedural controls used by the sender and recipient may be different. This could result in record falsification by signature transfer.

The agency agrees that the word "link" would offer persons greater flexibility in implementing the intent of this provision and in associating the names of individuals with their identification codes/passwords without actually recording the passwords themselves in electronic records. The agency has revised proposed § 11.70 to state that signatures shall be linked to their electronic records.

108. Several comments argued that proposed § 11.70 requires absolute protection of electronic records from falsification, an objective that is unrealistic to the extent that determined individuals could falsify records.

The agency acknowledges that, despite elaborate system controls, certain determined individuals may find a way to defeat antifalsification measures. FDA will pursue such illegal activities as vigorously as it does falsification of paper records. For purposes of part 11, the agency's intent is to require measures that prevent electronic records falsification by ordinary means. Therefore, FDA has revised § 11.70 by adding the phrase "by ordinary means" at the end of this section.

109. Several comments suggested changing the phrase "another electronic record" to "an electronic record" to clarify that the antifalsification provision applies to the current record as well as any other record.

The agency agrees and has revised § 11.70 accordingly.

110. Two comments argued that signature-to-record binding is unnecessary, in the context of PDMA, beyond the point of record creation (i.e., when records are transmitted to a point of receipt). The comments asserted that persons who might be in a position to separate a signature from a record (for purposes of falsification) are individuals responsible for record integrity and thus unlikely to falsify records. The comments also stated that signature-to-record binding is produced by software coding at the time the record is signed, and suggested that proposed § 11.70 clarify that binding would be necessary only up to the point of actual transmission of the electronic record to a central point of receipt.

The agency disagrees with the comment's premise that the need for binding to prevent falsification depends on the disposition of people to falsify records. The agency believes that reliance on individual tendencies is insufficient insurance against falsification. The agency also notes that in the traditional paper record, the signature remains bound to its corresponding record regardless of where the record may go.

111. One comment suggested that proposed § 11.70 be deleted because it appears to require that all records be kept on inalterable media. The comment also suggested that the phrase "otherwise transferred" be deleted on the basis that it should be permissible for copies of handwritten signatures (recorded electronically) to be made when used, in addition to another unique individual identification mechanism.

The agency advises that neither § 11.70, nor other sections in part 11, requires that records be kept on inalterable media. What is required is that whenever revisions to a record are made, the original entries must not be obscured. In addition, this section does not prohibit copies of handwritten signatures recorded electronically from being made for legitimate reasons that do not relate to record falsification. Section 11.70 merely states that such copies must not be made that falsify electronic records.

112. One comment suggested that proposed § 11.70 be revised to require application of response cryptographic methods because only those methods could be used to comply with the regulation. The comment noted that, for certificate based public key cryptographic methods, the agency should address verifiable binding between the signer's name and public key as well as binding between digital signatures and electronic records. The comment also suggested that the regulation should reference electronic signatures in the context of secure time and date stamping.

The agency intends to permit maximum flexibility in how organizations achieve the linking called for in § 11.70, and, as discussed above, has revised the regulation accordingly. Therefore, FDA does not believe that cryptographic and digital signature methods would be the only ways of linking an electronic signature to an electronic document. In fact, one firm commented that its system binds a person's handwritten signature to an electronic record. The agency agrees that use of digital signatures accomplishes the same objective because, if a digital signature were to be copied from one record to another, the second record would fail the digital signature verification procedure. Furthermore, FDA notes that concerns regarding binding a person's name with the person's public key would be addressed in the context of § 11.100(b) because an organization must establish an individual's identity before assigning or certifying an electronic signature (or any of the electronic signature components).

113. Two comments requested clarification of the types of technologies that could be used to meet the requirements of proposed § 11.70.

As discussed in comment 107 of this document, the agency is affording persons maximum flexibility in using any appropriate method to link electronic signatures to their respective electronic records to prevent record falsification. Use of digital signatures is one such method, as is use of software locks to prevent sections of codes representing signatures from being copied or removed. Because this is an area of developing technology, it is likely that other linking methods will emerge.

XI. ELECTRONIC SIGNATURES—GENERAL REQUIREMENTS (§ 11.100)

Proposed § 11.100(a) states that each electronic signature must be unique to one individual and not be reused or reassigned to anyone else.

114. One comment asserted that several people should be permitted to share a common identification code and password where access control is limited to inquiry only.

Part 11 does not prohibit the establishment of a common group identification code/password for read only access purposes. However, such commonly shared codes and passwords would not be regarded, and must not be used, as electronic signatures. Shared access to a common database may nonetheless be implemented by granting appropriate common record access privileges to groups of people, each of whom has a unique electronic signature.

115. Several comments said proposed § 11.100(a) should permit identification codes to be reused and reassigned from one employee to another, as long as an audit trail exists to associate an identification code with a given individual at any one time, and different passwords are used. Several comments said the section should indicate if the agency intends to restrict authority delegation by the nonreassignment or nonreuse provision, or by the provision in § 11.200(a)(2) requiring electronic signatures to be used only by their genuine owners. The comments questioned whether reuse means restricting one noncryptographic based signature to only one record and argued that passwords need not be unique if the combined identification code and password are unique to one individual. One comment recommended caution in using the term "ownership" because of possible confusion with intellectual property rights or ownership of the computer systems themselves.

The agency advises that, where an electronic signature consists of the combined identification code and password, § 11.100 would not prohibit the reassignment of the identification code provided the combined identification code and password remain unique to prevent record falsification. The agency believes that such reassignments are inadvisable, however, to the extent that they might be combined with an easily guessed password, thus increasing the chances that an individual might assume a signature belonging to someone else. The agency also advises that where people can read identification codes (e.g., printed numbers and letters that are typed at a keyboard or read from a card), the risks of someone obtaining that information as part of a falsification effort would be greatly increased as compared to an identification code that is not in human readable form (one that is, for example, encoded on a "secure card" or other device).

Regarding the delegation of authority to use electronic signatures, FDA does not intend to restrict the ability of one individual to sign a record or otherwise act on behalf of another individual. However, the applied electronic signature must be the assignee's and the record should clearly indicate the capacity in which the person is acting (e.g., on behalf of, or under the authority of, someone else). This is analogous to traditional paper records and handwritten signatures when person "A" signs his or her own name under the signature block of person "B," with appropriate explanatory notations such as "for" or "as representative of" person B. In such cases, person A does not simply sign the name of person B. The agency expects the same procedure to be used for electronic records and electronic signatures.

The agency intends the term "reuse" to refer to an electronic signature used by a different person. The agency does not regard as "reuse" the replicate application of a noncryptographic based electronic signature (such as an identification code and password) to different electronic records. For clarity, FDA has revised the phrase "not be reused or reassigned to" to state "not be reused by, or reassigned to," in § 11.100(a).

The reference in § 11.200(a) to ownership is made in the context of an individual owning or being assigned a particular electronic signature that no other individual may use. FDA believes this is clear and that concerns regarding ownership in the context of intellectual property rights or hardware are misplaced.

116. One comment suggested that proposed § 11.100(a) should accommodate electronic signatures assigned to organizations rather than individuals.

The agency advises that, for purposes of part 11, electronic signatures are those of individual human beings and not organizations. For example, FDA does not regard a corporate seal as an individual's signature. Humans may represent and obligate organizations by signing records, however. For clarification, the agency is substituting the word "individual" for "person" in the definition of electronic signature (§ 11.3(b)(7)) because the broader definition of person within the act includes organizations.

117. Proposed § 11.100(b) states that, before an electronic signature is assigned to a person, the identity of the individual must be verified by the assigning authority.

Two comments noted that where people use identification codes in combination with passwords only the identification code portion of the electronic signature is assigned, not the password. Another comment argued that the word "assigned" is inappropriate in the context of electronic signatures based upon public key cryptography because the appropriate authority certifies the bind between the individual's public key and identity, and not the electronic signature itself.

The agency acknowledges that, for certain types of electronic signatures, the authorizing or certifying organization issues or approves only a portion of what eventually becomes an individual's electronic signature. FDA wishes to accommodate a broad variety of electronic signatures and is therefore revising § 11.100(b) to require that an organization verify the identity of an individual before it establishes, assigns, certifies, or otherwise sanctions an individual's electronic signature or any element of such electronic signature.

118. One comment suggested that the word "verified" in proposed § 11.100(b) be changed to "confirmed." Other comments addressed the method of verifying a person's identity and suggested that the section specify acceptable verification methods, including high level procedures regarding the relative strength of that verification, and the need for personal appearances or supporting documentation such as birth certificates. Two comments said the verification provision should be deleted because normal internal controls are adequate, and that it was impractical for multinational companies whose employees are globally dispersed.

The agency does not believe that there is a sufficient difference between "verified" and "confirmed" to warrant a change in this section. Both words indicate that organizations substantiate a person's identity to prevent impersonations when an electronic signature, or any of its elements, is being established or certified. The agency disagrees with the assertion that this requirement is unnecessary. Without verifying someone's identity at the outset of establishing or certifying an individual's electronic signature, or a portion thereof, an imposter might easily access and compromise many records. Moreover, an imposter could continue this activity for a prolonged period of time despite other system controls, with potentially serious consequences.

The agency does not believe that the size of an organization, or global dispersion of its employees, is reason to abandon this vital control. Such dispersion may, in fact, make it easier for an impostor to pose as someone else in the absence of such verification. Further, the agency does not accept the implication that multinational firms would not verify the identity of their employees as part of other routine procedures, such as when individuals are first hired.

In addition, in cases where an organization is widely dispersed and electronic signatures are established or certified centrally, § 11.100(b) does not prohibit organizations from having their local units perform the verification and relaying this information to the central authority. Similarly, local units may conduct the electronic signature assignment or certification.

FDA does not believe it is necessary at this time to specify methods of identity verification and expects that organizations will consider risks attendant to sanctioning an erroneously assigned electronic signature.

119. Proposed § 11.100(c) states that persons using electronic signatures must certify to the agency that their electronic signature system guarantees the authenticity, validity, and binding nature of any electronic signature. Persons utilizing electronic signatures would, upon agency request, provide additional certification or testimony that a specific electronic signature is authentic, valid, and binding. Such certification would be submitted to the FDA district office in which territory the electronic signature system is in use.

Many comments objected to the proposed requirement that persons provide FDA with certification regarding their electronic signature systems. The comments asserted that the requirement was: (1) Unprecedented, (2) unrealistic, (3) unnecessary, (4) contradictory to the principles and intent of system validation, (5) too burdensome for FDA to manage logistically, (6) apparently intended only to simplify FDA litigation, (7) impossible to meet regarding "guarantees" of authenticity, and (8) an apparent substitute for FDA inspections.

FDA agrees in part with these comments. This final rule reduces the scope and burden of certification to a statement of intent that electronic signatures are the legally binding equivalent of handwritten signatures.

As noted previously, the agency believes it is important, within the context of its health protection activities, to ensure that persons who implement electronic signatures fully equate the legally binding nature of electronic signatures with the traditional handwritten paper-based signatures. The agency is concerned that individuals might disavow an electronic signature as something completely different from a traditional handwritten signature. Such contention could result in confusion and possibly extensive litigation.

Moreover, a limited certification as provided in this final rule is consistent with other legal, regulatory, and commercial practices. For example, electronic data exchange trading partner agreements are often written on paper and signed with traditional handwritten signatures to establish that certain electronic identifiers are recognized as equivalent to traditional handwritten signatures.

FDA does not expect electronic signature systems to be guaranteed foolproof. The agency does not intend, under § 11.100(c), to establish a requirement that is unattainable. Certification of an electronic signature system as the legally binding equivalent of a traditional handwritten signature is separate and distinct from system validation. This provision is not intended as a substitute for FDA inspection and such inspection alone may not be able to determine in a conclusive manner an organization's intent regarding electronic signature equivalency.

The agency has revised proposed § 11.100(c) to clarify its intent. The agency wishes to emphasize that the final rule dramatically curtails what FDA had proposed and is essential for the agency to be able to protect and promote the public health because FDA must be able to hold people to the commitments they make under their electronic signatures. The certification in the final rule is merely a statement of intent that electronic signatures are the legally binding equivalent of traditional handwritten signatures.

120. Several comments questioned the procedures necessary for submitting the certification to FDA, including: (1) The scheduling of the certification; (2) whether to submit certificates for each individual or for each electronic signature; (3) the meaning of "territory" in the context of wide area networks; (4) whether such certificates could be submitted electronically; and (5) whether organizations, after submitting a certificate, had to wait for a response from FDA before implementing their electronic signature systems. Two comments suggested revising proposed § 11.100(c) to require that all certifications be submitted to FDA only upon agency request. One comment suggested changing "should" to "shall" in the last sentence of § 11.100(c) if the agency's intent is to require certificates to be submitted to the respective FDA district office.

The agency intends that certificates be submitted once, in the form of a paper letter, bearing a traditional handwritten signature, at the time an organization first establishes an electronic signature system after the effective date of part 11, or, where such systems have been used before the effective date, upon continued use of the electronic signature system.

A separate certification is not needed for each electronic signature, although certification of a particular electronic signature is to be submitted if the agency requests it. The agency does not intend to establish certification as a review and approval function. In addition, organizations need not await FDA's response before putting electronic signature systems into effect, or before continuing to use an existing system.

A single certification may be stated in broad terms that encompass electronic signatures of all current and future employees, thus obviating the need for subsequent certifications submitted on a preestablished schedule.

To further simplify the process and to minimize the number of certifications that persons would have to provide, the agency has revised § 11.100(c) to permit submission of a single certification that covers all electronic signatures used by an organization. The revised rule also simplifies the process by providing a single agency receiving unit. The final rule instructs persons to send certifications to FDA's Office of Regional Operations (HFC–100), 5600 Fishers Lane, Rockville, MD 20857. Persons outside the United States may send their certifications to the same office.

The agency offers, as guidance, an example of an acceptable § 11.100(c) certification:

> Pursuant to Section 11.100 of Title 21 of the Code of Federal Regulations, this is to certify that [name of organization] intends that all electronic signatures executed by our employees, agents, or representatives, located anywhere in the world, are the legally binding equivalent of traditional handwritten signatures.

The agency has revised § 11.100 to clarify where and when certificates are to be submitted.

The agency does not agree that the initial certification be provided only upon agency request because FDA believes it is vital to have such certificates, as a matter of record, in advance of any possible litigation. This would clearly establish the intent of organizations to equate the legally binding nature of electronic signatures with traditional handwritten signatures. In addition, the agency believes that having the certification on file ahead of time will have the beneficial effect of reinforcing the gravity of electronic signatures by putting an organization's employees on notice that the organization has gone on record with FDA as equating electronic signatures with handwritten signatures.

121. One comment suggested that proposed § 11.100(c) be revised to exclude from certification instances in which the purported signer claims that he or she did not create or authorize the signature.

The agency declines to make this revision because a provision for nonrepudiation is already contained in § 11.10.

As a result of the considerations discussed in comments 119 and 120 of this document, the agency has revised proposed § 11.100(c) to state that:

> **(c)** Persons using electronic signatures shall, prior to or at the time of such use, certify to the agency that the electronic signatures in their system, used on or after August 20, 1997, are intended to be the legally binding equivalent of traditional handwritten signatures.
>
> > **(1)** The certification shall be submitted in paper form and signed with a traditional handwritten signature to the Office of Regional Operations (HFC–100), 5600 Fishers Lane, Rockville, MD 20857.

(2) Persons using electronic signatures shall, upon agency request, provide additional certification or testimony that a specific electronic signature is the legally binding equivalent of the signer's handwritten signature.

XII. ELECTRONIC SIGNATURE COMPONENTS AND CONTROLS (§ 11.200)

122. Proposed § 11.200 sets forth requirements for electronic signature identification mechanisms and controls. Two comments suggested that the term "identification code" should be defined. Several comments suggested that the term "identification mechanisms" should be changed to "identification components" because each component of an electronic signature need not be executed by a different mechanism.

The agency believes that the term "identification code" is sufficiently broad and generally understood and does not need to be defined in these regulations. FDA agrees that the word "component" more accurately reflects the agency's intent than the word "mechanism," and has substituted "component" for "mechanism" in revised § 11.200. The agency has also revised the section heading to read "Electronic signature components and controls" to be consistent with the wording of the section.

123. Proposed § 11.200(a) states that electronic signatures not based upon biometric/behavioral links must: (1) Employ at least two distinct identification mechanisms (such as an identification code and password), each of which is contemporaneously executed at each signing; (2) be used only by their genuine owners; and (3) be administered and executed to ensure that attempted use of an individual's electronic signature by anyone other than its genuine owner requires collaboration of two or more individuals.

Two comments said that proposed § 11.200(a) should acknowledge that passwords may be known not only to their genuine owners, but also to system administrators in case people forget their passwords.

The agency does not believe that system administrators would routinely need to know an individual's password because they would have sufficient privileges to assist those individuals who forget passwords.

124. Several comments argued that the agency should accept a single password alone as an electronic signature because: (1) Combining the password with an identification code adds little security, (2) administrative controls and passwords are sufficient, (3) authorized access is more difficult when two components are needed, (4) people would not want to gain unauthorized entry into a manufacturing environment, and (5) changing current systems that use only a password would be costly.

The comments generally addressed the need for two components in electronic signatures within the context of the requirement that all components be used each time an electronic signature is executed. Several comments suggested that, for purposes of system access, individuals should enter both a user identification code and password, but that, for subsequent signings during one period of access, a single element (such as a password) known only to, and usable by, the individual should be sufficient.

The agency believes that it is very important to distinguish between those (nonbiometric) electronic signatures that are executed repetitively during a single, continuous controlled period of time (access session or logged-on period) and those that are not. The agency is concerned, from statements made in comments, that people might use passwords that are not always unique and are frequently words that are easily associated with an individual. Accordingly, where nonbiometric electronic signatures are not executed repetitively during a single, continuous controlled period, it would be extremely bad practice to use a password alone as an electronic signature. The agency believes that using a password alone in such cases would clearly increase the likelihood that one individual, by chance or deduction, could enter a password that belonged to someone else and thereby easily and readily impersonate that individual. This action could falsify electronic records.

The agency acknowledges that there are some situations involving repetitive signings in which it may not be necessary for an individual to execute each component of a nonbiometric electronic signature for every signing. The agency is persuaded by the comments that such situations generally involve certain conditions. For example, an individual performs an initial system access or "log on," which is effectively the first signing, by executing all components of the electronic signature (typically both an identification code and a password). The individual then performs subsequent signings by executing at least one component of the electronic signature, under controlled conditions that prevent another person from impersonating the legitimate signer. The agency's concern here is the possibility that, if the person leaves the workstation, someone else could access the workstation (or other computer device used to execute the signing) and impersonate the legitimate signer by entering an identification code or password.

The agency believes that, in such situations, it is vital to have stringent controls in place to prevent the impersonation. Such controls include: (1) Requiring an individual to remain in close proximity to the workstation throughout the signing session; (2) use of automatic inactivity disconnect measures that would "de-log" the first individual if no entries or actions were taken within a fixed short timeframe; and (3) requiring that the single component needed for subsequent signings be known to, and usable only by, the authorized individual.

The agency's objective in accepting the execution of fewer than all the components of a nonbiometric electronic signature for repetitive signings is to make it impractical to falsify records. The agency believes that this would be attained by complying with all of the following procedures where nonbiometric electronic signatures are executed more than once during a single, continuous controlled session: (1) All electronic signature components are executed for the first signing; (2) at least one electronic signature component is executed at each subsequent signing; (3) the electronic signature component executed after the initial signing is only used by its genuine owner, and is designed to ensure it can only be used by its genuine owner; and (4) the electronic signatures are administered and executed to ensure that their attempted use by anyone other than their genuine owners requires collaboration of two or more individuals. Items 1 and 4 are already incorporated in proposed § 11.200(a). FDA has included items 2 and 3 in final § 11.200(a).

The agency cautions, however, that if its experience with enforcement of part 11 demonstrates that these controls are insufficient to deter falsifications, FDA may propose more stringent controls.

125. One comment asserted that, if the agency intends the term "identification code" to mean the typical user identification, it should not characterize the term as a distinct mechanism because such codes do not necessarily exhibit security attributes. The comment also suggested that proposed § 11.200(a) address the appropriate application of each possible combination of a two-factor authentication method.

The agency acknowledges that the identification code alone does not exhibit security attributes. Security derives from the totality of system controls used to prevent falsification. However, uniqueness of the identification code when combined with another electronic signature component, which may not be unique (such as a password), makes the combination unique and thereby enables a legitimate electronic signature. FDA does not now believe it necessary to address, in § 11.200(a), the application of all possible combinations of multifactored authentication methods.

126. One comment requested clarification of "each signing," noting that a laboratory employee may enter a group of test results under one signing.

The agency advises that each signing means each time an individual executes a signature. Particular requirements regarding what records need to be signed derive from other regulations, not part 11. For example, in the case of a laboratory employee who performs a number of analytical tests, within the context of drug CGMP regulations, it is permissible for one signature to indicate the performance of a group of tests (21 CFR 211.194(a)(7)). A separate signing is not required in this context for each separate test as long as the record clearly shows that the single signature means the signer performed all the tests.

127. One comment suggested that the proposed requirement, that collaboration of at least two individuals is needed to prevent attempts at electronic signature falsification, be deleted because a responsible person should be allowed to override the electronic signature of a subordinate. Several comments addressed the phrase "attempted use" and suggested that it be deleted or changed to "unauthorized use." The comments said that willful breaking or circumvention of any security measure does not require two or more people to execute, and that the central question is whether collaboration is required to use the electronic signature.

The agency advises that the intent of the collaboration provision is to require that the components of a nonbiometric electronic signature cannot be used by one individual without the prior knowledge of a second individual. One type of situation the agency seeks to prevent is the use of a component such as a card or token that a person may leave unattended. If an individual must collaborate with another individual by disclosing a password, the risks of betrayal and disclosure are greatly increased and this helps to deter such actions. Because the agency is not condoning such actions, § 11.200(a)(2) requires that electronic signatures be used only by the genuine owner. The agency disagrees with the comments that the term "attempted use" should be changed to "unauthorized uses," because "unauthorized uses" could infer that use of someone else's electronic signature is acceptable if it is authorized.

Regarding electronic signature "overrides," the agency would consider as falsification the act of substituting the signature of a supervisor for that of a subordinate. The electronic signature of the subordinate must remain inviolate for purposes of authentication and documentation. Although supervisors may overrule the actions of their staff, the electronic signatures of the subordinates must remain a permanent part of the record, and the supervisor's own electronic signature must appear separately. The agency believes that such an approach is fully consistent with procedures for paper records.

As a result of the revisions noted in comments 123 to 127 of this document, § 11.200(a) now reads as follows:

(a) Electronic signatures that are not based upon biometrics shall:

(1) Employ at least two distinct identification components such as an identification code and password.

(i) When an individual executes a series of signings during a single, continuous period of controlled system access, the first signing shall be executed using all electronic signature components; subsequent signings shall be executed using at least one electronic signature component that is only executable by, and designed to be used only by, the individual.

(ii) When an individual executes one or more signings not performed during a single, continuous period of controlled system access, each signing shall be executed using all of the electronic signature components.

(2) Be used only by their genuine owners; and

(3) Be administered and executed to ensure that attempted use of an individual's electronic signature by anyone other than its genuine owner requires collaboration of two or more individuals.

128. Proposed § 11.200(b) states that electronic signatures based upon biometric/behavioral links be designed to ensure that they could not be used by anyone other than their genuine owners.

One comment suggested that the agency make available, by public workshop or other means, any information it has regarding existing biometric systems so that industry can provide proper input. Another comment asserted that proposed § 11.200(b) placed too great an emphasis on biometrics, did not establish particular levels of assurance for biometrics, and did not provide for systems using mixtures of biometric and nonbiometric electronic signatures. The comment recommended revising the phrase "designed to ensure they cannot be used" to read "provide assurances that prevent their execution."

The agency's experience with biometric electronic signatures is contained in the administrative record for this rulemaking, under docket no. 92N–0251, and includes recommendations from public comments to the ANPRM and the proposed rule. The agency has also gathered, and continues to gather, additional information from literature reviews, general press reports, meetings, and the agency's experience with this technology. Interested persons have had extensive opportunity for input and comment regarding biometrics in part 11. In addition, interested persons may continue to contact the agency at any time regarding biometrics or any other relevant technologies. The agency notes that the rule does not require the use of biometric-based electronic signatures.

As the agency's experience with biometric electronic signatures increases, FDA will consider holding or participating in public workshops if that approach would be helpful to those wishing to adopt such technologies to comply with part 11.

The agency does not believe that proposed § 11.200(b) places too much emphasis on biometric electronic signatures. As discussed above, the regulation makes a clear distinction between electronic signatures that are and are not based on biometrics, but treats their acceptance equally.

The agency recognizes the inherent security advantages of biometrics, however, in that record falsification is more difficult to perform. System controls needed to make biometric-based electronic signatures reliable and trustworthy are thus different in certain respects from controls needed to make nonbiometric electronic signatures reliable and trustworthy. The requirements in part 11 reflect those differences.

The agency does not believe that it is necessary at this time to set numerical security assurance standards that any system would have to meet.

The regulation does not prohibit individuals from using combinations of biometric and nonbiometric-based electronic signatures. However, when combinations are used, FDA advises that requirements for each element in the combination would also apply. For example, if passwords are used in combination with biometrics, then the benefits of using passwords would only be realized, in the agency's view, by adhering to controls that ensure password integrity (see § 11.300).

In addition, the agency believes that the phrase "designed to ensure that they cannot be used" more accurately reflects the agency's intent than the suggested alternate wording, and is more consistent with the concept of systems validation. Under such validation, falsification preventive attributes would be designed into the biometric systems.

To be consistent with the revised definition of biometrics in § 11.3(b)(3), the agency has revised § 11.200(b) to read, "Electronic signatures based upon biometrics shall be designed to ensure that they cannot be used by anyone other than their genuine owners."

XIII. ELECTRONIC SIGNATURES—CONTROLS FOR IDENTIFICATION CODES/PASSWORDS (§ 11.300)

The introductory paragraph of proposed § 11.300 states that electronic signatures based upon use of identification codes in combination with passwords must employ controls to ensure their security and integrity.

To clarify the intent of this provision, the agency has added the words "[p]ersons who use" to the first sentence of § 11.300. This change is consistent with §§ 11.10 and 11.30. The introductory paragraph now reads, "Persons who use electronic signatures based upon use of identification codes in combination with passwords shall employ controls to ensure their security and integrity. Such controls shall include: * * *."

129. One comment suggested deletion of the phrase "in combination with passwords" from the first sentence of this section.

The agency disagrees with the suggested revision because the change is inconsistent with FDA's intent to address controls for electronic signatures based on combinations of identification codes and passwords, and would, in effect, permit a single component nonbiometric-based electronic signature.

130. Proposed § 11.300(a) states that controls for identification codes/ passwords must include maintaining the uniqueness of each issuance of identification code and password.

One comment alleged that most passwords are commonly used words, such as a child's name, a State, city, street, month, holiday, or date, that are significant to the person who creates the password. Another stated that the rule should explain uniqueness and distinguish between issuance and use because identification code/password combinations generally do not change for each use.

FDA does not intend to require that individuals use a completely different identification code/password combination each time they execute an electronic signature. For reasons explained in the response to comment 16, what is required to be unique is each combined password and identification code and FDA has revised the wording of § 11.300(a) to clarify this provision. The agency is aware, however, of identification devices that generate new passwords on a continuous basis in synchronization with a "host" computer. This results in unique passwords for each system access. Thus, it is possible in theory to generate a unique nonbiometric electronic signature for each signing.

The agency cautions against using passwords that are common words easily associated with their originators because such a practice would make it relatively easy for someone to impersonate someone else by guessing the password and combining it with an unsecured (or even commonly known) identification code.

131. Proposed § 11.300(b) states that controls for identification codes/ passwords must ensure that code/ password issuances are periodically checked, recalled, or revised.

Several comments objected to this proposed requirement because: (1) It is unnecessary, (2) it excessively prescribes "how to," (3) it duplicates the requirements in § 11.300(c), and (4) it is administratively impractical for larger organizations. However, the comments said individuals should be encouraged to change their passwords periodically. Several comments suggested that proposed § 11.300(b) include a clarifying example such as "to cover events such as password aging." One comment said that the section should indicate who is to perform the periodic checking, recalling, or revising.

The agency disagrees with the objections to this provision. FDA does not view the provision as a "how to" because organizations have full flexibility in determining the frequency and methods of checking, recalling, or revising their code/password issuances. The agency does not believe that this paragraph duplicates the regulation in § 11.300(c) because paragraph (c) specifically addresses followup to losses of electronic signature issuances, whereas § 11.300(b) addresses periodic issuance changes to ensure against their having been unknowingly compromised. This provision would be met by ensuring that people change their passwords periodically.

FDA disagrees that this system control is unnecessary or impractical in large organizations because the presence of more people may increase the opportunities for compromising identification codes/passwords. The agency is confident that larger organizations will be fully capable of handling periodic issuance checks, revisions, or recalls.

FDA agrees with the comments that suggested a clarifying example and has revised § 11.300(b) to include password aging as such an example. The agency cautions, however, that the example should not be taken to mean that password expiration would be the only rationale for revising, recalling, and checking issuances. If, for example, identification codes and passwords have been copied or compromised, they should be changed.

FDA does not believe it necessary at this time to specify who in an organization is to carry out this system control, although the agency expects that units that issue electronic signatures would likely have this duty.

132. Proposed § 11.300(c) states that controls for identification codes/ passwords must include the following of loss management procedures to electronically deauthorize lost tokens, cards, etc., and to issue temporary or permanent replacements using suitable, rigorous controls for substitutes.

One comment suggested that this section be deleted because it excessively prescribes "how to." Another comment argued that the proposal was not detailed enough and should distinguish among fundamental types of cards (e.g., magstripe, integrated circuit, and optical) and include separate sections that address their respective use. Two comments questioned why the proposal called for "rigorous controls" in this section as opposed to other sections. One of the comments recommended that this section should also apply to cards or devices that are stolen as well as lost.

The agency believes that the requirement that organizations institute loss management procedures is neither too detailed nor too general. Organizations retain full flexibility in establishing the details of such procedures. The agency does not believe it necessary at this time to offer specific provisions relating to different types of cards or tokens. Organizations that use such devices retain full flexibility to establish appropriate controls for their operations. To clarify the agency's broad intent to cover all types of devices that contain or generate identification code or password information, FDA has revised § 11.300(c) to replace "etc." with "and other devices that bear or generate identification code or password information."

The agency agrees that § 11.300(c) should cover loss management procedures regardless of how devices become potentially compromised, and has revised this section by adding, after the word "lost," the phrase "stolen, missing, or otherwise potentially compromised." FDA uses the term "rigorous" because device disappearance may be the result of inadequate controls over the issuance and management of the original cards or devices, thus necessitating more stringent measures to prevent problem recurrence. For example, personnel training on device safekeeping may need to be strengthened.

133. Proposed § 11.300(d) states that controls for identification codes/ passwords must include the use of transaction safeguards to prevent unauthorized use of passwords and/or identification codes, and, detecting and reporting to the system security unit and organizational management in an emergent manner any attempts at their unauthorized use.

Several comments suggested that the term "emergent" in proposed § 11.300(d) be replaced with "timely" to describe reports regarding attempted unauthorized use of identification codes/passwords because: (1) A timely report would be sufficient, (2) technology to report emergently is not available, and (3) timely is a more recognizable and common term.

FDA agrees in part. The agency considers attempts at unauthorized use of identification codes and passwords to be extremely serious because such attempts signal potential electronic signature and electronic record falsification, data corruption, or worse— consequences that could also ultimately be very costly to organizations. In FDA's view, the significance of such attempts requires the immediate and urgent attention of appropriate security personnel in the same manner that individuals would respond to a fire alarm. To clarify its intent with a more widely recognized term, the agency is replacing "emergent" with "immediate and urgent" in the final rule. The agency believes that the same technology that accepts or rejects an identification code and password can be used to relay to security personnel an appropriate message regarding attempted misuse.

134. One comment suggested that the word "any" be deleted from the phrase "any attempts" in proposed § 11.300(d) because it is excessive. Another comment, noting that the question of attempts to enter a system or access a file by unauthorized personnel is very serious, urged the agency to substitute "all" for "any." This comment added that there are devices on the market that can be used by unauthorized individuals to locate personal identification codes and passwords.

The agency believes the word "any" is sufficiently broad to cover all attempts at misuse of identification codes and passwords, and rejects the suggestion to delete the word. If the word "any" were deleted, laxity could result from any inference that persons are less likely to be caught in an essentially permissive, nonvigilant system. FDA is aware of the "sniffing" devices referred to by one comment and cautions persons to establish suitable countermeasures against them.

135. One comment suggested that proposed § 11.300(d) be deleted because it is impractical, especially when simple typing errors are made. Another suggested that this section pertain to access to electronic records, not just the system, on the basis that simple miskeys may be typed when accessing a system.

As discussed in comments 133 and 134 of this document, the agency believes this provision is necessary and reasonable. The agency's security concerns extend to system as well as record access. Once having gained unauthorized system access, an individual could conceivably alter passwords to mask further intrusion and misdeeds. If this section were removed, falsifications would be more probable to the extent that some establishments would not alert security personnel.

However, the agency advises that a simple typing error may not indicate an unauthorized use attempt, although a pattern of such errors, especially in short succession, or such an apparent error executed when the individual who "owns" that identification code or password is deceased, absent, or otherwise known to be unavailable, could signal a security problem that should not be ignored. FDA notes that this section offers organizations maximum latitude in deciding what they perceive to be attempts at unauthorized use.

136. One comment suggested substituting the phrase "electronic signature" for "passwords and/or identification codes."

The agency disagrees with this comment because the net effect of the revision might be to ignore attempted misuse of important elements of an electronic signature such as a "password" attack on a system.

137. Several comments argued that: (1) It is not necessary to report misuse attempts simultaneously to management when reporting to the appropriate security unit, (2) security units would respond to management in accordance with their established procedures and lines of authority, and (3) management would not always be involved.

The agency agrees that not every misuse attempt would have to be reported simultaneously to an organization's management if the security unit that was alerted responded appropriately. FDA notes, however, that some apparent security breaches could be serious enough to warrant management's immediate and urgent attention. The agency has revised proposed § 11.300(d) to give organizations maximum flexibility in establishing criteria for management notification. Accordingly, § 11.300(d) now states that controls for identification codes/passwords must include:

> Use of transaction safeguards to prevent unauthorized use of passwords and/or identification codes, and to detect and report in an immediate and urgent manner any attempts at their unauthorized use to the system security unit, and, as appropriate, to organizational management.

138. Proposed § 11.300(e) states that controls for identification codes/ passwords must include initial and periodic testing of devices, such as tokens or cards, bearing identifying information, for proper function.

Many comments objected to this proposed device testing requirement as unnecessary because it is part of system validation and because devices are access fail-safe in that non-working devices would deny rather than permit system access. The comments suggested revising this section to require that failed devices deny user access. One comment stated that § 11.300(e) is unclear on the meaning of "identifying information" and that the phrase "tokens or cards" is redundant because cards are a form of tokens.

FDA wishes to clarify the reason for this proposed requirement, and to emphasize that proper device functioning includes, in addition to system access, the correctness of the identifying information and security performance attributes. Testing for system access alone could fail to discern significant unauthorized device alterations. If, for example, a device has been modified to change the identifying information, system access may still be allowed, which would enable someone to assume the identity of another person. In addition, devices may have been changed to grant individuals additional system privileges and action authorizations beyond those granted by the organization. Of lesser significance would be simple wear and tear on such devices, which result in reduced performance. For instance, a bar code may not be read with the same consistent accuracy as intended if the code becomes marred, stained, or otherwise disfigured. Access may be granted, but only after many more scannings than desired. The agency expects that device testing would detect such defects.

Because validation of electronic signature systems would not cover unauthorized device modifications, or subsequent wear and tear, validation would not obviate the need for periodic testing.

The agency notes that § 11.300(e) does not limit the types of devices organizations may use. In addition, not all tokens may be cards, and identifying information is intended to include identification codes and passwords. Therefore, FDA has revised proposed § 11.300(e) to clarify the agency's intent and to be consistent with § 11.300(c). Revised § 11.300(e) requires initial and periodic testing of devices, such as tokens or cards, that bear or generate identification code or password information to ensure that they function properly and have not been altered in an unauthorized manner.

XIV.   PAPERWORK REDUCTION ACT OF 1995

This final rule contains information collection provisions that are subject to review by the Office of Management and Budget (OMB) under the Paperwork Reduction Act of 1995 (44 U.S.C. 3501–3520). Therefore, in accordance with 5 CFR 1320, the title, description, and description of respondents of the collection of information requirements are shown below with an estimate of the annual reporting and recordkeeping burdens. Included in the estimate is the time for reviewing instructions, searching existing data sources, gathering and maintaining the data needed, and completing and reviewing the collection of information.

Most of the burden created by the information collection provision of this final rule will be a one-time burden associated with the creation of standard operating procedures, validation, and certification. The agency anticipates the use of electronic media will substantially reduce the paperwork burden associated with maintaining FDA-required records.

Title: Electronic records; Electronic signatures.

Description: FDA is issuing regulations that provide criteria for acceptance of electronic records, electronic signatures, and handwritten signatures executed to electronic records

TABLE 1. Estimated Annual Recordkeeping Burden

21 CFR Section	Annual No. of Recordkeepers	Hours per Recordkeeper	Total Hours
11.10	50	40	2,000
11.30	50	40	2,000
11.50	50	40	2,000
11.300	50	40	2,000
Total annual burden hours			8,000

TABLE 2. Estimated Annual Reporting Burden

21 CFR Section	Annual No. of Respondents	Hours per Response	Total Burden Hours
11.100	1,000	1	1,000
Total annual burden hours			1,000

as equivalent to paper records. Rules apply to any FDA records requirements unless specific restrictions are issued in the future. Records required to be submitted to FDA may be submitted electronically, provided the agency has stated its ability to accept the records electronically in an agency established public docket.

Description of Respondents: Businesses and other for-profit organizations, state or local governments, Federal agencies, and nonprofit institutions.

Although the August 31, 1994, proposed rule (59 FR 45160) provided a 90-day comment period under the Paperwork Reduction Act of 1980, FDA is providing an additional opportunity for public comment under the Paperwork Reduction Act of 1995, which was enacted after the expiration of the comment period and applies to this final rule. Therefore, FDA now invites comments on: (1) Whether the proposed collection of information is necessary for the proper performance of FDA's functions, including whether the information will have practical utility; (2) the accuracy of FDA's estimate of the burden of the proposed collection of information, including the validity of the methodology and assumptions used; (3) ways to enhance the quality, utility, and clarity of the information to be collected; and (4) ways to minimize the burden of the collection of information on respondents, including through the use of automated collection techniques, when appropriate, and other forms of information technology. Individuals and organizations may submit comments on the information collection provisions of this final rule by May 19, 1997. Comments should be directed to the Dockets Management Branch (address above).

At the close of the 60-day comment period, FDA will review the comments received, revise the information collection provisions as necessary, and submit these provisions to OMB for review and approval. FDA will publish a notice in the Federal Register when the information collection provisions are submitted to OMB, and an opportunity for public comment to OMB will be provided at that time. Prior to the effective date of this final rule, FDA will publish a notice in the Federal Register of OMB's decision to approve, modify, or disapprove the information collection provisions. An agency may not conduct or sponsor, and a person is not required to respond to, a collection of information unless it displays a currently valid OMB control number.

XV. ENVIRONMENTAL IMPACT

The agency has determined under 21 CFR 25.24(a)(8) that this action is of a type that does not individually or cumulatively have a significant effect on the human environment. Therefore, neither an environmental assessment nor an environmental impact statement is required.

XVI. ANALYSIS OF IMPACTS

FDA has examined the impacts of the final rule under Executive Order 12866, under the Regulatory Flexibility Act (5 U.S.C. 601–612), and under the Unfunded Mandates Reform Act (Pub. L. 104–4). Executive Order 12866 directs agencies to assess all costs and benefits of available regulatory alternatives and, when regulation is necessary, to select regulatory approaches that maximize net benefits (including potential economic, environmental, public health and safety, and other advantages; and distributive impacts and equity). Unless an agency certifies that a rule will not have a significant economic impact on a substantial number of small entities, the Regulatory Flexibility Act requires an analysis of regulatory options that would minimize any significant impact of a rule on small entities. The Unfunded Mandates Reform Act requires that agencies prepare an assessment of anticipated costs and benefits before proposing any rule that may result in an annual expenditure by State, local and tribal governments, in the aggregate, or by the private sector, of $100 million (adjusted annually for inflation).

The agency believes that this final rule is consistent with the regulatory philosophy and principles identified in the Executive Order. This rule permits persons to maintain any FDA required record or report in electronic format. It also permits FDA to accept electronic records, electronic signatures, and handwritten signatures executed to electronic records as equivalent to paper records and handwritten signatures executed on paper. The rule applies to any paper records required by statute or agency regulations. The rule was substantially influenced by comments to the ANPRM and the proposed rule. The provisions of this rule permit the use of electronic technology under conditions that the agency believes are necessary to ensure the integrity of electronic systems, records, and signatures, and the ability of the agency to protect and promote the public health.

This rule is a significant regulatory action as defined by the Executive Order and is subject to review under the Executive Order. This rule does not impose any mandates on State, local, or tribal governments, nor is it a significant regulatory action under the Unfunded Mandates Reform Act.

The activities regulated by this rule are voluntary; no entity is required by this rule to maintain or submit records electronically if it does not wish to do so. Presumably, no firm (or other regulated entity) will implement electronic recordkeeping unless the benefits to that firm are expected to exceed any costs (including capital and maintenance costs). Thus, the industry will incur no net costs as a result of this rule.

Based on the fact that the activities regulated by this rule are entirely voluntary and will not have any net adverse effects on small entities, the Commissioner of Food and Drugs certifies that this rule will not have a significant economic impact on a substantial number of small entities. Therefore, under the Regulatory Flexibility Act, no further regulatory flexibility analysis is required.

Although no further analysis is required, in developing this rule, FDA has considered the impact of the rule on small entities. The agency has also considered various regulatory options to maximize the net benefits of the rule to small entities without compromising the integrity of electronic systems, records, and signatures, or the agency's ability to protect and promote the public health. The following analysis briefly examines the potential impact of this rule on small businesses and other small entities, and describes the measures that FDA incorporated in this final rule to reduce the costs of applying electronic record/signature systems consistent with the objectives of the rule. This analysis includes each of the elements required for a final regulatory flexibility analysis under 5 U.S.C. 604(a).

A. Objectives

The purpose of this rule is to permit the use of a technology that was not contemplated when most existing FDA regulations were written, without undermining in any way the integrity of records and reports or the ability of FDA to carry out its statutory health protection mandate. The rule will permit regulated industry and FDA to operate with greater flexibility, in ways that will improve both the efficiency and the speed of industry's operations and the regulatory process. At the same time, it ensures that individuals will assign the same level of importance to affixing an electronic signature, and the records to which that signature attests, as they currently do to a handwritten signature.

B. Small Entities Affected

This rule potentially affects all large and small entities that are required by any statute administered by FDA, or any FDA regulation, to keep records or make reports or other submissions to FDA, including small businesses, nonprofit organizations, and small government entities. Because the rule affects such a broad range of industries, no data currently exist to estimate precisely the total number of small entities that will potentially benefit from the rule, but the number is substantial. For example, within the medical devices industry alone, the Small Business Administration (SBA) estimates that over 3,221 firms are small businesses (i.e., have fewer than 500 employees). SBA also estimates that 504 pharmaceutical firms are small businesses with fewer than 500 employees. Of the approximately 2,204 registered blood and plasma establishments that are neither government-owned nor part of the American Red Cross, most are nonprofit establishments that are not nationally dominant and thus may be small entities as defined by the Regulatory Flexibility Act.

Not all submissions will immediately be acceptable electronically, even if the submission and the electronic record conform to the criteria set forth in this rule. A particular required submission will be acceptable in electronic form only after it has been identified to this effect in public docket 92S–0251. (The agency unit that can receive that electronic submission will also be identified in the docket.) Thus, although all small entities subject to FDA regulations are potentially affected by this rule, the rule will actually only benefit those that: (1) Are required to submit records or other documents that have been identified in the public docket as acceptable if submitted electronically, and (2) choose this method of submission, instead of traditional paper record submissions. The potential range of submissions includes such records as new drug applications, medical device premarket notifications, food additive petitions, and medicated feed applications. These, and all other required submissions, will be considered by FDA as candidates for optional electronic format.

Although the benefits of making electronic submissions to FDA will be phased in over time, as the agency accepts more submissions in electronic form, firms can, upon the rule's effective date, immediately benefit from using electronic records/signatures for records they are required to keep, but not submit to FDA. Such records include, but are not limited to: Pharmaceutical and medical device batch production records, complaint records, and food processing records.

Some small entities will be affected by this rule even if they are not among the industries regulated by FDA. Because it will increase the market demand for certain types of software (e.g., document management, signature, and encryption software) and services (e.g., digital notaries and digital signature certification authorities), this rule will benefit some small firms engaged in developing and providing those products and services.

C. Description of the Impact

For any paper record that an entity is required to keep under existing statutes or FDA regulations, FDA will now accept an electronic record instead of a paper one, as long as the electronic record conforms to the requirements of this rule. FDA will also consider an electronic signature to be equivalent to a handwritten signature if it meets the requirements of this rule. Thus, entities regulated by FDA may, if they choose, submit required records and authorizations to the agency electronically once those records have been listed in the docket as acceptable in electronic form. This action is voluntary; paper records and handwritten signatures are still fully acceptable. No entity will be required to change the way it is currently allowed to submit paper records to the agency.

1. Benefits and Costs.   For any firm choosing to convert to electronic recordkeeping, the direct benefits are expected to include:

(1) Improved ability for the firm to analyze trends, problems, etc., enhancing internal evaluation and quality control;

(2) Reduced data entry errors, due to automated checks;

(3) Reduced costs of storage space;

(4) Reduced shipping costs for data transmission to FDA; and

(5) More efficient FDA reviews and approvals of FDA-regulated products.

No small entity will be required to convert to electronic submissions. Furthermore, it is expected that no individual firm, or other entity, will choose the electronic option unless that firm finds that the benefits to the firm from conversion will exceed any conversion costs.

There may be some small entities that currently submit records on paper, but archive records electronically. These entities will need to ensure that their existing electronic systems conform to the requirements for electronic recordkeeping described in this rule. Once they have done so, however, they may also take advantage of all the other benefits of electronic recordkeeping. Therefore, no individual small entity is expected to experience direct costs that exceed benefits as a result of this rule.

Furthermore, because almost all of the rule's provisions reflect contemporary security measures and controls that respondents to the ANPRM identified, most firms should have to make few, if any, modifications to their systems.

For entities that do choose electronic recordkeeping, the magnitude of the costs associated with doing so will depend on several factors, such as the level of appropriate computer hardware and software already in place in a given firm, the types of conforming technologies selected, and the size and dispersion of the firm. For example, biometric signature technologies may be more expensive than nonbiometric technologies; firms that choose the former technology may encounter relatively higher costs. Large, geographically dispersed firms may need some institutional security procedures that smaller firms, with fewer persons in more geographically concentrated areas, may not need. Firms that require wholesale technology replacements in order to adopt electronic record/signature technology may face much higher costs than those that require only minor modifications (e.g., because they already have similar technology for internal security and quality control purposes). Among the firms that must undertake major changes to implement electronic recordkeeping, costs will be lower for those able to undertake these changes simultaneously with other planned computer and security upgrades. New firms entering the market may have a slight advantage in implementing technologies that conform with this rule, because the technologies and associated procedures can be put in place as part of the general startup.

2. Compliance Requirements. If a small entity chooses to keep electronic records and/ or make electronic submissions, it must do so in ways that conform to the requirements for electronic records and electronic signatures set forth in this rule. These requirements, described previously in section II. of this document, involve measures designed to ensure the integrity of system operations, of information stored in the system, and of the authorized signatures affixed to electronic records. The requirements apply to all small (and large) entities in all industry sectors regulated by FDA.

The agency believes that because the rule is flexible and reflects contemporary standards, firms should have no difficulty in putting in place the needed systems and controls. However, to assist firms in meeting the provisions of this rule, FDA may hold public meetings and publish more detailed guidance. Firms may contact FDA's Industry and Small Business Liaison Staff, HF–50, at 5600 Fishers Lane, Rockville, MD 20857 (301–827–3430) for more information.

3. Professional skills required. If a firm elects electronic recordkeeping and submissions, it must take steps to ensure that all persons involved in developing, maintaining, and using electronic records and electronic signature systems have the education, training, and experience to perform the tasks involved. The level of training and experience that will be required depends on the tasks that the person performs. For example, an individual whose sole involvement with electronic records is infrequent might only need sufficient training to understand and use the required procedures. On the other hand, an individual involved in developing an electronic record system for a firm wishing to convert from a paper recordkeeping system would probably need more education or training in computer systems and software design and implementation. In addition, FDA expects that such a person would also have specific on-the-job training and experience related to the particular type of records kept by that firm.

The relevant education, training, and experience of each individual involved in developing, maintaining, or using electronic records/submissions must be documented. However, no specific examinations or credentials for these individuals are required by the rule.

D. Minimizing the Burden on Small Entities

This rule includes several conditions that an electronic record or signature must meet in order to be acceptable as an alternative to a paper record or handwritten signature. These conditions are necessary to permit the agency to protect and promote the public health. For example, FDA must retain the ability to audit records to detect unauthorized modifications, simple errors, and to deter falsification. Whereas there are many scientific techniques to show changes in paper records (e.g., analysis of the paper, signs of erasures, and handwriting analysis), these methods do not apply to electronic records. For electronic records and submissions to have the same integrity as paper records, they must be developed, maintained, and used under circumstances that make it difficult for them to be inappropriately modified. Without these assurances, FDA's objective of enabling electronic records and signatures to have standing equal to paper records and handwritten signatures, and to satisfy the requirements of existing statutes and regulations, cannot be met.

Within these constraints, FDA has attempted to select alternatives that provide as much flexibility as practicable without endangering the integrity of the electronic records. The agency decided not to make the required extent and stringency of controls dependent on the type of record or transactions, so that firms can decide for themselves what level of controls are worthwhile in each case. For example, FDA chose to give firms maximum flexibility in determining: (1) The circumstances under which management would have to be notified of security problems, (2) the means by which firms achieve the required link between an electronic signature and an electronic record, (3) the circumstances under which extra security and authentication measures are warranted in open systems, (4) when to use operational system checks to ensure proper event sequencing, and (5) when to use terminal checks to ensure that data and instructions originate from a valid source.

Numerous other specific considerations were addressed in the public comments to the proposed rule. A summary of the issues raised by those comments, the agency's assessment of these issues, and any changes made in the proposed rule as a result of these comments is presented earlier in this preamble.

FDA rejected alternatives for limiting potentially acceptable electronic submissions to a particular category, and for issuing different electronic submissions standards for small and large entities. The former alternative would unnecessarily limit the potential benefits of this rule; whereas the latter alternative would threaten the integrity of electronic records and submissions from small entities.

As discussed previously in this preamble, FDA rejected comments that suggested a total of 17 additional more stringent controls that might be more expensive to implement. These include: (1) Examination and certification of individuals who perform certain important tasks, (2) exclusive use of cryptographic methods to link electronic signatures to electronic records, (3) controls for each possible combination of a two factored authentication method, (4) controls for each different type of identification card, and (5) recording in audit trails the reason why records were changed.

List of Subjects in 21 CFR Part 11:   Administrative practice and procedure, Electronic records, Electronic signatures, Reporting and recordkeeping requirements.

Therefore, under the Federal Food, Drug, and Cosmetic Act, the Public Health Service Act, and under authority delegated to the Commissioner of Food and Drugs, Title 21, Chapter I of the Code of Federal Regulations is amended by adding part 11 to read as follows:

PART 11—ELECTRONIC RECORDS; ELECTRONIC SIGNATURES

Subpart A—General Provisions

Subpart C—Electronic Signatures

    11.100   General requirements.

    11.200   Electronic signature components and controls.

    11.300   Controls for identification codes/passwords.

Authority: Secs. 201–903 of the Federal Food, Drug, and Cosmetic Act (21 U.S.C. 321–393); sec. 351 of the Public Health Service Act (42 U.S.C. 262).

Subpart A—General Provisions

§ 11.1 Scope

(a) The regulations in this part set forth the criteria under which the agency considers electronic records, electronic signatures, and handwritten signatures executed to electronic records to be trustworthy, reliable, and generally equivalent to paper records and handwritten signatures executed on paper.

(b) This part applies to records in electronic form that are created, modified, maintained, archived, retrieved, or transmitted, under any records requirements set forth in agency regulations. This part also applies to electronic records submitted to the agency under requirements of the Federal Food, Drug, and Cosmetic Act and the Public Health Service Act, even if such records are not specifically identified in agency regulations. However, this part does not apply to paper records that are, or have been, transmitted by electronic means.

(c) Where electronic signatures and their associated electronic records meet the requirements of this part, the agency will consider the electronic signatures to be equivalent to full handwritten signatures, initials, and other general signings as required by agency regulations, unless specifically excepted by regulation(s) effective on or after August 20, 1997.

(d) Electronic records that meet the requirements of this part may be used in lieu of paper records, in accordance with § 11.2, unless paper records are specifically required.

(e) Computer systems (including hardware and software), controls, and attendant documentation maintained under this part shall be readily available for, and subject to, FDA inspection.

§ 11.2 Implementation

(a) For records required to be maintained but not submitted to the agency, persons may use electronic records in lieu of paper records or electronic signatures in lieu of traditional signatures, in whole or in part, provided that the requirements of this part are met.

(b) For records submitted to the agency, persons may use electronic records in lieu of paper records or electronic signatures in lieu of traditional signatures, in whole or in part, provided that:

    (1) The requirements of this part are met; and

    (2) The document or parts of a document to be submitted have been identified in public docket No. 92S– 0251 as being the type of submission the agency accepts in electronic form. This docket will identify specifically what types of documents or parts of documents are acceptable for submission in electronic form without paper records and the agency receiving unit(s) (e.g., specific center, office, division, branch) to which such submissions may be made. Documents to agency receiving

unit(s) not specified in the public docket will not be considered as official if they are submitted in electronic form; paper forms of such documents will be considered as official and must accompany any electronic records. Persons are expected to consult with the intended agency receiving unit for details on how (e.g., method of transmission, media, file formats, and technical protocols) and whether to proceed with the electronic submission.

§ 11.3 Definitions

(a) The definitions and interpretations of terms contained in section 201 of the act apply to those terms when used in this part.

(b) The following definitions of terms also apply to this part:

(1) *Act* means the Federal Food, Drug, and Cosmetic Act (secs. 201–903 (21 U.S.C. 321–393)).

(2) *Agency* means the Food and Drug Administration.

(3) *Biometrics* means a method of verifying an individual's identity based on measurement of the individual's physical feature(s) or repeatable action(s) where those features and/or actions are both unique to that individual and measurable.

(4) *Closed system* means an environment in which system access is controlled by persons who are responsible for the content of electronic records that are on the system.

(5) *Digital signature* means an electronic signature based upon cryptographic methods of originator authentication, computed by using a set of rules and a set of parameters such that the identity of the signer and the integrity of the data can be verified.

(6) *Electronic record* means any combination of text, graphics, data, audio, pictorial, or other information representation in digital form that is created, modified, maintained, archived, retrieved, or distributed by a computer system.

(7) *Electronic signature* means a computer data compilation of any symbol or series of symbols executed, adopted, or authorized by an individual to be the legally binding equivalent of the individual's handwritten signature.

(8) *Handwritten signature* means the scripted name or legal mark of an individual handwritten by that individual and executed or adopted with the present intention to authenticate a writing in a permanent form. The act of signing with a writing or marking instrument such as a pen or stylus is preserved. The scripted name or legal mark, while conventionally applied to paper, may also be applied to other devices that capture the name or mark.

(9) *Open system* means an environment in which system access is not controlled by persons who are responsible for the content of electronic records that are on the system.

Subpart B—Electronic Records

§ 11.10   Controls for closed systems.   Persons who use closed systems to create, modify, maintain, or transmit electronic records shall employ procedures and controls designed to ensure the authenticity, integrity, and, when appropriate, the confidentiality of electronic records, and to ensure that the signer cannot readily repudiate the signed record as not genuine. Such procedures and controls shall include the following:

(a) Validation of systems to ensure accuracy, reliability, consistent intended performance, and the ability to discern invalid or altered records.

(b) The ability to generate accurate and complete copies of records in both human readable and electronic form suitable for inspection, review, and copying by the agency. Persons should contact the agency if there are any questions regarding the ability of the agency to perform such review and copying of the electronic records.

(c) Protection of records to enable their accurate and ready retrieval throughout the records retention period.

(d) Limiting system access to authorized individuals.

(e) Use of secure, computer-generated, time-stamped audit trails to independently record the date and time of operator entries and actions that create, modify, or delete electronic records. Record changes shall not obscure previously recorded information. Such audit trail documentation shall be retained for a period at least as long as that required for the subject electronic records and shall be available for agency review and copying.

(f) Use of operational system checks to enforce permitted sequencing of steps and events, as appropriate.

(g) Use of authority checks to ensure that only authorized individuals can use the system, electronically sign a record, access the operation or computer system input or output device, alter a record, or perform the operation at hand.

(h) Use of device (e.g., terminal) checks to determine, as appropriate, the validity of the source of data input or operational instruction.

(i) Determination that persons who develop, maintain, or use electronic record/electronic signature systems have the education, training, and experience to perform their assigned tasks.

(j) The establishment of, and adherence to, written policies that hold individuals accountable and responsible for actions initiated under their electronic signatures, in order to deter record and signature falsification.

(k) Use of appropriate controls over systems documentation including:

    (1) Adequate controls over the distribution of, access to, and use of documentation for system operation and maintenance.

    (2) Revision and change control procedures to maintain an audit trail that documents time-sequenced development and modification of systems documentation.

§ 11.30   Controls for open systems.   Persons who use open systems to create, modify, maintain, or transmit electronic records shall employ procedures and controls designed to ensure the authenticity, integrity, and, as appropriate, the confidentiality of electronic records from the point of their creation to the point of their receipt. Such procedures and controls shall include those identified in § 11.10, as appropriate, and additional measures such as document encryption and use of appropriate digital signature standards to ensure, as necessary under the circumstances, record authenticity, integrity, and confidentiality.

§ 11.50   Signature manifestations

(a) Signed electronic records shall contain information associated with the signing that clearly indicates all of the following:

    (1) The printed name of the signer;

(2) The date and time when the signature was executed; and

(3) The meaning (such as review, approval, responsibility, or authorship) associated with the signature.

(b) The items identified in paragraphs (a)(1), (a)(2), and (a)(3) of this section shall be subject to the same controls as for electronic records and shall be included as part of any human readable form of the electronic record (such as electronic display or printout).

§ 11.70 *Signature/record linking.*

Electronic signatures and handwritten signatures executed to electronic records shall be linked to their respective electronic records to ensure that the signatures cannot be excised, copied, or otherwise transferred to falsify an electronic record by ordinary means.

Subpart C—Electronic Signatures

§ 11.100 *General requirements*

(a) Each electronic signature shall be unique to one individual and shall not be reused by, or reassigned to, anyone else.

(b) Before an organization establishes, assigns, certifies, or otherwise sanctions an individual's electronic signature, or any element of such electronic signature, the organization shall verify the identity of the individual.

(c) Persons using electronic signatures shall, prior to or at the time of such use, certify to the agency that the electronic signatures in their system, used on or after August 20, 1997, are intended to be the legally binding equivalent of traditional handwritten signatures.

(1) The certification shall be submitted in paper form and signed with a traditional handwritten signature, to the Office of Regional Operations (HFC–100), 5600 Fishers Lane, Rockville, MD 20857.

(2) Persons using electronic signatures shall, upon agency request, provide additional certification or testimony that a specific electronic signature is the legally binding equivalent of the signer's handwritten signature.

§ 11.200 *Electronic signature components and controls*

(a) Electronic signatures that are not based upon biometrics shall:

(1) Employ at least two distinct identification components such as an identification code and password.

(i) When an individual executes a series of signings during a single, continuous period of controlled system access, the first signing shall be executed using all electronic signature components; subsequent signings shall be executed using at least one electronic signature component that is only executable by, and designed to be used only by, the individual.

(ii) When an individual executes one or more signings not performed during a single, continuous period of controlled system access, each signing shall be executed using all of the electronic signature components.

(2) Be used only by their genuine owners; and

(**3**) Be administered and executed to ensure that attempted use of an individual's electronic signature by anyone other than its genuine owner requires collaboration of two or more individuals.

(**b**) Electronic signatures based upon biometrics shall be designed to ensure that they cannot be used by anyone other than their genuine owners.

§ 11.300   Controls for identification codes/passwords.   Persons who use electronic signatures based upon use of identification codes in combination with passwords shall employ controls to ensure their security and integrity. Such controls shall include:

(**a**) Maintaining the uniqueness of each combined identification code and password, such that no two individuals have the same combination of identification code and password.

(**b**) Ensuring that identification code and password issuances are periodically checked, recalled, or revised (e.g., to cover such events as password aging).

(**c**) Following loss management procedures to electronically deauthorize lost, stolen, missing, or otherwise potentially compromised tokens, cards, and other devices that bear or generate identification code or password information, and to issue temporary or permanent replacements using suitable, rigorous controls.

(**d**) Use of transaction safeguards to prevent unauthorized use of passwords and/or identification codes, and to detect and report in an immediate and urgent manner any attempts at their unauthorized use to the system security unit, and, as appropriate, to organizational management.

(**e**) Initial and periodic testing of devices, such as tokens or cards, that bear or generate identification code or password information to ensure that they function properly and have not been altered in an unauthorized manner.

Dated: March 11, 1997.
William B. Schultz,
Deputy Commissioner for Policy.
[FR Doc. 97–6833 Filed 3–20–97; 8:45 am]
BILLING CODE 4160–01–F

FEDERAL REGISTER

PART II   Thursday, February 20, 2003

Department of Health and Human Services

Office of the Secretary

45 CFR Parts 160, 162, and 164
Health Insurance Reform: Security Standards; Final Rule

DEPARTMENT OF HEALTH AND HUMAN SERVICES

Office of the Secretary

45 CFR Parts 160, 162, and 164

[CMS–0049–F]

RIN 0938–AI57

Health Insurance Reform: Security Standards

AGENCY: Centers for Medicare & Medicaid Services (CMS), HHS.

ACTION: Final rule.

SUMMARY: This final rule adopts standards for the security of electronic protected health information to be implemented by health plans, health care clearinghouses, and certain health care providers. The use of the security standards will improve the Medicare and Medicaid programs, and other Federal health programs and private health programs, and the effectiveness and efficiency of the health care industry in general by establishing a level of protection for certain electronic health information. This final rule implements some of the requirements of the Administrative Simplification subtitle of the Health Insurance Portability and Accountability Act of 1996 (HIPAA).

DATES: *Effective Date:* These regulations are effective on April 21, 2003.

Compliance Date: Covered entities, with the exception of small health plans, must comply with the requirements of this final rule by April 21, 2005. Small health plans must comply with the requirements of this final rule by April 21, 2006.

FOR FURTHER INFORMATION CONTACT:
William Schooler, (410) 786–0089.

SUPPLEMENTARY INFORMATION:

Availability of Copies and Electronic Access

To order copies of the **Federal Register** containing this document, send your request to: New Orders, Superintendent of Documents, P.O. Box 371954, Pittsburgh, PA 15250–7954. Specify the date of the issue requested and enclose a check or money order payable to the Superintendent of Documents, or enclose your Visa or Master Card number and expiration date. Credit card orders can also be placed by calling the order desk at (202) 512–1800 or by faxing to (202) 512–2250. The cost for each copy is $10. As an alternative, you can view and photocopy the **Federal Register** document at most libraries designated as Federal Depository Libraries and at many other public and academic libraries throughout the country that receive the **Federal Register**.

This **Federal Register** document is also available from the **Federal Register** online database through GPO access, a service of the U.S. Government Printing Office. The Web site address is http://www.access.gpo.gov/nara/index.html.

I. BACKGROUND

The Department of Health and Human Services (HHS) Medicare Program, other Federal agencies operating health plans or providing health care, State Medicaid agencies, private health plans, health care providers, and health care clearinghouses must assure their customers (for example, patients, insured individuals, providers, and health plans) that the integrity, confidentiality, and availability of electronic protected health information they collect, maintain, use, or transmit is protected. The confidentiality of health information is threatened not only by the risk of improper access to stored information, but also by the risk of interception during electronic transmission of the information. The purpose of this final rule is to adopt national standards for safeguards to protect the confidentiality, integrity, and availability of electronic protected health information. Currently, no standard measures exist in the health care industry that address all aspects of the security of electronic health information while it is being stored or during the exchange of that information between entities.

This final rule adopts standards as required under title II, subtitle F, sections 261 through 264 of the Health Insurance Portability and Accountability Act of 1996 (HIPAA), Pub. L. 104–191. These standards require measures to be taken to secure this information while in the custody of entities covered by HIPAA (covered entities) as well as in transit between covered entities and from covered entities to others.

The Congress included provisions to address the need for safeguarding electronic health information and other administrative simplification issues in HIPAA. In subtitle F of title II of that law, the Congress added to title XI of the Social Security Act a new part C, entitled "Administrative Simplification" (hereafter, we refer to the Social Security Act as "the Act"; we refer to the other laws cited in this document by their names). The purpose of subtitle F is to improve the Medicare program under title XVIII of the Act, the Medicaid program under title XIX of the Act, and the efficiency and effectiveness of the health care system, by encouraging the development of a health information system through the establishment of standards and requirements to enable the electronic exchange of certain health information.

Part C of title XI consists of sections 1171 through 1179 of the Act. These sections define various terms and impose requirements on HHS, health plans, health care clearinghouses, and certain health care providers. These statutory sections are discussed in the Transactions Rule, at 65 FR 50312, on pages 50312 through 50313, and in the final rules adopting Standards for Privacy of Individually Identifiable Health Information, published on December 28, 2000 at 65 FR 82462 (Privacy Rules), on pages 82470 through 82471, and on August 14, 2002 at 67 FR 53182. The reader is referred to those discussions.

Section 1173(d) of the Act requires the Secretary of HHS to adopt security standards that take into account the technical capabilities of record systems used to maintain health information, the costs of security measures, the need to train persons who have access to health information, the value of audit trails in computerized record systems, and the needs and capabilities of small health care providers and rural health care providers. Section 1173(d) of the Act also requires that the standards ensure that a health care clearinghouse, if part of a larger organization, has policies and security procedures that isolate the activities of the clearinghouse with respect to processing information so as to prevent unauthorized access to health information by the larger organization. Section 1173(d) of the Act provides that covered entities that maintain or transmit health information are required to maintain reasonable and appropriate administrative, physical, and technical safeguards to ensure the integrity and confidentiality of the information and to protect against any reasonably anticipated threats or hazards to the security or integrity of the information and unauthorized use or disclosure of the information. These safeguards must also otherwise ensure compliance with the statute by the officers and employees of the covered entities.

II.   GENERAL OVERVIEW OF THE PROVISIONS OF THE PROPOSED RULE

On August 12, 1998, we published a proposed rule (63 FR 43242) to establish a minimum standard for security of electronic health information. We proposed that the standard would require the safeguarding of all electronic health information by covered entities. The proposed rule also proposed a standard for electronic signatures. This final rule adopts only security standards. All comments concerning the proposed electronic signature standard, responses to these comments, and a final rule for electronic signatures will be published at a later date. A detailed discussion of the provisions of the August 12, 1998 proposed rule can be found at 63 FR 43245 through 43259.

We originally proposed to add part 142, entitled "Administrative Requirements," to title 45 of the Code of Federal Regulations (CFR). It has now been determined that this material will reside in subchapter C of title 45, consisting of parts 160, 162, and 164. Subpart A of part 160 contains the general provisions applicable to all the Administrative Simplification rules; other subparts of part 160 will contain other requirements applicable to all standards. Part 162 contains the standards for transactions and code sets and will contain the identifier standards. Part 164 contains the standards relating to privacy and security. Subpart A of part 164 contains general provisions applicable to part 164; subpart E contains the privacy standards. Subpart C of part 164, which is adopted in this final rule, adopts standards for the security of electronic protected health information.

III. ANALYSIS OF, AND RESPONSES TO, PUBLIC COMMENTS ON THE PROPOSED RULE

We received approximately 2,350 timely public comments on the August 12, 1998 proposed rule. The comments came from professional associations and societies, health care workers, law firms, health insurers, hospitals, and private individuals. We reviewed each commenter's letter and grouped related comments. Some comments were identical. After associating like comments, we placed them in categories based on subject matter or based on the section(s) of the regulations affected and then reviewed the comments.

In this section of the preamble, we summarize the provisions of the proposed regulations, summarize the related provisions in this final rule, and respond to comments received concerning each area.

It should be noted that the proposed Security Rule contained multiple proposed "requirements" and "implementation features." In this final rule, we replace the term "requirement" with "standard." We also replace the phrase "implementation feature" with "implementation specification." We do this to maintain consistency with the use of those terms as they appear in the statute, the Transactions Rule, and the Privacy Rule. Within the comment and response portion of this final rule, for purposes of continuity, however, we use "requirement" and "implementation feature" when we are referring specifically to matters from the proposed rule. In all other instances, we use "standard" and "implementation specification."

The proposed rule would require that each covered entity (as now described in § 160.102) engaged in the electronic maintenance or transmission of health information pertaining to individuals assess potential risks and vulnerabilities to such information in its possession in electronic form, and develop, implement, and maintain appropriate security measures to protect that information. Importantly, these measures would be required to be documented and kept current.

The proposed security standard was based on three basic concepts that were derived from the Administrative Simplification provisions of HIPAA. First, the standard should be comprehensive and coordinated to address all aspects of security. Second, it should be scalable, so that it can be effectively implemented by covered entities of all types and sizes. Third, it should not be linked to specific technologies, allowing covered entities to make use of future technology advancements.

The proposed standard consisted of four categories of requirements that a covered entity would have to address in order to safeguard the integrity, confidentiality, and availability of its electronic health information pertaining to individuals: administrative procedures, physical safeguards, technical security services, and technical mechanisms. The implementation features described the requirements in greater detail when that detail was needed. Within the four categories, the requirements and implementation features were presented in alphabetical order to convey that no one item was considered to be more important than another.

The four proposed categories of requirements and implementation features were depicted in tabular form along with the electronic signature standard in a combined matrix located at Addendum 1. We also provided a glossary of terms, at Addendum 2, to facilitate a common understanding of the matrix entries, and at Addendum 3, we mapped available existing industry standards and guidelines to the proposed security requirements.

A. General Issues

The comment process overwhelmingly validated our basic assumptions that the entities affected by this regulation are so varied in terms of installed technology, size, resources, and relative risk, that it would be impossible to dictate a specific solution or set of solutions that would be useable by all covered entities. Many commenters also supported the concept of technological neutrality, which would afford them the flexibility to select appropriate technology solutions and to adopt new technology over time.

1. Security Rule and Privacy Rule Distinctions. As many commenters recognized, security and privacy are inextricably linked. The protection of the privacy of information depends in large part on the existence of security measures to protect that information. It is important that we note several distinct differences between the Privacy Rule and the Security Rule.

The security standards below define administrative, physical, and technical safeguards to protect the confidentiality, integrity, and availability of electronic protected health information. The standards require covered entities to implement basic safeguards to protect electronic protected health information from unauthorized access, alteration, deletion, and transmission. The Privacy Rule, by contrast, sets standards for how protected health information should be controlled by setting forth what uses and disclosures are authorized or required and what rights patients have with respect to their health information.

As is discussed more fully below, this rule narrows the scope of the information to which the safeguards must be applied from that proposed in the proposed rule, electronic health information pertaining to individuals, to protected health information in electronic form. Thus, the scope of information covered in this rule is consistent with the Privacy Rule, which addresses privacy protections for "protected health information." However, the scope of the Security Rule is more limited than that of the Privacy Rule. The Privacy Rule applies to protected health information in any form, whereas this rule applies only to protected health information in electronic form. It is true that, under section 1173(d) of the Act, the Secretary has authority to cover "health information," which, by statute, includes information in other than electronic form. However, because the proposed rule proposed to cover only health information in electronic form, we do not include security standards for health information in non-electronic form in this final rule.

We received a number of comments that pertained to privacy issues. These issues were considered in the development of the Privacy Rule and many of these comments were addressed in the preamble of the Privacy Rule. Therefore, we are referring the reader to that document for a discussion of those issues.

2. *Level of Detail.* We solicited comments as to the level of detail expressed in the required implementation features; that is, we specifically wanted to know whether commenters believe the level of detail of any proposed requirement went beyond what is necessary or appropriate. We received numerous comments expressing the view that the security standards should not be overly prescriptive because the speed with which technology is evolving could make specific requirements obsolete and might in fact deter technological progress. We have accordingly written the final rule to frame the standards in terms that are as generic as possible and which, generally speaking, may be met through various approaches or technologies.

3. *Implementation Specifications.* In addition to adopting standards, this rule adopts implementation specifications that provide instructions for implementing those standards.

However, in some cases, the standard itself includes all the necessary instructions for implementation. In these instances, there may be no corresponding implementation specification for the standard specifically set forth in the regulations text. In those instances, the standards themselves also serve as the implementation specification. In other words, in those instances, we are adopting one set of instructions as both the standard and the implementation specification. The implementation specification would, accordingly, in those instances be required.

In this final rule, we adopt both "required" and "addressable" implementation specifications. We introduce the concept of "addressable implementation specifications" to provide covered entities additional flexibility with respect to compliance with the security standards.

In meeting standards that contain addressable implementation specifications, a covered entity will ultimately do one of the following: (a) Implement one or more of the addressable implementation specifications; (b) implement one or more alternative security measures; (c) implement a combination of both; or (d) not implement either an addressable implementation specification or an alternative security measure. In all cases, the covered entity must meet the standards, as explained below.

The entity must decide whether a given addressable implementation specification is a reasonable and appropriate security measure to apply within its particular security framework. This decision will depend on a variety of factors, such as, among others, the entity's risk analysis, risk mitigation strategy, what security measures are already in place, and the cost of implementation. Based upon this decision the following applies:

(a) If a given addressable implementation specification is determined to be reasonable and appropriate, the covered entity must implement it.

(b) If a given addressable implementation specification is determined to be an inappropriate and/or unreasonable security measure for the covered entity, but the standard cannot be met without implementation of an additional security safeguard, the covered entity may implement an alternate measure that accomplishes the same end as the addressable implementation specification. An entity that meets a given standard through alternative measures must document the decision not to implement the addressable implementation specification, the rationale behind that decision, and the alternative safeguard implemented to meet the standard. For example, the addressable implementation specification for the integrity standard calls for electronic mechanisms to corroborate that data have not been altered or destroyed in an unauthorized manner (see 45 CFR 164.312(c)(2)). In a small provider's office environment, it might well be unreasonable and inappropriate to make electronic copies of the data in question. Rather, it might well be more practical and afford a sufficient safeguard to make paper copies of the data.

(c) A covered entity may also decide that a given implementation specification is simply not applicable (that is, neither reasonable nor appropriate) to its situation and that the standard can be met without implementation of an alternative measure in place of the addressable implementation specification. In this scenario, the covered entity must document the decision not to implement the addressable specification, the rationale behind that decision, and how the standard is being met. For example, under the information access management standard, an access establishment and modification implementation specification reads: "implement policies and procedures that, based upon the entity's access authorization policies, establish, document, review, and modify a user's right of access to a workstation, transaction, program, or process" (45 CFR 164.308(a) (4)(ii)(c)). It is possible that a small practice, with one or more individuals equally responsible for establishing and maintaining all automated patient records, will not need to establish policies and procedures for granting access to that electronic protected health information because the access rights are equal for all of the individuals.

a. Comment: A large number of commenters indicated that mandating 69 implementation features would result in a regulation that is too burdensome, intrusive, and difficult to implement. These commenters requested that the implementation features be made optional to meet the requirements. A number of other commenters requested that all implementation features be removed from the regulation.

Response: Deleting the implementation specifications would result in the standards being too general to understand, apply effectively, and enforce consistently. Moreover, a number of implementation specifications are so basic that no covered entity could effectively protect electronic protected health information without implementing them. We selected 13 of these mandatory implementation specifications based on (1) the expertise of Federal security experts and generally accepted industry practices and, (2) the recommendation for immediate implementation of certain technical and organizational practices and procedures described in Chapter 6 of *For The Record: Protecting Electronic Health Information,* a 1997 report by the National Research Council (NRC). These mandatory implementation specifications are referred to as required implementation specifications and are reflected in the NRC report's recommendations. Risk Analysis and Risk management are found in the NRC recommendation title System Assessment; Sanction Policy is required in the Sanctions recommendation; Information system Activity Review is discussed in Audit Trails; Response and Reporting circumstances.

In addition, a number of voluntary national and regional organizations have been formed to address HIPAA implementation issues and to facilitate communication among trading partners. These include the Strategic National Implementation Process (SNIP) developed under the auspices of the Workgroup for Electronic Data Interchange (WEDI), an organization named in the HIPAA statute to consult with the Secretary of HHS on HIPAA issues. Some of these organizations have developed white papers, tools, and recommended best practices addressing a number of HIPAA issues, including security. Covered entities may wish to examine these products to determine if they are relevant and useful in their own implementation efforts. A partial list of these organizations can be found at *http://www.wedi/ snip./org*. We believe that these and other future industry-developed guidelines and/or models may provide valuable assistance to covered entities implementing these standards but must caution that HHS does not rate or endorse any such guidelines and/or models and the value of its content must be determine by the user.

b. Comment: Many commenters asked us to develop guidelines and models to aid in complying with the Security Rule. Several commenters either offered to participate in the development of guidelines and models or suggested entities that should be invited to participate.

Response: We agree that creation of compliance tools and guidelines for different business environments could assist covered entities to implement the HIPAA Security Rule. We plan to issue guidance documents after the publication of this final rule. However, it is critical for each covered entity to establish policies and procedures that address its own unique risks and circumstances.

In addition, a number of voluntary national and regional organizations have been formed to address HIPAA implementation issues and to facilitate communication among trading partners. These include the Strategic National Implementation Process (SNIP) developed under the auspices of the Workgroup for Electronic Data Interchange (WEDI), an organization named in the HIPAA statute to consult with the Secretary of HHS on HIPAA issues. Some of these organizations have developed white papers, tools, and recommended best practices addressing a number of HIPAA issues, including security.

Covered entities may wish to examine these products to determine if they are relevant and useful in their own implementation efforts. A partial list of these organizations can be found at *http://www.snip.wedi.org*. We believe that these and other future industry-developed guidelines and/or models may provide valuable assistance to covered entities implementing these standards but must caution that HHS does not rate or endorse any such guidelines and/ or models and the value of its content must be determined by the user.

4. Examples

Comment: We received a number of comments that demonstrated confusion regarding the purpose of the examples of security solutions that were included throughout the proposed rule. Commenters stated that they could not, or did not wish to, adopt various security measures suggested in examples. Other commenters asked that we include additional options within the examples. Some commenters referred specifically to the example provided in the proposed rule demonstrating how a small or rural provider might comply with the standards. One commenter asked for clarification that the examples are not mandatory measures that are required to demonstrate compliance, but are merely meant as a guide when implementing the security standards. Another commenter expressed support for the use of examples to clarify the intent of text descriptions.

Response: We wish to clarify that examples are used only as illustrations of possible approaches, and are included to serve as a springboard for ideas. The steps that a covered entity will actually need to take to comply with these regulations will be dependent upon its own particular environment and circumstances and risk assessment. The examples do not describe mandatory measures, nor do they represent the only, or even the best, way of achieving compliance. The most appropriate means of compliance for any covered entity can only be determined by that entity assessing its own risks and deciding upon the measures that would best mitigate those risks.

B. Applicability (§ 164.302)

We proposed that the security standards would apply to health plans, health care clearinghouses, and to health care providers that maintain or transmit health information electronically. The proposed security standards would apply to all electronic health information maintained or transmitted, regardless of format (standard transaction or a proprietary format). No distinction would be made between internal corporate entity communication or communication external to the corporate entity. Electronic transmissions would include transactions using all media, even when the information is physically moved from one location to another using magnetic tape, disk, or other machine readable media. Transmissions over the Internet (wide-open), extranet (using Internet technology to link a business with information only accessible to collaborating parties), leased lines, dialup lines, and private networks would be included. We proposed that telephone voice response and "faxback" systems (a request for information made via voice using a fax machine and requested information returned via that same machine as a fax) would not be included but we solicited comments on this proposed exclusion.

This final rule simplifies the applicability statement greatly. Section 164.302 provides that the security standards apply to covered entities; the scope of the information covered is specified in § 164.306 (see the discussion under that section below regarding the changes and revisions to the scope of information covered).

1. Comment: A number of commenters requested clarification of who must comply with the standards. The preamble and proposed § 142.102 and § 142.302 stated: "Each person described in section 1172(a) of the Act who maintains or transmits health information shall maintain reasonable and appropriate administrative, technical, and physical safeguards." Commenters suggested that this statement is in conflict with the law, which defines a covered entity as a health plan, a clearinghouse, or a health care provider that conducts certain transactions electronically. The commentors apparently did not realize that section 1172(a) of the Act contains the definition of covered entities.

Response: Section 164.302 below makes the security standards applicable to "covered entities." The term "covered entity" is defined at § 160.103 as one of the following: (1) A health plan; (2) a health care clearinghouse; (3) a health care provider who transmits any health information in electronic form in connection with a transaction covered by part 162 of title 45 of the Code of Federal Regulations (CFR). The rationale for the use and the meaning of the term "covered entity" is discussed in the preamble to the Privacy Rule (65 FR 82476 through 82477).

As that discussion makes clear, the standards only apply to health care providers who engage electronically in the transactions for which standards have been adopted.

2. Comment: Several commenters recommended expansion of applicability, either to other specific entities, or to all entities involved in health care. Others wanted to know whether the standards apply to entities such as employers, public health organizations, medical schools, universities, research organizations, plan brokers, or non-EDI providers. One commenter asked whether the standards apply to State data organizations operating in capacities other than as plans, clearinghouses, or providers. Still other commenters stated that it was inappropriate to include physicians and other health care professionals in the same category as plans and clearinghouses, arguing that providers should be subject to different, less burdensome requirements because they already protect health information.

Response: The statute does not cover all health care entities that transmit or maintain individually identifiable health information. Section 1172(a) of the Act provides that only health plans, health care clearinghouses, and certain health care providers (as discussed above) are covered. With respect to the comments regarding the difference between providers and plans/clearinghouses, we have structured the Security Rule to be scalable and flexible enough to allow different entities to implement the standards in a manner that is appropriate for their circumstances. Regarding the coverage of entities not within the jurisdiction of HIPAA, *see* the Privacy Rule at 82567 through 82571.

3. Comment: One commenter asked whether the standards would apply to research organizations, both to those affiliated with health care providers and those that are not.

Response: Only health plans, health care clearinghouses, and certain health care providers are required to comply with the security standards. Researchers who are members of a covered entity's work force may be covered by the security standards as part of the covered entity. *See* the definition of "workforce" at 45 CFR 160.103. Note, however, that a covered entity could, under appropriate circumstances, exclude a researcher or research division from its health care component or components (*see* § 164.105(a)). Researchers who are not part of the covered entity's workforce and are not themselves covered entities are not subject to the standards.

4. Comment: Several commenters stated that internal networks and external networks should be treated differently. One commenter asked for further clarification of the difference between what needs to be secured external to a corporation versus the security of data movement within an organization. Another stated that complying with the security standards for internal communications may prove difficult and costly to monitor and control. In contrast, one commenter stated that the existence of requirements should not depend on whether use of information is for internal or external purposes.

Another commenter argued that the regulation goes beyond the intent of the law, and while communication of electronic information between entities should be covered, the law was never intended to mandate changes to an entity's internal automated systems. One commenter requested that raw data that are only for the internal use of a facility be excluded, provided that reasonable safeguards are in place to keep the raw data under the control of the facility.

Response: Section 1173(d)(2) of the Act states: Each person described in section 1172(a) who maintains or transmits health information shall maintain reasonable and appropriate administrative, technical, and physical safeguards—(A) to ensure the integrity and confidentiality of the information; (B) to protect against any reasonably anticipated—(i) threats or hazards to the security or integrity of the information; and (ii) unauthorized uses or disclosures of the information; and (C) otherwise to ensure compliance with this part by the officers and employees of such person.

This language draws no distinction between internal and external data movement. Therefore, this final rule covers electronic protected health information at rest (that is, in storage) as well as during transmission. Appropriate protections must be applied, regardless of whether the data are at rest or being transmitted. However, because each entity's security needs are unique, the specific protections determined appropriate to adequately protect information will vary and will be determined by each entity in complying with the standards (*see* the discussion below).

5. Comment: Several commenters found the following statement in the proposed rule (63 FR 43245) at section II.A. confusing and asked for clarification: "With the exception of the security standard, transmission within a corporate entity would not be required to comply with the standards."

Response: In the final Transactions Rule, we revised our approach concerning the transaction and code set exemptions, replacing this concept with other tests that determine whether a particular transaction is subject to those standards (*see* the discussion in the Transactions Rule at 65 FR 50316 through 50318). We also note that the Privacy Rule regulates a covered entity's use, as well as disclosure, of protected health information.

6. *Comment:* One commenter stated that research would be hampered if proposed § 142.306(a) applied. The commenter believes that research uses of health information should be excluded or the standard should be revised to allow appropriate flexibility for research depending on the risk to patients or subjects (for example, if the information is anonymous, there is no risk, and it would not be necessary to meet the security standards).

Response: If electronic protected health information is de-identified (as truly anonymous information would be), it is not covered by this rule because it is no longer electronic protected health information (see 45 CFR 164.502(d) and 164.514(a)). Electronic protected health information received, created, or maintained by a covered entity, or that is transmitted by covered entities, is covered by the security standards and must be protected. To the extent a researcher is a covered entity, the researcher must comply with these standards with respect to electronic protected health information. Otherwise, the conditions for release of such information to researchers is governed by the Privacy Rule. *See,* for example, 45 CFR 164.512(i), 164.514(e) and 164.502(d). These standards would not apply to the researchers as such in the latter circumstances.

7. *Comment:* One commenter asked to what extent individual patients are subject to the standards. For example, some telemedicine practices support the use of diagnostic systems in the patient's home, which can be used to conduct tests and send results to a remote physician. In other cases, patients may be responsible for the filing of insurance claims directly and will need the ability to verify facts, confirm receipt of claims, and so on. The commenter asked if it is the intent of the rule to include electronic transmission to or from the patient.

Response: Patients are not covered entities and, thus, are not subject to these standards. With respect to transmissions from covered entities, covered entities must protect electronic protected health information when they transmit that information. *See* also the discussion of encryption in section III.G.

C. Transition to the Final Rule

The proposed rule included definitions for a number of terms that have now already been promulgated as part of the Transactions Rule or the Privacy Rule. Comments related to the definitions of "code set," "health care" clearinghouse," "health plan," "health care provider," "small health plan," "standard" and "transaction," are addressed in the Transactions Rule at 65 FR 50319 through 50320. Comments concerning the definition of "individually identifiable health information" are discussed below, but are also addressed in the Privacy Rule at 65 FR 82611 through 82613. In addition, a few terms were redefined in the final Standards for Privacy of Individually Identifiable Health Information (67 FR 53182), issued on August 14, 2002 (Privacy Modifications). Certain terms that were defined in the proposed rule are not used in the final rule because they are no longer necessary. Other terms defined in the proposed rule are defined within the explanation of the standards in the final rule and are discussed in the preamble discussions in § 164.308 through § 164.312.

Definitions of terms relevant to the security standards now appear in the regulations text provisions as indicated below:

§ 160.103: Definitions of the following terms relevant to this rule appear in § 160.103: "business associate," "covered entity," "disclosure," "electronic media," "electronic protected health information," "health care," "health care clearinghouse," "health care provider," "health information," "health plan," "individual," "individually identifiable health information," "implementation specification," "organized health care arrangement," "protected health

information," "standard," "use," and "workforce." These terms were discussed in connection with the Transaction and Privacy Rules and with the exception of the terms "covered entity" "disclosure" "electronic protected health information," "health information," "individual," "organized health care arrangement," "protected health information," and "use," we will not discuss them in this document. We note that the definition of those terms are not changed in the final rule.

§ 162.103: We have moved the definition of "electronic media" at § 162.103 to § 160.103 and have modified it to clarify that the term includes storage of information. The term "electronic media" is used in the definition of "protected health information." Both the privacy and security standards apply to information "at rest" as well as to information being transmitted.

We note that we have deleted the reference to § 162.103 in paragraph (1)(ii) of the definition of "protected health information," since both definitions, "electronic media" and "protected health information," have been moved to this section. Also, it is unnecessary, because the definitions of § 160.103 apply to all of the rule in parts 160, 162, and 164.

We have also clarified that the physical movement of electronic media from place to place is not limited to magnetic tape, disk, or compact disk. This clarification removes a restriction as to what is considered to be physical electronic media, thereby allowing for future technological innovation. We further clarified that transmission of information not in electronic form before the transmission, for example, paper or voice, is not covered by this definition.

§ 164.103: The following term "plan sponsor" now appears in the new § 164.103, which consists of definitions of terms common to both subpart C and subpart E (the privacy standards). This definition was moved, without substantive change, from § 164.501 and has the meaning given to it in that section, and comments relating to this definition are discussed in connection with that section in the Privacy Rule at 65 FR 82607, 82611 through 82613, 82618 through 82622, and 82629.

§ 164.304: Definitions specifically applicable to the Security Rule appear in § 164.304, and these are discussed below. These definitions are from, or derived from, currently accepted definitions in industry publications, such as, the International Organization for Standards (ISO) 7498–2 and the American Society for Testing and Materials (ASTM) E1762–95.

The following terms in § 164.304 are taken from the proposed rule text or the glossary in Addendum 2 of the proposed rule (63 FR 43271), were not commented on, and/or are unchanged or have only minor technical changes for purposes of clarification and are not discussed below: "access," "authentication," "availability," "confidentiality," "encryption," "password," and "security."

§ 164.314: Four terms were defined in § 164.504(a) of the Privacy Rule ("common control," "common ownership," "health care component," and "hybrid entity"). Because these terms apply to both security and privacy, their definitions have been moved to § 164.103 without change.

Those terms are discussed in the Privacy Rule at 65 FR 82502 through 82503 and at 67 FR 53203 through 53207.

1. Covered Entity (§ 160.103)

Comment: One commenter asked if transcription services were covered entities. The question arose because transcription is often the first electronic or printed source of clinical information. Concern was expressed about the application of physical safeguard standards to the transcribers working for transcription companies or health care providers, either as employees or as independent contractors.

Another commenter expressed concern that scalability was limited to only small providers. The commenter explained that Third Party Administrators (TPAs) allow claim processors to work at home. Some TPAs have noted that it would be impossible to comply with the security standards for home-based claims processors.

Response: A covered entity's responsibility to implement security standards extends to the members of its workforce, whether they work at home or on-site. Because a covered entity is responsible for ensuring the security of the information in its care, the covered entity must include "at home" functions in its security process. While an independent transcription company or a TPA may not be covered entities, they will be a business associate of the covered entity because their activities fall under paragraph (1)(i)(a) of the definition of that term. For business associate provisions see proposed preamble section III.E.8. and § 164.308(b)(1) and § 164.314(c) of this final rule.

2. Health Care and Medical Care (§ 160.103)

Comment: One commenter asked whether "medical care," which is defined in the proposed rule, and "health care," which is not, are synonymous.

Response: The term "medical care," as used in the proposed rule (63 FR 43242), was intended to be synonymous with "health care." The term ldquo;medical care" is not included in this final rule. It is, however, included in the definition of "health plan," where its meaning is not synonymous with "health care." For a full discussion of this issue and its resolution, *see* the Privacy Rule (65 FR 82578).

3. Health Information and Individually Identifiable Health Information 160.103). We note that the definitions of "health information" and "individually identifiable health information" remain unchanged from those published in the Transactions and Privacy Rules.

a. *Comment:* A number of commenters asked that the definition of "health information" be expanded to include information collected by additional entities. Several commenters wanted the definition to include health information collected, maintained, or transmitted by any entity, and one commenter suggested the inclusion of aggregated information not identifiable to an individual. Several commenters asked that eligibility information be excluded from the definition of information. Several commenters wanted the definition broadened to include demographics.

Response: Our definition of health information is taken from the definition in section 1171(4) of the Act, which provides that health information relates to the health or condition of an individual, the provision of health care to an individual, or payment for the provision of health care to an individual. The statutory definition also specifies the entities by which health information is created or received. We note that, because "individually identifiable health information" is a subset of "health information" and by statute includes demographic information, "health information" necessarily includes demographic information. We think this is clear as a matter of statutory construction and does not require further regulatory change.

b. *Comment:* Several commenters asked that we clarify the difference between "health information" and "individually identifiable" and "health information pertaining to an individual" as used in the August 12, 1998 proposed rule (63 FR 43242). Additionally, commenters asked that we be more consistent in the use of these terms and recommended use of the term "individually identifiable health information."

Two commenters stated that it is important to distinguish between "health information pertaining to an individual" and "individually identifiable health information," as in reporting statistics at various levels there will always be a need to bring forth information pertaining to an individual.

One commenter recommended that the standards apply only to individually identifiable health information. Another stated that in § 142.306(b) of the proposed rule, "health information pertaining to an individual" should be changed to "individually identifiable health information," as nonidentifiable information can be used for utilization review and other purposes. As written, the regulation text could limit the ability to use data, for example, from a clearinghouse for compliance monitoring.

Response: In general, we agree with these commenters, and note that these comments are largely mooted by the decision, reflected in § 164.306 below and discussed in section III.D.1. of this final rule, to cover only electronic protected health information in this final rule.

c. *Comment:*   Several commenters stated that the definition of "individually identifiable health information" is not in the regulations and should be added.

Response: We note that the definition of "individually identifiable health information" appears at § 160.103, which applies to this final rule.

4.   *Protected Health Information (§ 160.103).*   This term is moved from § 164.501 to § 160.103 because it applies to both subparts C (security) and E (privacy). See 67 FR 53192 through 531936 regarding the definition of "protected health information."

Also, the term "electronic media" is included in paragraphs (1)(i) and (ii) of the definition of "protected health information," as specified in this section.

In addition, we added the definitions of "covered functions," "plan sponsor," and "Required by law" to § 164.103.

5.   *Breach (§ 164.304)*

Comment: One commenter asked that "breach" be defined.

Response: The term "breach" has been deleted and therefore not defined. Instead, we define the term "security incident," which better describes the types of situations we were referring to as breaches.

6.   *Facility (§ 164.304).*   This new term has been added as a result of changing the name of the "physical access control" standard to "facility access control." This change was made based on comments indicating that the original term was not descriptive. We have defined the term "facility" as the physical premises and interior and exterior of a building.

7.   *Security Incident (§ 164.304)*

Comment: We received comments asking that this term be defined.

Response. This final rule defines "Security incident" in § 164.304 as "the attempted or successful unauthorized access, use, disclosure, modification, or destruction of information or interference with system operations in an information system."

8. System (§ 164.304)

Comment: One commenter asked that "system" be defined.

Response: This final rule defines "system," in the context of an information system, in § 164.304 as "an interconnected set of information resources under the same direct management control that shares common functionality. A system normally includes hardware, software, information, data, applications, communications, and people."

9. Workstation (§ 164.304)

Comment: One commenter expressed concern that the use of the term "workstation" implied limited applicability to fixed devices (such as terminals), excluding laptops and other portable devices.

Response: We have added a definition of the term "workstation" to clarify that portable devices are also included. This final rule defines workstation as "an electronic computing device, for example, a laptop or desktop computer, or any other device that performs similar functions, and electronic media stored in its immediate environment."

10. Definitions Not Adopted.
Several definitions in the proposed regulations text and glossary are not adopted as definitions in the final rule: "participant," "contingency plan," "risk," "role-based access control," and "user-based access control." The terms "participant," "role-based access control," and "user-based access control" are not used in this final rule and thus are not defined. "Risk" is not defined as its meaning is generally understood. While we do not define the term, we address "contingency plan" as a standard in § 164.308(a)(7) below.

a. *Comment:* We received comments requesting that we define the following terms: "token" and "documentation."

Response: These terms were defined in Addendum 2 of the proposed rule. In this final rule, we do not adopt a definition for "token" because it is not used in the final rule. "Documentation" is discussed in § 164.316 below.

b. *Comment:* We received several comments that "small" and "rural" should be defined as those terms apply to providers. We received an equal number of comments stating that there is no need to define these terms. One commenter stated that definitions for these terms would be necessary only if special exemptions existed for small and rural providers. Several commenters suggested initiation of a study to determine limitations and potential barriers small and rural providers will have in implementing these regulations.

Response: The statute requires that we address the needs of small and rural providers. We believe that we have done this through the provisions, which require the risk assessment and the response to be assessment based on the needs and capabilities of the entity. This scalability concept takes the needs of those providers into account and eliminates any need to define those terms.

c. Comment: In the proposed rule, we proposed the following definition for the term "Access control": "A method of restricting access to resources, allowing only privileged entities access. Types of access control include, among others, mandatory access control, discretionary access control, time-of day, classification, and subject-object separation." One commenter believed the proposed definition is too restrictive and requested revision of the definition to read: "Access control refers to a method of restricting access to resources, allowing access to only those entities which have been specifically granted the desired access rights." Another commenter wanted the definition expanded to include partitioned rule-based access control (PRBAC).

Response: We agree with the commenter who suggested that the definition as proposed seemed too restrictive. In this case, as in many others, a number of commenters believed the examples given in the proposed rule provided the only acceptable compliance actions. As previously noted, in order to clarify that the examples listed were not to be considered all-inclusive, we have generalized the proposed requirements in this final rule. In this case, we have also generalized the requirements and placed the substantive provisions governing access control at § 164.308(a)(4), § 164.310(a)(1), and § 164.312(a)(1). With respect to PRBAC, the access control standard does not exclude this control, and entities should adopt it if appropriate to their circumstances.

D. General Rules (§ 164.306)

In the proposed rule, we proposed to cover all health information maintained or transmitted in electronic form by a covered entity. We proposed to adopt, in § 142.308, a nation-wide security standard that would require covered entities to implement security measures that would be technology-neutral and scalable, and yet integrate all the components of security (administrative procedures, physical safeguards, technical security services, and technical security mechanisms) that must be in place to preserve health information confidentiality, integrity, and availability (three basic elements of security). Since no comprehensive, scalable, and technology-neutral set of standards currently exists, we proposed to designate a new standard, which would define the security requirements to be fulfilled.

The proposed rule proposed to define the security standard as a set of scalable, technology-neutral requirements with implementation features that providers, plans, and clearinghouses would have to include in their operations to ensure that health information pertaining to an individual that is electronically maintained or electronically transmitted remains safeguarded. The proposed rule would have required that each affected entity assess its own security needs and risks and devise, implement, and maintain appropriate security to address its own unique security needs. How individual security requirements would be satisfied and which technology to use would be business decisions that each entity would have to make.

In the final rule we adopt this basic framework. In § 164.306, we set forth general rules pertaining to the security standards. In paragraph (a), we describe the general requirements. Paragraph (a) generally reflects section 1173(d)(2) of the Act, but makes explicit the connection between the security standards and the privacy standards (see § 164.306(a)(3)). In § 164.306(a)(1), we provide that the security standards apply to all electronic protected health information the covered entity creates, receives, maintains, or transmits. In paragraph (b)(1), we provide explicitly for the scalability of this rule by discussing the flexibility of the standards, and paragraph (b)(2) of § 164.306 discusses various factors covered entities must consider in complying with the standards.

The provisions of § 164.306(c) provide the framework for the security standards, and establish the requirement that covered entities must comply with the standards. The administrative, physical, and technical safeguards a covered entity employs must be reasonable and appropriate to accomplish the tasks outlined in paragraphs (1) through (4) of § 164.306(a). Thus, an entity's risk analysis and risk management measures required by § 164.308(a)(1) must be designed to lead to the implementation of security measures that will comply with § 164.306(a).

It should be noted that the implementation of reasonable and appropriate security measures also supports compliance with the privacy standards, just as the lack of adequate security can increase the risk of violation of the privacy standards. If, for example, a particular safeguard is inadequate because it routinely permits reasonably anticipated uses or disclosures of electronic protected health information that are not permitted by the Privacy Rule, and that could have been prevented by implementation of one or more security measures appropriate to the scale of the covered entity, the covered entity would not only be violating the Privacy Rule, but would also not be in compliance with § 164.306(a)(3) of this rule.

Paragraph (d) of § 164.306 establishes two types of implementation specifications, required and addressable. It provides that required implementation specifications must be met. However, with respect to implementation specifications that are addressable, § 164.306(d)(3) specifies that covered entities must assess whether an implementation specification is a reasonable and appropriate safeguard in its environment, which may include consideration of factors such as the size and capability of the organization as well as the risk. If the organization determines it is a reasonable and appropriate safeguard, it must implement the specification. If an addressable implementation specification is determined not to be a reasonable and appropriate answer to a covered entity's security needs, the covered entity must do one of two things: implement another equivalent measure if reasonable and appropriate; or if the standard can otherwise be met, the covered entity may choose to not implement the implementation specification or any equivalent alternative measure at all. The covered entity must document the rationale behind not implementing the implementation specification. See the detailed discussion in section II.A.3.

Paragraph (e) of § 164.306 addresses the requirement for covered entities to maintain the security measures implemented by reviewing and modifying the measures as needed to continue the provision of reasonable and appropriate protections, for example, as technology moves forward, and as new threats or vulnerabilities are discovered.

1. Scope of Health Information Covered by the Rule (§ 164.306(a)).

We proposed to cover health information maintained or transmitted by a covered entity in electronic form. We have modified, by narrowing, the scope of health information to be safeguarded under this rule from that which was proposed. The statute requires the privacy standards to cover individually identifiable health information. The Privacy Rule covers all individually identifiable information except for: (1) Education records covered by the Family and Educational Rights and Privacy Act (FERPA); (2) records described in 20 U.S.C. 1232g(a)(4)(B)(iv); and (3) employment records. (see the Privacy Rule at 65 FR 82496. See also 67 FR 53191 through 53193). The scope of information covered in the Privacy Rule is referred to as "protected health information." Based upon the comments we received, we align the requirements of the Security and Privacy Rules with regard to the scope of information covered, in order to eliminate confusion and ease implementation. Thus, this final rule requires protection of the same scope of information as that covered by the Privacy Rule, except that it only covers that information if it is in electronic form.

We note that standards for the security of all health information or protected health information in nonelectronic form may be proposed at a later date.

a.   Comment: One commenter stated that the rule should apply to aggregate information that is not identifiable to an individual. In contrast, another commenter asked that health information used for statistical analysis be exempted if the covered entity may reasonably expect that the removed information cannot be used to reidentify an individual.

Response: As a general proposition, any electronic protected health information received, created, maintained, or transmitted by a covered entity is covered by this final rule. We agree with the second commenter that certain information, from which identifiers have been stripped, does not come within the purview of this final rule. Information that is de-identified, as defined in the Privacy Rule at § 164.502(d) and § 164.514(a), is not "individually identifiable" within the meaning of these rules and, thus, does not come within the definition of "protected health information." It accordingly is not covered by this final rule. For a full discussion of the issues of de-identification and re-identification of individually identifiable health information see 65 FR 82499 and 82708 through 82712 and 67 FR 53232 through 53234.

b.   Comment: Several commenters asked whether systems that determine eligibility of clients for insurance coverage under broad categories such as medical coverage groups are considered health information. One commenter asked that we specifically exclude eligibility information from the standards.

Response: We cannot accept the latter suggestion. Eligibility information will typically be individually identifiable, and much eligibility information will also contain health information. If the information is "individually identifiable" and is "health information," (with three very specific exceptions noted in the general discussion above) and it is in electronic form, it is covered by the security standards if maintained or transmitted by a covered entity.

c.   Comment: Several commenters requested clarification as to whether the standards apply to identifiable health information in paper form. Some commenters believed the rule should be applicable to paper; others argued that it should apply to all confidential, identifiable health information.

Response: While we agree that protected health information in paper or other form also should have appropriate security protections, the proposed rule proposing the security standards proposed to apply those standards to health information in electronic form only. We are, accordingly, not extending the scope in this final rule.
We may establish standards to secure protected health information in other media in a future rule, in accordance with our statutory authority to do so. See discussion, supra, responding to a comment on the definition of "health information" and "individually identifiable health information."

d.   Comment: The proposed rule would have excluded "telephone voice response" and "faxback" systems from the security standards, and we specifically solicited comments on that issue. A number of commenters agreed that telephone voice response and faxback should be excluded from the regulation, suggesting that the privacy standards rather than the security standards should apply. Others wanted those systems included, on the grounds that inclusion is necessary for consistency and in keeping with the intent of the Act. Still others specifically wanted personal computer-fax transmissions included. One commenter asked for clarification of when we would cover faxes, and another commenter asked why we were excluding them. Several commenters suggested that the other security requirements provide for adequate security of these systems.

Response: In light of these comments, we have decided that telephone voice response and "faxback" (that is, a request for information from a computer made via voice or telephone keypad input with the requested information returned as a fax) systems fall under this rule because they are used as input and output devices for computers, not because they have computers in them. Excluding these features would provide a huge loophole in any system concerned with security of the information contained and/or processed therein. It should be noted that employment of telephone voice response and/or faxback systems will generally require security protection by only one of the parties involved, and not the other. Information being transmitted via a telephone (either by voice or a DTMP tone pad) is not in electronic form (as defined in the first paragraph of the definition of "electronic media") before transmission and therefore is not subject to the Security Rule. Information being returned via a telephone voice response system in response to a telephone request is data that is already in electronic form and stored in a computer. This latter transmission does require protection under the Security Rule.

Although most recently made electronic devices contain microprocessors (a form of computer) controlled by firmware (an unchangeable form of computer program), we intend the term "computer" to include only software programmable computers, for example, personal computers, minicomputers, and mainframes. Copy machines, fax machines, and telephones, even those that contain memory and can produce multiple copies for multiple people are not intended to be included in the term "computer." Therefore, because "paper-to-paper" faxes, person-to-person telephone calls, video teleconferencing, or messages left on voice-mail were not in electronic form before the transmission, those activities are not covered by this rule. See also the definition of "electronic media" at § 160.103.

We note that this guidance differs from the guidance regarding the applicability of the Transactions Rule to faxback and voice response systems. HHS has stated that faxback and voice response systems are not required to follow the standards mandated in the Transactions Rule. This new guidance refers only to this rule.

e. *Comment:*   One commenter asked whether there is a need to implement special security practices to address the shipping and receiving of health information and asked that we more fully explain our expectations and solutions in the final rules.

Response: If the handling of electronic protected health information involves shipping and receiving, appropriate measures must be taken to protect the information. However, specific solutions are not provided within this rule, as discussed in section III.A.3 of this final rule. The device and media controls standard under § 164.310(d)(1) addresses this situation.

f. *Comment:*   One commenter wanted the "HTML" statement reworded to eliminate a specific exemption for HTML from the regulation.

Response: The Transactions Rule did not adopt the proposed exemption for HTML. The use of HTML or any other electronic protocol is not exempt from the security standards. Generally, if protected health information is contained in any form of electronic transmission, it must be appropriately safeguarded.

g. *Comment:*   One commenter asked to what degree "family history" is considered health information under this rule and what protections apply to family members included in a patient's family history.

Response: Any health-related "family history" contained in a patient's record that identifies a patient, including a person other than the patient, is individually identifiable health information and, to the extent it is also electronic protected health information, must be afforded the security protections.

h. Comment: Two commenters asked that the rule prohibit re-identification of de-identified data. In contrast, several commenters asked that we identify a minimum list or threshold of specific re-identification data elements (for example, name, city, and ZIP) that would fall under this final rule so that, for example, the rule would not affect numerous systems, for example, network adequacy and population-based clinical analysis databases. One commenter asked that we establish a means to use re-identified information if the entity already has access to the information or is authorized to have access.

Response: The issue of re-identification is addressed in the Privacy Rule at § 164.502(d) and § 164.514(c). The reader is referred to those sections and the related discussion in the preamble to the Privacy Rule (65 FR 82712) and the preamble to the Privacy Modifications (67 FR 53232 through 53234) for a full discussion of the issues of re-identification. We note that once information in the possession (or constructive possession) of a covered entity is re-identified and meets the definition of electronic protected health information, the security standards apply.

2. Technology-Neutral Standards

Comment: Many commenters expressed support for our efforts to develop standards for the security of health information. A number of comments were made in support of the technology-neutral approach of the proposed rule. For example, one commenter stated, "By avoiding prescription of the specific technologies health care entities should use to meet the law's requirements, you are opening the door for industry to apply innovation. Technologies that don't currently exist or are impractical today could, in the near future, enhance health information security while minimizing the overall cost." Several other commenters stated that the requirements should be general enough to withstand changes to technology without becoming obsolete. One commenter anticipates no problems with meeting the standards.

In contrast, one commenter suggested that whenever possible, specific technology recommendations should provide sufficient detail to promote systems interoperability and decrease the tendency toward adoption of multiple divergent standards. Several commenters stated that by letting each organization determine its own rules, the rules impose procedural burdens without any substantive benefit to security.

Response: The overwhelming majority of comments supported our position. We do not believe it is appropriate to make the standards technology-specific because technology is simply moving too fast, for example, the increased use and sophistication of internet-enabled hand held devices. We believe that the implementation of these rules will promote the security of electronic protected health information by (1) providing integrity and confidentiality; (2) allowing only authorized individuals access to that information; and (3) ensuring its availability to those authorized to access the information. The standards do not allow organizations to make their own rules, only their own technology choices.

3. Miscellaneous Comments

a. Comment: Some commenters stated that the requirements and implementation features set out in the proposed rule were not specific enough to be considered standards, and that the actual standards are delegated to the discretion of the covered entities, at the expense of medical record privacy. Several commenters stated that it was inappropriate to balance the interests of those seeking to use identifiable medical information without patient consent against the interest of patients. Several other commenters believe that allowing covered entities to make their own decisions about the adequacy and balance of security measures undermined patient confidentiality interests, and stated that the proposed rule did not appear to adequately consider patient concerns and viewpoints.

Response: Again, the overwhelming majority of commenters supported our approach. This final rule sets forth requirements with which covered entities must comply and labels those requirements as standards and implementation specifications. Adequate implementation of this final rule by covered entities will ensure that the electronic protected health information in a covered entity's care will be as protected as is feasible for that entity.

We disagree that covered entities are given complete discretion to determine their security polices under this rule, resulting in effect, in no standards. While cost is one factor a covered identity may consider in determining whether to implement a particular implementation specification, there is nonetheless a clear requirement that adequate security measures be implemented, see 45 CFR 164.306(b). Cost is not meant to free covered entities from this responsibility.

b. Comment: Several commenters requested we withdraw the regulations, citing resource shortages due to Y2K preparation, upcoming privacy legislation, and/or the "excessive micro-management" contained in the rules. One commenter stated that, to insurers, these rules were onerous, not necessary, and not justified as cost-effective, as they already have effective practices for computer security and are subject to rigorous State laws for the safeguarding of health information. Another commenter stated that these rules would adversely affect a provider's practice environment.

Response: The HIPAA statute requires us to promulgate a rule adopting security standards for health information. Resource concerns due to Y2K should no longer be an issue. Covered entities will have 2 years (or, in the case of small health plans, 3 years) from the adoption of this final rule in which to comply. Concerns relative to effective and compliance dates and the Privacy Rule are discussed under § 164.318, Compliance dates for initial implementation, below and at 65 FR 82751 through 82752.

We disagree that these standards will adversely affect a provider's practice environment. The scalability of the standards allows each covered entity to implement security protections that are appropriate to its specific needs, risks, and environments. These protections are necessary to maintain the confidentiality, integrity, and availability of patient data. A covered entity that lacks adequate protections risks inadvertent disclosure of patient data, with resulting loss of public trust, and potential legal action. For example, a covered entity with poor facility access controls and procedures would be susceptible to hacking of its databases. A provider with appropriate security protections already in place would only need to ensure that the protections are documented and are reassessed periodically to ensure that they continue to be appropriate and are actually being implemented. Our decision to classify many implementation specifications as addressable, rather than mandatory, provides even more flexibility to covered entities to develop cost-effective solutions. We believe that insurers who already have effective security programs in place will have met many of the requirements of this regulation.

c. Comment: One commenter believes the rule is arbitrary and capricious in its requirements without any justification that they will significantly improve the security of medical records and with the likelihood that their implementation may actually increase the vulnerability of the data. The commenter noted that the data backup requirements increase access to data and that security awareness training provides more information to employees.

Response: The standards are based on generally accepted security procedures, existing industry standards and guidelines, and recommendations contained in the National Research Council's 1997 report *For The Record: Protecting Electronic Health Information,* Chapter 6. We also consulted extensively with experts in the field of security throughout the health care industry. The standards are consistent with generally accepted security principles and practices that are already in widespread use.

Data backup need not result in increased access to that data. Backups should be stored in a secure location with controlled access. The appropriate secure location and access control will vary, based upon the security needs of the covered entity. For example, a procedure as simple as locking backup diskettes in a safe place and restricting who has access to the key may be suitable for one entity, whereas another may need to store backed-up information off-site in a secure computer facility. The information provided in security awareness training heightens awareness of security anomalies and helps to prevent security incidents.

d. Comment: Several commenters suggested that the proposed rule appears to reflect the Medicare program's perspective on security risks and solutions, and that it should be noted that not all industry segments share all the same risks as Medicare. One commenter stated that as future proposed rules are drafted, we should solicit input from those most significantly affected, for example, providers, plans, and clearinghouses.

Others stated that Medicaid agencies were not sufficiently involved in the discussions and debate. Still another stated that States would be unable to perform some basic business functions if all the standards are not designed to meet their needs.

Response: We believe that the standards are consistent with common industry practices and equitable, and that there has been adequate consultation with interested parties in the development of the standards. These standards are the result of an intensive process of public consultation. We consulted with the National Uniform Billing Committee, the National Uniform Claim Committee, the American Dental Association, and the Workgroup for Electronic Data Interchange, in the course of developing the proposed rule. Those organizations were specifically named in the Act to advise the Secretary, and their membership is drawn from the full spectrum of industry segments. In addition, the National Committee on Vital and Health Statistics (NCVHS), an independent advisory group to the Secretary, held numerous public hearings to obtain the views of interested parties. Again, many segments of the health care industry, including provider groups, health plans, clearinghouses, vendors, and government programs participated actively. The NCVHS developed recommendations to the Secretary, which were relied upon as we developed the proposed rule. Finally, we note that the opportunity to comment was available to all during the public comment period.

e. Comment: One commenter stated that there is a need to ensure the confidentiality of risk analysis information that may contain sensitive information.

Response: The information included in a risk analysis would not be subject to the security standards if it does not include electronic protected health information. We agree that risk analysis data could contain sensitive information, just as other business information can be sensitive. Covered entities may wish to develop their own business rules regarding access to and protections for risk analysis data.

f. Comment: One commenter expressed concern over the statement in the preamble of the proposed rule (63 FR 43250) that read: "No one item is considered to be more important than another." The commenter suggested that security management should be viewed as most critical and perhaps what forms the foundation for all other security actions.

Response: The majority of comments received on this subject requested that we prioritize the standards. In response, we have regrouped the standards and implementation specifications in what we believe is a logical order within each of three categories: "Administrative safeguards," "Physical safeguards," and "Technical safeguards." In this final rule, we order the standards in such a way that the "Security management process" is listed first under the "Administrative safeguards" section, as we believe this forms the foundation on which all of the other standards depend. The determination of the specific security measures to be implemented to comply with the standards will, in large part, be dependent upon completion of the implementation specifications within the security management process standard (see § 164.308(a)(1)). We emphasize, however, that an entity implementing these standards may choose to implement them in any order, as long as the standards are met.

g. Comment: One commenter stated that there is a need for requirements concerning organizational practices (for example, education, training, and security and confidentiality policies), as well as technical practices and procedures.

Response: We agree. Section 164.308 of this final rule describes administrative safeguards that address these topics. Section 164.308 requires covered entities to implement standards and required implementation specifications, as well as consider and implement, when appropriate and reasonable, addressable implementation specifications. For example, the security management process standard requires implementation of a risk analysis, risk management, a sanction policy, and an information system activity review. The information access management standard requires consideration, and implementation where appropriate and reasonable, of access authorization and access establishment and modification policies and procedures. Other areas addressed are assigned security responsibility, workforce security, security awareness and training, security incident procedures, contingency planning, business associate contracts, and evaluation.

h. Comment: One commenter stated that internal and external security requirements should be separated and dealt with independently.

Response: The presentation of the standards within this final rule could have been structured in numerous ways, including by addressing separate internal and external security standards. We chose the current structure as we considered it a logical breakout for purposes of display within this final rule. Under our structure a covered entity may apply a given standard to internal activities and to external activities. Had we displayed separately the standards for internal security and the standards for external security, we would have needed to describe a number of the standards twice, as many apply to both internal and external security. However, a given entity may address the standards in whatever order it chooses, as long as the standards are met.

i. Comment: Two commenters stated that the standards identified in Addendum 3 of the proposed rule may not all have matured to implementation readiness.

Response: Addendum 3 of the proposed rule cross-referred individual requirements on the matrix to existing industry standards of varying levels of maturity. Addendum 3 was intended to show what we evaluated in searching for existing industry standards that could be adopted on a national level. No one standard was found to be comprehensive enough to be adopted, and none were proposed as the standards to be met under the Security Rule.

j. Comment: One commenter suggested we include a revised preamble in the final publication. Another questioned how clarification of points in the preamble will be handled if the preamble is not part of the final regulation.

Response: Preambles to proposed rules are not republished in the final rule. The preamble in this final rule contains summaries of the information presented in the preamble of the proposed rule, summaries of the comments received during the public comment period, and responses to questions and concerns raised in those comments and a summary of changes made. Additional clarification will be provided by HHS on an ongoing basis through written documents and postings on HHS's websites.

k. Comment: One commenter asked that we clarify that no third party can require implementation of more security features than are required in the final rule, for example, a third party could not require encryption but may choose to accept it if the other party so desires.

Response: The security standards establish a minimum level of security to be met by covered entities. It is not our intent to limit the level of security that may be agreed to between trading partners or others above this floor.

l. Comment: One commenter asked how privacy legislation would affect these rules. The commenter inquired whether covered entities will have to reassess and revise actions already taken in the spirit of compliance with the security regulations.

Response: We cannot predict if or how future legislation may affect the rules below. At present, the privacy standards at subpart E of 42 CFR part 164 have been adopted, and this final rule is compatible with them.

m. Comment: One commenter stated that a data classification policy, that is a method of assigning sensitivity ratings to specific pieces of data, should be part of the final regulations.

Response: We did not adopt such a policy because this final rule requires a floor of protection of all electronic protected health information. A covered entity has the option to exceed this floor. The sensitivity of information, the risks to and vulnerabilities of electronic protected health information and the means that should be employed to protect it are business determinations and decisions to be made by each covered entity.

n. Comment: One commenter stated that this proposed rule conflicts with previously stated rules that acceptable "standards" must have been developed by ANSI-recognized Standards Development Organizations (SDOs).

Response: In general, HHS is required to adopt standards developed by ANSI-accredited SDOs when such standards exist. The currently existing security standards developed by ANSI-recognized SDOs are targeted to specific technologies and/or activities. No existing security standard, or group of standards, is technology-neutral, scaleable to the extent required by HIPAA, and broad enough to be adopted in this final rule. Therefore, this final rule adopts standards under section 1172(c)(2)(B) of the Act, which permits us to develop standards when no industry standards exist.

o. Comment: One commenter stated that this regulation goes beyond the scope of the law, unjustifiably extending into business practices, employee policies, and facility security.

Response: We do not believe that this regulation goes beyond the scope of the law. The law requires HHS to adopt standards for reasonable and appropriate security safeguards concerning such matters as compliance by the officers and employees of covered entities, protection against reasonably anticipated unauthorized uses and disclosures of health information, and so on. Such standards will inevitably address the areas the commenter pointed to.

The intent of this regulation is to provide standards for the protection of electronic protected health information in accordance with the Act. In order to do this, covered entities are required to implement administrative, physical, and technical safeguards. Those entities must ensure that data are protected, to the extent feasible, from inappropriate access, modification, dissemination, and destruction. As noted above, however, this final rule has been modified to increase flexibility as to how this protection is accomplished.

p. Comment: One commenter stated that all sections regarding confidentiality and privacy should be removed, since they do not belong in this regulation.

Response: As the discussion in section III.A above of this final rule makes clear, the privacy and security standards are very closely related. Section 1173(d)(2) of the Act specifically mentions "confidentiality" and authorizes uses and disclosures of information as part of what security safeguards must address. Thus, we cannot omit all references to confidentiality and privacy in discussions of the security standards.

However, we have relocated material that relates to both security and privacy (including definitions) to the general section of part 164.

q. Comment: One commenter asked that data retention be addressed more specifically, since this will become a significant issue over time. It is recommended that a national work group be convened to address this issue.

Response: The commenter's concern is noted. While the documentation relating to Security Rule implementation must be retained for a period of 6 years (see § 164.316(b)(2)), it is not within the scope of this final rule to address data retention time frames for administrative or clinical records.

r. Comment: One commenter stated that requiring provider practices to develop policies, procedures, and training programs and to implement record keeping and documentation systems would be tremendously resource-intensive and increase the costs of health care.

Response: We expect that many of the standards of this final rule are already being met in one form or another by covered entities. For example, as part of normal business operations, health care providers already take measures to protect the health information in their keeping. Health care providers already keep records, train their employees, and require employees to follow office policies and procedures. Similarly, health plans are already frequently required by State law to keep information confidential. While revisions to a practice's or plan's current activities may be necessary, the development of entirely new systems or procedures may not be necessary.

s. Comment: One commenter stated that there is no system for which risk has been eliminated and expressed concern over phrases such as covered entities must "assure that electronic health information pertaining to an individual remains secure."

Response: We agree with the commenter that there is no such thing as a totally secure system that carries no risks to security. Furthermore, we believe the Congress' intent in the use of the word "ensure" in section 1173(d) of the Act was to set an exceptionally high goal for the security of electronic protected health information. However, we note that the Congress also recognized that some trade-offs would be necessary, and that "ensuring" protection did not mean providing protection, no matter how expensive. See section 1173(d)(1)(A)(ii) of the Act. Therefore, when we state that a covered entity must ensure the safety of the information in its keeping, we intend that a covered entity take steps, to the best of its ability, to protect that information. This will involve establishing a balance between the information's identifiable risks and vulnerabilities, and the cost of various protective measures, and will also be dependent upon the size, complexity, and capabilities of the covered entity, as provided in § 164.306(b).

E. Administrative Safeguards (§ 164.308)

We proposed that measures taken to comply with the rule be appropriate to protect the health information in a covered entity's care. Most importantly, we proposed to require that both the measures taken and documentation of those measures be kept current, that is, reviewed and updated periodically to continue appropriately to protect the health information in the care of covered entities. We would have required the documentation to be made available to those individuals responsible for implementing the procedure.

We proposed a number of administrative requirements and supporting implementation features, and required documentation for those administrative requirements and implementation features.

In this final rule, we have placed these administrative standards in § 164.308. We have reordered them, deleted much of the detail of the proposed requirements, as discussed below, and omitted two of the proposed sets of requirements (system configuration requirements and a requirement for a formal mechanism for processing records) as discussed in paragraph 10 of the discussion of § 164.308 of section III.E. of this preamble. Otherwise, the basic elements of the administrative safeguards are adopted in this final rule as proposed.

1. Security Management Process (§ 164.308(a)(1)(i)). We proposed the establishment of a formal security management process to involve the creation, administration, and oversight of policies to address the full range of security issues and to ensure the prevention, detection, containment, and correction of security violations. This process would include implementation features consisting of a risk analysis, risk management, and sanction and security policies.

We also proposed, in a separate requirement under administrative procedures, an internal audit, which would be an in-house review of the records of system activity (for example, logins, file accesses, and security incidents) maintained by an entity.

In this final rule, risk analysis, risk management, and sanction policy are adopted as required implementation specifications although some of the details are changed, and the proposed internal audit requirement has been renamed as "information system activity review" and incorporated here as an additional implementation specification.

a. Comment: Three commenters asked that this requirement be deleted. Two commenters cited this requirement as a possible burden. Several commenters asked that the implementation features be made optional.

Response: This standard and its component implementation specifications form the foundation upon which an entity's necessary security activities are built. See NIST SP 800–30, "Risk Management Guide for Information Technology Systems," chapters 3 and 4, January 2002. An entity must identify the risks to and vulnerabilities of the information in its care before it can take effective steps to eliminate or minimize those risks and vulnerabilities. Some form of sanction or punishment activity must be instituted for noncompliance. Indeed, we question how the statutory requirement for safeguards "to ensure compliance * * * by a [covered entity's] officers and employees" could be met without a requirement for a sanction policy. See section 1176(d)(2)(C) of the Act. Accordingly, implementation of these specifications remains mandatory. However, it is important to note that covered entities have the flexibility to implement the standard in a manner consistent with numerous factors, including such things as, but not limited to, their size, degree of risk, and environment. We have deleted the implementation specification calling for an organizational security policy, as it duplicated requirements of the security management and training standard.

We note that the implementation specification for a risk analysis at § 164.308(a)(1)(ii) (A) does not specifically require that a covered entity perform a risk analysis often enough to ensure that its security measures are adequate to provide the level of security required by § 164.306(a). In the proposed rule, an assurance of adequate security was framed as a requirement to keep security measures "current." We continue to believe that security measures must remain current, and have added regulatory language in § 164.306(e) as a more precise way of communicating that security measures in general that must be periodically reassessed and updated as needed.

The risk analysis implementation specification contains other terms that merit explanation. Under § 164.308(a)(1)(ii)(A), the risk analysis must look at risks to the covered entity's electronic protected health information. A thorough and accurate risk analysis would consider "all relevant losses" that would be expected if the security measures were not in place. "Relevant losses" would include losses caused by unauthorized uses and disclosures and loss of data integrity that would be expected to occur absent the security measures.

b. Comment: Relative to the development of an entity's sanction policy, one commenter asked that we describe the sanction penalties for breach of security. Another suggested establishment of a standard to which one's conduct could be held and adoption of mitigating circumstances so that the fact that a person acted in good faith would be a factor that could be used to reduce or otherwise minimize any sanction imposed. Another commenter suggested sanction activities not be implemented before the full implementation and testing of all electronic transaction standards.

Response: The sanction policy is a required implementation specification because— (1) the statute requires covered entities to have safeguards to ensure compliance by officers and employees; (2) a negative consequence to noncompliance enhances the likelihood of compliance; and (3) sanction policies are recognized as a usual and necessary component of an adequate security program. The type and severity of sanctions imposed, and for what causes, must be determined by each covered entity based upon its security policy and the relative severity of the violation.

c. *Comment:* Commenters requested the definitions of "risk analysis" and "breach."

Response: "Risk analysis" is defined and described in the specification of the security management process standard, and is discussed in the preamble discussion of § 164.308(a) (1)(ii)(A) of this final rule. The term breach is no longer used and is, therefore, not defined.

d. *Comment:* One commenter asked whether all health information is considered equally "sensitive," the thought being that, in determining risk, an entity may consider the loss of a smaller amount of extraordinarily sensitive data to be more significant than the loss of a larger amount of routinely collected data. The commenter stated that common reasoning would suggest that the smaller amount of data would be considered more sensitive.

Response: All electronic protected health information must be protected at least to the degree provided by these standards. If an entity desires to protect the information to a greater degree than the risk analysis would indicate, it is free to do so.

e. *Comment:* One commenter asked that we add "threat assessment" to this requirement.

Response: We have not done this because we view threat assessment as an inherent part of a risk analysis; adding it would be redundant.

f. *Comment:* We proposed a requirement for internal audit, the inhouse review of the records of system activity (for example, logins, file accesses, and security incidents) maintained by an entity. Several commenters wanted this requirement deleted. One suggested the audit trail requirement should not be mandatory, while another stated that internal audits would be unnecessary if physical security requirements are implemented.

A number of commenters asked that we clarify the nature and scope of what an internal audit covers and what the audit time frame should be. Several commenters offered further detail concerning what should and should not be required in an internal audit for security purposes. One commenter stated that ongoing intrusion detection should be included in this requirement. Another wanted us to specify the retention times for archived audit logs.

Several commenters had difficulty with the term "audit" and suggested we change the title of the requirement to "logging and violation monitoring."

A number of commenters stated this requirement could result in an undue burden and would be economically unfeasible.

Response: Our intent for this requirement was to promote the periodic review of an entity's internal security controls, for example, logs, access reports, and incident tracking. The extent, frequency, and nature of the reviews would be determined by the covered entity's security environment. The term "internal audit" apparently, based on the comments received, has certain rigid formal connotations we did not intend. We agree that the implementation of formal internal audits could prove burdensome or even unfeasible, to some covered entities due to the cost and effort involved. However, we do not want to overlook the value of internal reviews. Based on our review of the comments and the text to which they refer, it is clear that this requirement should be renamed for clarity and that it should actually be an implementation specification of the security management process rather than an independent standard. We accordingly remove "internal audit" as a separate requirement and add "information system activity review" under the security management process standard as a mandatory implementation specification.

2. *Assigned Security Responsibility (§ 164.308(a)(2)).* We proposed that the responsibility for security be assigned to a specific individual or organization to provide an organizational focus and importance to security, and that the assignment be documented. Responsibilities would include the management and supervision of (1) the use of security measures to protect data, and (2) the conduct of personnel in relation to the protection of data.

In this final rule, we clarify that the final responsibility for a covered entity's security must be assigned to one official. The requirement for documentation is retained, but is made part of § 164.316 below. This policy is consistent with the analogous policy in the Privacy Rule, at 45 CFR 164.530(a), and the same considerations apply. See 65 FR 82744 through 87445. The same person could fill the role for both security and privacy.

a. Comment: Commenters were concerned that delegation of assigned security responsibility, especially in large organizations, needs to be to more than a single individual. Commenters believe that a large health organization's security concerns would likely cross many departmental boundaries requiring group responsibility.

Response: The assigned security responsibility standard adopted in this final rule specifies that final security responsibility must rest with one individual to ensure accountability within each covered entity. More than one individual may be given specific security responsibilities, especially within a large organization, but a single individual must be designated as having the overall final responsibility for the security of the entity's electronic protected health information. This decision also aligns this rule with the final Privacy Rule provisions concerning the Privacy Official.

b. Comment: One commenter disagreed with placing assigned security responsibility as part of physical safeguards. The commenter suggested that assigned security responsibility should be included under the Administrative Procedures.

Response: Upon review of the matrix and regulations text, we agree with the commenter, because this requirement involves an administrative decision at the highest levels of who should be responsible for ensuring security measures are implemented and maintained. Assigned security responsibility has been removed from "Physical safeguards" and is now located under "Administrative safeguards" at § 164.308.

3. *Workforce Security (§ 164.308(a)(3)(i)).* We proposed implementation of a number of features for personnel security, including ensuring that maintenance personnel are supervised by a knowledgeable person, maintaining a record of access authorizations, ensuring that operating and maintenance personnel have proper access authorization, establishing personnel clearance procedures, establishing and maintaining personnel security policies and procedures, and ensuring that system users have proper training.

In this final rule, to provide clarification and reduce duplication, we have combined the "Assure supervision of maintenance personnel by authorized, knowledgeable person" implementation feature and the "Operating, and in some cases, maintenance personnel have proper access authorization" feature into one addressable implementation specification titled "Authorization and/or supervision."

In a related, but separate, requirement entitled "Termination procedures," we proposed implementation features for the ending of an employee's employment or an internal or external user's access. These features would include things such as changing combination locks, removal from access lists, removal of user account(s), and the turning in of keys, tokens, or cards that allow access.

In this final rule, "Termination procedures" has been made an addressable implementation specification under "Workforce security." This is addressable because in certain circumstances, for example, a solo physician practice whose staff consists only of the physician's spouse, formal procedures may not be necessary.

The proposed "Personnel security policy/procedure" and "record of access authorizations" implementation features have been removed from this final rule, as they have been determined to be redundant. Implementation of the balance of the "Workforce security" implementation specifications and the other standards contained within this final rule will result in assurance that all personnel with access to electronic protected health information have the required access authority as well as appropriate clearances.

a. Comment: The majority of comments concerned the supervision of maintenance personnel by an authorized knowledgeable person. Commenters stated this would not be feasible in smaller settings. For example, the availability of technically knowledgeable persons to ensure this supervision would be an issue. We were asked to either reword this implementation feature or delete it.

Response: We agree that a "knowledgeable" person may not be available to supervise maintenance personnel. We have accordingly modified this implementation specification so that, in this final rule, we are adopting an addressable implementation specification titled, "Authorization and/or supervision," requiring that workforce members, for example, operations and maintenance personnel, must either be supervised or have authorization when working with electronic protected health information or in locations where it resides (see § 164.308(a)(3)(ii)(A)). Entities can decide on the feasibility of meeting this specification based on their risk analysis.

b. Comment: The second largest group of comments requested assurance that, with regard to the proposed "Personnel clearance procedure" implementation feature, having appropriate clearances does not mean performing background checks on everyone. We were asked to delete references to "clearance" and use the term "authorization" in its place.

Response: We agree with the commenters concerning background checks. This feature was not intended to be interpreted as an absolute requirement for background checks. We retain the use of the term "clearance," however, because we believe that it more accurately conveys the screening process intended than does the term "authorization." We have attempted to clarify our intent in the language of § 164.308(a)(3)(ii)(B), which now reads, "Implement procedures to determine that the access of a workforce member to electronic protected health information is appropriate." The need for and extent of a screening process is normally based on an assessment of risk, cost, benefit, and feasibility as well as other protective measures in place. Effective personnel screening processes may be applied in a way to allow a range of implementation, from minimal procedures to more stringent procedures based on the risk analysis performed by the covered entity. So long as the standard is met and the underlying standard of § 164.306(a) is met, covered entities have choices in how they meet these standards. To clarify the intent of this provision, we retitle the implementation specification "Workforce clearance procedure."

c. Comment: One commenter asked that we expand the implementation features to include the identification of the restrictions that should be placed on members of the workforce and others.

Response: We have not adopted this comment in the interest of maintaining flexibility as discussed in § 164.306. Restrictions would be dependent upon job responsibilities, the amount and type of supervision required and other factors. We note that a covered entity should consider in this regard the applicable requirements of the Privacy Rule (*see,* for example, § 164.514(d)(2) (relating to minimum necessary requirements), and § 164.530(c) (relating to safeguards).

d. Comment: One commenter believes that the proposed "Personnel security" requirement was reasonable, since an administrative determination of trustworthiness is needed before allowing access to sensitive information. Two commenters asked that we delete the requirement entirely. A number of commenters requested that we delete the implementation features. Another commenter stated that all the implementation features may not be applicable or even appropriate to a given entity and should be so qualified.

Response: While we do not believe this requirement should be eliminated, we agree that all the implementation specifications may not be applicable or even appropriate to a given entity. For example, a personal clearance may not be reasonable or appropriate for a small provider whose only assistant is his or her spouse. The implementation specifications are not mandatory, but must be addressed. This final rule has been changed to reflect this approach (see § 164.308(a)(3)(ii)(B)).

e. Comment: The majority of commenters on the "Termination procedures" requirement asked that it be made optional, stating that it may not be applicable or even appropriate in all circumstances and should be so qualified or posed as guidelines. A number of commenters stated that the requirement should be deleted. One commenter stated that much of the material covered under the "Termination procedures" requirement is already covered in "Information access control." A number of commenters stated that this requirement was too detailed and some of the requirements excessive.

Response: Based upon the comments received, we agree that termination procedures should not be a separate standard; however, consideration of termination procedures remains relevant for any covered entity with employees, because of the risks associated with the potential for unauthorized acts by former employees, such as acts of retribution or use of proprietary information for personal gain. We further agree with the reasoning of the commenters who asked that these procedures be made optional; therefore, "Termination procedures" is now reflected in this final rule as an addressable implementation specification. We also removed reference to all specific termination activities, for example, changing locks, because, although the activities may be considered appropriate for some covered entities, they may not be reasonable for others.

f. Comment: One commenter asked whether human resource employee termination policies and procedures must be documented to show the types of security breaches that would result in termination.

Response: Policies and procedures implemented to adhere to this standard must be documented (see § 164.316 below). The purpose of termination procedure documentation under this implementation specification is not to detail when or under which circumstances an employee should be terminated. This information would more appropriately be part of the entity's sanction policy. The purpose of termination procedure documentation is to ensure that termination procedures include security-unique actions to be followed, for example, revoking passwords and retrieving keys when a termination occurs.

4.   *Information Access Management (§ 164.308(a)(4))*.   We proposed an "information access control" requirement for establishment and maintenance of formal, documented policies and procedures defining levels of access for all personnel authorized to access health information, and how access is granted and modified. In § 164.308(a)(4)(ii)(B) and (C) below, the proposed implementation features are made addressable specifications. We have added in § 164.308(a)(4)(ii)(A), a required implementation specification to isolate health care clearinghouse functions to address the provisions of section 1173(d)(1)(B) of the Act which related to this area.

   a.   Comment: One commenter asked that the requirement be deleted, expressing the opinion that this requirement goes beyond "reasonable boundaries" into regulating common business practices. In contrast, another asked that we expand this requirement to identify participating parties and access privileges relative to specific data elements.

   Response: We disagree that this requirement improperly imposes upon business functions. Restricting access to those persons and entities with a need for access is a basic tenet of security. By this mechanism, the risk of inappropriate disclosure, alteration, or destruction of information is minimized. We cannot, however, specifically identify participating parties and access privileges relative to data elements within this regulation. These will vary depending upon the entity, the needs within the user community, the system in which the data resides, and the specific data being accessed. This standard is consistent with § 164.514(d) in the Privacy Rule (minimum necessary requirements for use and disclosure of protected health information), and is, therefore, being retained.

   b.   Comment: Several commenters asked that we not mandate the implementation features, but leave them as optional, a suggested means of compliance. The commenters noted that this might make the rules more scalable and flexible, since this approach would allow providers to implement safeguards that best addressed their needs. Along this line, one commenter expressed the belief that each organization should implement features deemed necessary based on its own risk assessment.

   Response: While the information access management standard in this final rule must be met, we agree that the implementation specifications at § 164.308(a)(4)(ii)(B) and (C) should not be mandated but posed as a suggested means of compliance, which must be addressed. These specifications may not be applicable to all entities based on their size and degree of automation. A fully automated covered entity spanning multiple locations and involving hundreds of employees may determine it has a need to adopt a formal policy for access authorization, while a small provider may decide that a desktop standard operating procedure will meet the specifications. The final rule has been revised accordingly.

   c.   Comment: Clarification was requested concerning the meaning of "formal."

   Response: The word "formal" has caused considerable concern among commenters, as it was thought "formal" carried the connotation of a rigidly defined structure similar to what might be found in the Department of Defense instructions. As used in the proposed rule, this word was not intended to convey such a strict structure. Rather, it was meant to convey that documentation should be an official organizational statement as opposed to word-of-mouth or cryptic notes scratched on a notepad. While documentation is still required (see § 164.316), to alleviate confusion, the word "formal" has been deleted.

d. *Comment:* One commenter asked that we clarify that this requirement relates to both the establishment of policies for the access control function and to access control (the implementation of those policies).

Response: "Information access management" does address both the establishment of access control policies and their implementation. We use the term "implement" to clarify that the procedures must be in use, and we believe that the requirement to implement policies and procedures requires, as an antecedent condition, the establishment or adaptation of those policies and procedures.

5. *Security Awareness and Training (§ 164.308(a)(5)(i)).* We proposed, under the requirement "Training," that security training be required for all staff, including management. Training would include awareness training for all personnel, periodic security reminders, user education concerning virus protection, user education in the importance of monitoring login success/failure, and how to report discrepancies, and user education in password management.

In this final rule, we adopt this proposed requirement in modified form. For the standard "Security awareness and training," in § 164.308(a)(5), we require training of the workforce as reasonable and appropriate to carry out their functions in the facility. All proposed training features have been combined as implementation specifications under this standard. Specific implementation specifications relative to content are addressable. The "Virus protection" implementation feature has been renamed "protection from malicious software," because we did not intend by the nomenclature to exclude coverage of malicious acts that might not come within the prior term, such as worms.

a. *Comment:* One commenter believes that security awareness training for all system users would be too difficult to do in a large organization.

Response: We disagree with the commenter. Security awareness training is a critical activity, regardless of an organization's size. This feature would typically become part of an entity's overall training program (which would include privacy and other information technology items as well). For example, the Government Information Systems Reform ACT (GISRA) of 2000 requires security awareness training as part of Federal agencies' information security programs, including Federal covered entities, such as the Medicare program. In addition, National Institute of Standards and Technology (NIST) SP 800–16, *Information Technology Security Training Requirements, A role and performance base model, April 1998,* provides an excellent source of information and guidance on this subject and is targeted at industry as well as government activities. We also note that covered entities must have discretion in how they implement the requirement, so they can incorporate this training in other existing activities. One approach would be to require this training as part of employee orientation.

b. *Comment:* A number of commenters asked that this requirement be made optional or used as a guideline only. Several commenters stated that this requirement is too specific and is burdensome. Several asked that the implementation features be removed.

Several others stated that this requirement is not appropriate for agents or contractors. One commenter asked how to apply this requirement to outsiders having access to data. Another asked if this requirement included all subcontractor staff. Others stated that contracts, signed by entities such as consultants, that address training should be sufficient.

Response: Security training remains a requirement because of its criticality; however, we have revised the implementation specifications to indicate that the amount and type of training needed will be dependent upon an entity's configuration and security risks. Business associates must be made aware of security policies and procedures, whether through contract language or other means. Covered entities are not required to provide training to business associates or anyone else that is not a member of their workforce.

c. Comment: Several commenters questioned why security awareness training appeared in two places, under "Physical safeguards" as well as "Administrative safeguards." Others questioned the appropriateness of security awareness training under "Physical safeguards."

Response: We reviewed the definitions of the proposed "Awareness training for all personnel" ("Administrative safeguards") implementation feature and the proposed "Security awareness training" ("Physical safeguards") requirement. We agree that, to avoid confusion and eliminate redundancy, security awareness and training should appear in only one place. We believe the appropriate location for it is under "Administrative safeguards," as such training is essentially an administrative function.

d. Comment: Several commenters objected to the blanket requirement for security awareness training of individuals who may be on site for a limited time period (for example, a single day).

Response: Each individual who has access to electronic protected health information must be aware of the appropriate security measures to reduce the risk of improper access, uses, and disclosures. This requirement does not mean lengthy training is appropriate in every instance; there are alternative methods to inform individuals of security responsibilities (for example, provisions of pamphlets or copies of security policies, and procedures).

e. Comment: One commenter asked that "training" be changed to "orientation."

Response: We believe the term "training," as presented within this rule is the more appropriate term. The rule does not contemplate a one-time type of activity as connoted by "orientation," but rather an on-going, evolving process as an entity's security needs and procedures change.

f. Comment: Several commenters asked how often training should be conducted and asked for a definition of "periodic," as it appears in the proposed implementation feature "Periodic security reminders." One asked if the training should be tailored to job need.

Response: Amount and timing of training should be determined by each covered entity; training should be an ongoing, evolving process in response to environmental and operational changes affecting the security of electronic protected health information. While initial training must be carried out by the compliance date, we provide flexibility for covered entities to construct training programs. Training can be tailored to job need if the covered entity so desires.

6. ***Security Incident Procedures (§ 164.308(a)(6)).*** We proposed a requirement for implementation of accurate and current security incident procedures: formal, documented report and response procedures so that security violations would be reported and handled promptly. We adopt this standard in the final rule, along with an implementation specification for response and reporting, since documenting and reporting incidents, as well as responding to incidents are an integral part of a security program.

a. *Comment:* Several commenters asked that we further define the scope of a breach of security. Along this same line, another commenter stated that the proposed security incident procedures were too vague as stated. We were asked to specify what a security incident would be, what the internal chain for reporting procedures would be, and what should be included in the documentation (for example, hardware/software, personnel responses).

Response: We define a security incident in § 164.304. Whether a specific action would be considered a security incident, the specific process of documenting incidents, what information should be contained in the documentation, and what the appropriate response should be will be dependent upon an entity's environment and the information involved. An entity should be able to rely upon the information gathered in complying with the other security standards, for example, its risk assessment and risk management procedures and the privacy standards, to determine what constitutes a security incident in the context of its business operations.

b. *Comment:* One commenter asked what types of incidents must be reported to outside entities. Another commented that we clarify that incident reporting is internal.

Response: Internal reporting is an inherent part of security incident procedures. This regulation does not specifically require any incident reporting to outside entities. External incident reporting is dependent upon business and legal considerations.

c. *Comment:* One commenter stated that network activity should be included here.

Response: We see no reason to exclude network activity under this requirement. Improper network activity should be treated as a security incident, because, by definition, it represents an improper instance of access to or use of information.

d. *Comment:* One commenter stated that this requirement should address suspected misuse also.

Response: We agree that security incidents include misuse of data; therefore, this requirement is addressed.

e. *Comment:* Several commenters asked that this requirement be deleted. One commenter asked that we delete the implementation features.

Response: As indicated above, we have adopted the proposed standard and combined the implementation specifications.

*7.   **Contingency Plan (§ 164.308(a)(7)(i)).***   We proposed that a contingency plan must be in effect for responding to system emergencies. The plan would include an applications and data criticality analysis, a data backup plan, a disaster recovery plan, an emergency mode operation plan, and testing and revision procedures.

In this final rule, we make the implementation specifications for testing and revision procedures and an applications and data criticality analysis addressable, but otherwise require that the contingency features proposed be met.

a.   Comment: Several commenters suggested the contingency plan requirement be deleted. Several thought that this aspect of the proposed regulation went beyond its intended scope. Another believed that more discussion and development is needed before developing regulatory guidance on contingency plans. Others wanted this to be an optional requirement. In contrast, one commenter requested more guidance concerning contingency planning. Still others wanted to require that a contingency plan be in place but stated that we should not regulate its contents. One comment stated that data backup, disaster recovery, and emergency mode operation should not be part of this requirement.

Response: A contingency plan is the only way to protect the availability, integrity, and security of data during unexpected negative events. Data are often most exposed in these events, since the usual security measures may be disabled, ignored, or not observed.

Each entity needs to determine its own risk in the event of an emergency that would result in a loss of operations. A contingency plan may involve highly complex processes in one processing site, or simple manual processes in another. The contents of any given contingency plan will depend upon the nature and configuration of the entity devising it.

While the contingency plan standard must be met, we agree that the proposed testing and revision implementation feature should be an addressable implementation specification in this final rule. Dependent upon the size, configuration, and environment of a given covered entity, the entity should decide if testing and revision of all parts of a contingency plan should be done or if there are more reasonable alternatives. The same is true for the proposed applications and data criticality analysis implementation feature. We have revised the final rule to reflect this approach.

b.   Comment: One commenter believed that adhering to this requirement could prove burdensome. Another stated that testing of certain parts of a contingency plan would be burdensome, and even infeasible, for smaller entities.

Response: Without contingency planning, a covered entity has no assurance that its critical data could survive an emergency situation. Recent events, such as September 11, 2001, illustrate the importance of such planning. Contingency planning will be scalable based upon, among other factors, office configuration, and risk assessment. However, in response to the scalability issue raised by the commenter, we have made the testing and revision implementation specification addressable (see § 164.308(a)(7)(ii)).

c.   Comment: Two commenters considered a 2-year implementation time frame for this requirement inadequate for large health plans. Another commenter stated that implementation of measures against natural disaster would be too big an issue for this regulation.

Response: The statute sets forth the compliance dates for the initial standards. The statute requires that compliance with initial standards is not later than 2 years after adoption of the standards for all covered entities except small health plans for which the compliance date is not later than 3 years after adoption.

The final rule calls for covered entities to consider how natural disasters could damage systems that contain electronic protected health information and develop policies and procedures for responding to such situations. We consider this to be a reasonable precautionary step to take since in many cases the risk would be deemed to be low.

d. Comment: A commenter requested clarification of the term "Emergency mode" with regard to the proposed "Emergency mode operation plan" implementation feature.

Response: We have clarified the "Emergency mode operations plan" to show that it only involves those critical business processes that must occur to protect the security of electronic protected health information during and immediately after a crisis situation.

8. Evaluation (§ 164.308(a)(8)). We proposed that certification would be required and could be performed internally or by an external accrediting agency. We solicited input on appropriate mechanisms to permit an independent assessment of compliance. We were particularly interested in input from those engaging in health care electronic data interchange (EDI), as well as independent certification and auditing organizations addressing issues of documentary evidence of steps taken for compliance; need for, or desirability of, independent verification, validation, and testing of system changes; and certifications required for off-the-shelf products used to meet the requirements of this regulation. We also solicited comments on the extent to which obtaining external certification would create an undue burden on small or rural providers.

In this final rule, we require covered entities to periodically conduct an evaluation of their security safeguards to demonstrate and document their compliance with the entity's security policy and the requirements of this subpart. Covered entities must assess the need for a new evaluation based on changes to their security environment since their last evaluation, for example, new technology adopted or responses to newly recognized risks to the security of their information.

a. Comment: We received several comments that certification should be performed externally. A larger group of commenters preferred self-certification. The majority of the comments, however, were to the effect that external certification should be encouraged but not mandated.

A number of commenters thought that mandating external certification would create an undue financial burden, regardless of the size of the entity being certified. One commenter stated that external certification would not place an undue burden on a small or rural provider.

Response: Evaluation by an external entity is a business decision to be left to each covered entity. Evaluation is required under § 164.308(a)(8), but a covered entity may comply with this standard either by using its own workforce or an external accreditation agency, which would be acting as a business associate. External evaluation may be too costly an option for small entities.

b. Comment: Several commenters stated that the certification should cover all components of the proposed rule, not just the information systems.

Response: We agree. We have revised this section to reflect that evaluation would be both technical and nontechnical components of security.

c. Comment: A number of commenters expressed a desire for the creation of certification guides or models to complement the rule.

Response: We agree that creation of compliance guidelines or models for different business environments would help in the implementation and evaluation of HIPAA security requirements and we encourage professional associations and others to do so. We may develop technical assistance materials, but do not intend to create certification criteria because we do not have the resources to address the large number of different business environments.

d.   Comment: Some commenters asked how certification is possible without specifying the level of risk that is permissible.

Response: The level of risk that is permissible is specified by § 164.306(a). How such risk is managed will be determined by a covered entity through its security risk analysis and the risk mitigation activities it implements in order to ensure that the level of security required by § 164.306 is provided.

e.   Comment: Several commenters requested creation of a list of Federally "certified" security software and off-the-shelf products. Several others stated that this request was not feasible. Regarding certification of off-the-shelf products, one commenter thought this should be encouraged, but not mandated; several thought this would be an impractical endeavor.

Response: While we will not assume the task of certifying software and off-the-shelf products for the reason described above, we have noted with interest that other Government agencies such as the National Institute of Standards and Technology (NIST) are working towards that end. The health care industry is encouraged to monitor the activity of NIST and provide comments and suggestions when requested (see *http:// www.niap.nist.gov.*).

f.   Comment: One commenter stated, "With HCFA's publishing of these HIPAA standards, and their desire to retain the final responsibility for determining violations and imposing penalties of the statute, it also seems appropriate for HCFA to also provide certifying services to ensure security compliance."

Response: In view of the enormous number and variety of covered entities, we believe that evaluation can best be handled through the marketplace, which can develop more usable and targeted evaluation instruments and processes.

8.   Business Associate Contracts or Other Arrangements (§ 164.308(b)(1)).   In the proposed rule § 142.308(a)(2) "Chain of trust" requirement, we proposed that covered entities be required to enter into a chain of trust partner agreement with their business partners, in which the partners would agree to electronically exchange data and protect the integrity, confidentiality, and availability of the data exchanged. This standard has been modified from the proposed requirement to reflect, in § 164.308(b)(1) "Business associate contracts and other arrangements," the business associate structure put in place by the Privacy Rule.

In this final rule, covered entities must enter into a contract or other arrangement with persons that meet the definition of business associate in § 160.103. The covered entity must obtain satisfactory assurances from the business associate that it will appropriately safeguard the information in accordance with these standards (see § 164.314(a)(1)).

The comments received on the proposed chain of trust partner agreements are discussed in section 2 "Business associate contracts and other arrangements" of the discussion of § 164.314 below.

9. *Proposed Requirements Not Adopted in This Final Rule*

a. Security Configuration Management. We proposed that an organization would be required to implement measures, practices, and procedures regarding security configuration management. They would be coordinated and integrated with other system configuration management practices for the security of information systems. These would include documentation, hardware and/or software installation and maintenance review and testing for security features, inventory procedures, security testing, and virus checking.

Comment: Several commenters asked that the entire requirement be deleted. Several others asked that the inventory and virus checking implementation features be removed as they believe those features are not germane to security configuration management. A number of commenters requested that security testing be deleted because this implementation feature is too detailed, unreasonable, impractical, and beyond the scope of the legislation. Others stated that the testing would be very complex and expensive. Others wanted more clarification of what we intend by security testing, and how much would be enough. A number of commenters asked that all of the implementation features be deleted. Others asked that the implementation features be made optional. Several commenters wanted to know the scope of organizational integration required. Several others asked if what we meant by Security Configuration Management was change or version control.

Response: Upon review, this requirement appears unnecessary because it is redundant of other requirements we are adopting in this rule. A covered entity will have addressed the activities described by the features under this proposed requirement by virtue of having implemented the risk analysis, risk management measures, sanction policies, and information systems criticality review called for under the security management process. The proposed documentation implementation feature has been made a separate standard (see § 164.316). As a result, the Security Configuration Management requirement is not adopted in this final rule.

b. Formal Mechanism for Processing Records. The proposed rule proposed requiring a formal mechanism for processing records, and documented policies and procedures for the routine and nonroutine receipt, manipulation, storage, dissemination, transmission, and/or disposal of health information. This requirement has not been adopted in the final rule.

Comment: Several commenters thought this requirement concerned the regulation of formal procedures for how an entity does business and stated that such procedures should not be regulated. Others asked for additional clarification of what is meant by this requirement. One commenter thought the requirement too ambiguous and asked for clarification as to whether we meant such things as "the proper handling of storage media, databases, transmissions," or "the clinical realm of processes."

Two commenters asked how extensive this requirement would be and whether systems' user manuals and policies and procedures for handling health information would suffice and what level of detail would be expected.

Several thought this requirement could result in a significant resource and monetary burden to develop and maintain formal procedures. Two asked for an explanation of the benefit to be derived from this requirement.

One asked that covered entities be required to document processes that create a security risk only and suggested that a risk assessment would determine the need for this documentation.

Response: We agree with the commenters that the standard is ambiguous, and upon review, is unnecessary because the remaining standards, for example, device and media controls, provide adequate safeguards. Accordingly, this requirement is not adopted in this final rule.

F.   Physical Safeguards (§ 164.310)

We proposed requirements and implementation features for documented physical safeguards to guard data integrity, confidentiality, and availability. We proposed to require safeguards in the following areas: Assigned security responsibility; media controls; physical access controls; policies and guidelines on workstation use; a secure workstation location; and security awareness training. A number of specific implementation features were proposed under the media controls and physical access controls requirements.

In § 164.310 of this final rule, most of the proposed implementation features are adopted as addressable implementation specifications. The proposed requirements for the assigned security responsibility and security awareness training requirements are relocated in § 164.308.

1.   General Comments

a.   Comment: Several commenters made suggestions to modify the language to more clearly describe "Physical safeguards."

Response: In response to comments, we have revised the definition of "Physical safeguards" to read as follows: "Physical safeguards are security measures to protect a covered entity's electronic information systems and related buildings and equipment, from natural and environmental hazards, and unauthorized intrusion."

b.   Comment: One commenter was concerned that electronic security systems could not be used in lieu of physical security systems.

Response: This final rule does not preclude the use of electronic security systems in lieu of, or in combination with, physical security systems to meet a "Physical safeguard" standard.

2.   Facility Access Controls (§ 164.310(a)(1)).   We proposed, under the "Physical access controls" requirement, formal, documented policies and procedures for limiting physical access to an entity while ensuring that properly authorized access is allowed. These controls would include the following implementation features: disaster recovery, emergency mode operation, equipment control (into and out of site), a facility security plan, procedures for verifying access authorizations before physical access, maintenance records, need-to-know procedures for personnel access, sign-in for visitors and escort, if appropriate, and testing and revision.

In § 164.310(a)(2) below, we combine and restate these as addressable implementation specifications. These are contingency operations, facility security plan, access control and validation procedures, and maintenance records.

a.   Comment: Many commenters were concerned because the proposed language would require implementation of all physical access control features. Other commenters were concerned that the language did not allow entities to use the results of their risk assessment and risk management process to arrive at the appropriate solutions for them.

Response: We agree that implementation of all implementation specifications may not be appropriate in all situations. While the facility access controls standard must be met, we agree that the implementation specifications should not be required in all circumstances, but should be addressable. In this final rule, all four implementation specifications are addressable.

We have also determined, based on "level of detail" comments requesting consolidation of the list of implementation features, that the proposed implementation feature "Equipment control (into and out of site)" was redundant. "Equipment control" is already covered under the "Device and media controls" standard at § 164.310(d)(1). Accordingly, we have eliminated it as a separate implementation specification.

b. Comment: One commenter raised the issue of a potential conflict of authority between those having access to the data and those responsible for checking and maintaining access controls.

Response: Any potential conflicts should be identified, addressed, and resolved in the policies and procedures developed according to the standards under § 164.308.

c. Comment: Several commenters questioned whether "Physical Access Controls" was a descriptive phrase to describe a technology to be used, or whether the phrase referred to a facility.

Response: We agree that the term "Physical" may be misleading; to remove any confusion, the requirement is reflected in this final rule as a standard titled "Facility access controls." We believe this is a more precise term to describe that the standard, and its associated implementation specifications, is applicable to an entity's business location or locations.

d. Comment: Several commenters requested that the disaster recovery and emergency mode operations features be moved to "Administrative safeguards." Other commenters recommended that disaster recovery and emergency mode operations should be replaced by, and included in, a "Contingency Operations" implementation feature.

Response: The "Administrative safeguards" section addresses the contingency planning that must be done to contend with emergency situations. The placement of the disaster recovery and emergency mode operations implementation specifications in the "Physical safeguards" section is also appropriate, however, because "Physical safeguards" defines the physical operations (processes) that provide access to the facility to implement the associated plans, developed under § 164.308. We agree, however, that the term "contingency operations" better describes, and would include, disaster recovery and emergency mode operations, and have modified the regulation text accordingly (see § 164.310(a)(1)).

e. Comment: Commenters were concerned about having to address in their facility security plan the exterior/ interior security of a building when they are one of many occupants rather than the sole occupant. Additional commenters were concerned that the responsibility for physical security of the building could not be delegated to a third party when the covered entity shares the building with other offices.

Response: The facility security plan is an addressable implementation specification. However, the covered entity retains responsibility for considering facility security even where it shares space within a building with other organizations. Facility security measures taken by a third party must be considered and documented in the covered entity's facility security plan, when appropriate.

3.   Workstation Use (§ 164.310(b)).   We proposed policy and guidelines on workstation use that included documented instructions/procedures delineating the proper functions to be performed and the manner in which those functions are to be performed (for example, logging off before leaving a workstation unattended) to maximize the security of health information. In this final rule, we adopt this standard.

Comment: One commenter was concerned most people may be misled by the use of "terminal" as an example in the definition of workstation. The concern was that the standard only addresses "fixed location devices," while in many instances the workstation has become a laptop computer.

Response: For clarity, we have added the definition of "workstation" to § 164.304 and deleted the word "terminal" from the description of workstation use in § 164.310(b).

4.   Workstation Security (§ 164.310(c)).   We proposed that each organization would be required to put in place physical safeguards to restrict access to information. In this final rule, we retain the general requirement for a secure workstation.

Comment: Comments were directed toward the example profiled in the definition of a secure workstation location. It was believed that what constitutes a secure workstation location must be dependent upon the entity's risk management process.

Response: We agree that what constitutes an appropriate solution to a covered entity's workstation security issues is dependent on the entity's risk analysis and risk management process. Because many commenters incorrectly interpreted the examples as the required and only solution for securing the workstation location, we have modified the regulations text description to generalize the requirement (see § 164.310(c)). Also, for clarity, the title "Secure workstation location" has been changed to "Workstation security" (see also the definition of "Workstation" at § 164.304).

5.   Device and Media Controls (§ 164.310(d)(1)).   We proposed that covered entities have media controls in the form of formal, documented policies and procedures that govern the receipt and removal of hardware and/or software (for example, diskettes and tapes) into and out of a facility. Implementation features would have included "Access control," "Accountability" (tracking mechanism), "Data backup," "Data storage," and "Disposal."
In this final rule, we adopt most of these provisions as addressable implementation specifications and add a specification for media re-use. We change the name from "Media controls" to "Device and media controls" to more clearly reflect that this standard concerns hardware as well as electronic media. The proposed "Access control" implementation feature has been removed, as it is addressed as part of other standards (see section III.C.12.c of this preamble).

a.   Comment: One commenter was concerned about the exclusion of removable media devices from examples of physical types of hardware and/or software.

Response: The media examples used were not intended to represent all possible physical types of hardware and/or software. Removable media devices, although not specifically listed, are not intended to be excluded.

b. Comment: Comments were made that the issue of equipment re-use or recycling of media containing mass storage was not addressed in "Media controls."

Response: We agree that equipment re-use or recycling should be addressed, since this equipment may contain electronic protected health information. The "Device and media controls" standard is accordingly expanded to include a required implementation specification that addresses the re-use of media (see § 164.310(d)(2)(ii)).

c. Comment: Several commenters asked for a definition of the term "facility," as used in the proposed "Media controls" requirement description. Commenters were unclear whether we were talking about a corporate entity or the physical plant.

Response: The term "facility" refers to the physical premises and the interior and exterior of a building(s). We have added this definition to § 164.304.

d. Comment: Several commenters believe the "Media controls" implementation features are too onerous and should be deleted.

Response: While the "Device and media controls" standard must be met, we believe, based upon further review, that implementation of all specifications would not be necessary in every situation, and might even be counter-productive in some situations. For example, small providers would be unlikely to be involved in large-scale moves of equipment that would require systematic tracking, unlike, for example, large health care providers or health plans. We have, therefore, reclassified the "Accountability and data backup" implementation specification as addressable to provide more flexibility in meeting the standard.

e. Comment: One commenter was concerned about the accountability impact of audit trails on system resources and the pace of system services.

Response: The proposed audit trail implementation feature appears as the addressable "Accountability" implementation specification. The name change better reflects the purpose and intended scope of the implementation specification. This implementation specification does not address audit trails within systems and/or software. Rather it requires a record of the actions of a person relative to the receipt and removal of hardware and/or software into and out of a facility that are traceable to that person. The impact of maintaining accountability on system resources and services will depend upon the complexity of the mechanism to establish accountability. For example, the appropriate mechanism for a given entity may be manual, such as receipt and removal restricted to specific persons, with logs kept. Maintaining accountability in such a fashion should have a minimal, if any, effect on system resources and services.

f. Comment: A commenter was concerned about the resource expenditure (system and fiscal) for total e-mail backup and wanted a clarification of the extensiveness of data backup.

Response: The data an entity needs to backup, and which operations should be used to carry out the backup, should be determined by the entity's risk analysis and risk management process. The data backup plan, which is part of the required contingency plan (see § 164.308(a)(7)(ii)(A)), should define exactly what information is needed to be retrievable to allow the entity to continue business "as usual" in the face of damage or destruction of data, hardware, or software. The extent to which e-mail backup would be needed would be determined through that analysis.

G. Technical Safeguards (§ 164.312)

We proposed five technical security services requirements with supporting implementation features: Access control; Audit controls; Authorization control; Data authentication; and Entity authentication. We also proposed specific technical security mechanisms for data transmitted over a communications network, Communications/network controls with supporting implementation features; Integrity controls; Message authentication; Access controls; Encryption; Alarm; Audit trails; Entity authentication; and Event reporting.

In this final rule, we consolidate these provisions into § 164.312. That section now includes standards regarding access controls, audit controls, integrity (previously titled data authentication), person or entity authentication, and transmission security. As discussed below, while certain implementation specifications are required, many of the proposed security implementation features are now addressable implementation specifications. The function of authorization control has been incorporated into the information access management standard under § 164.308, Administrative safeguards.

1. Access Control (§ 164.312(a)(1)). In the proposed rule, we proposed to require that the access controls requirement include features for emergency access procedures and provisions for context-based, role-based, and/or user-based access; we also proposed the optional use of encryption as a means of providing access control. In this final rule, we require unique user identification and provision for emergency access procedures, and retain encryption as an addressable implementation specification. We also make "Automatic logoff" an addressable implementation specification. "Automatic logoff" and "Unique user identification" were formerly implementation features under the proposed "Entity authentication" (see § 164.312(d)).

a. Comment: Some commenters believe that in specifying "Context," "Role," and "User" based controls, use of other controls would effectively be excluded, for example, "Partition rule-based access controls," and the development of new access control technology.

Response: We agree with the commenters that other types of access controls should be allowed. There was no intent to limit the implementation features to the named technologies and this final rule has been reworded to make it clear that use of any appropriate access control mechanism is allowed. Proposed implementation features titled "Context-based access," "Role-based access," and "User-based access" have been deleted and the access control standard at § 164.312(a)(1) states the general requirement.

b. Comment: A large number of comments were received objecting to the identification of "Automatic logoff" as a mandatory implementation feature. Generally the comments asked that we not be so specific and allow other forms of inactivity lockout, and that this type of feature be made optional, based more on the particular configuration in use and a risk assessment/analysis.

Response: We agree with the comments that mandating an automatic logoff is too specific. This final rule has been written to clarify that the proposed implementation feature of automatic logoff now appears as an addressable access control implementation specification and also permits the use of an equivalent measure.

c. Comment: We received comments asking that encryption be deleted as an implementation feature and stating that encryption is not required for "data at rest."

Response: The use of file encryption is an acceptable method of denying access to information in that file. Encryption provides confidentiality, which is a form of control. The use of encryption, for the purpose of access control of data at rest, should be based upon an entity's risk analysis. Therefore, encryption has been adopted as an addressable implementation specification in this final rule.

d. Comment: We received one comment stating that the proposed implementation feature "Procedure for emergency access," is not access control and recommending that emergency access be made a separate requirement.

Response: We believe that emergency access is a necessary part of access controls and, therefore, is properly a required implementation specification of the "Access controls" standard. Access controls will still be necessary under emergency conditions, although they may be very different from those used in normal operational circumstances. For example, in a situation when normal environmental systems, including electrical power, have been severely damaged or rendered inoperative due to a natural or man-made disaster, procedures should be established beforehand to provide guidance on possible ways to gain access to needed electronic protected health information.

2. *Audit Controls (§ 164.312(b)).* We proposed that audit control mechanisms be put in place to record and examine system activity. We adopt this requirement in this final rule.

a. Comment: We received a comment stating that "Audit controls" should be an implementation feature rather than the standard, and suggesting that we change the title of the standard to "Accountability," and provide additional detail to the audit control implementation feature.

Response: We do not adopt the term "Accountability" in this final rule because it is not descriptive of the requirement, which is to have the capability to record and examine system activity. We believe that it is appropriate to specify audit controls as a type of technical safeguard. Entities have flexibility to implement the standard in a manner appropriate to their needs as deemed necessary by their own risk analyses. For example, see NIST Special Publication 800–14, *Generally Accepted Principles and Practices for Securing Information Technology Systems* and NIST Special Publication 800–33, *Underlying Technical Models for Information Technology Security.*

b. Comment: One commenter recommended that this final rule state that audit control mechanisms should be implemented based on the findings of an entity's risk assessment and risk analysis. The commenter asserted that audit control mechanisms should be utilized only when appropriate and necessary and should not adversely affect system performance.

Response: We support the use of a risk assessment and risk analysis to determine how intensive any audit control function should be. We believe that the audit control requirement should remain mandatory, however, since it provides a means to assess activities regarding the electronic protected health information in an entity's care.

c. Comment: One commenter was concerned about the interplay of State and Federal requirements for auditing of privacy data and requested additional guidance on the interplay of privacy rights, laws, and the expectation for audits under the rule.

Response: In general, the security standards will supercede any contrary provision of State law. Security standards in this final rule establish a minimum level of security that covered entities must meet. We note that covered entities may be required by other Federal law to adhere to additional, or more stringent security measures. Section 1178(a)(2) of the statute provides several exceptions to this general rule. With regard to protected health information, the preemption of State laws and the relationship of the Privacy Rule to other Federal laws is discussed in the Privacy Rule beginning at 65 FR 82480; the preemption provisions of the rule are set out at 45 CFR part 160, subpart B.

It should be noted that although the Privacy Rule does not incorporate a requirement for an "audit trail" function, it does call for providing an accounting of certain disclosures of protected health information to an individual upon request. There has been a tendency to assume that this Privacy Rule requirement would be satisfied via some sort of process involving audit trails. We caution against assuming that the Security Rule's requirement for an audit capability will satisfy the Privacy Rule's requirement regarding accounting for disclosures of protected health information. The two rules cover overlapping, but not identical information. Further, audit trails are typically used to record uses within an electronic information system, while the Privacy Rule requirement for accounting applies to certain disclosures outside of the covered entity (for example, to public health authorities).

3. Integrity (§ 164.312(c)(1)). We proposed under the "Data authentication" requirement, that each organization be required to corroborate that data in its possession have not been altered or destroyed in an unauthorized manner and provided examples of mechanisms that could be used to accomplish this task. We adopt the proposed requirement for data authentication in the final rule as an addressable implementation specification "Mechanism to authenticate data," under the "Integrity" standard.

a. Comment: We received a large number of comments requesting clarification of the "Data authentication" requirement. Many of these comments suggested that the requirement be called "Data integrity" instead of "Data authentication." Others asked for guidance regarding just what "data" must be authenticated. A significant number of commenters indicated that this requirement would put an extraordinary burden on large segments of the health care industry, particularly when legacy systems are in use. Requests were received to make this an "optional" requirement, based on an entity's risk assessment and analysis.

Response: We adopt the suggested "integrity" terminology because it more clearly describes the intent of the standard. We retain the meaning of the term "Data authentication" under the addressable implementation specification "Mechanism to authenticate data," and provide an example of a potential means to achieve data integrity.

Error-correcting memory and magnetic disc storage are examples of the built-in data authentication mechanisms that are ubiquitous in hardware and operating systems today. The risk analysis process will address what data must be authenticated and should provide answers appropriate to the different situations faced by the various health care entities implementing this regulation.

Further, we believe that this standard will not prove difficult to implement, since there are numerous techniques available, such as processes that employ digital signature or check sum technology to accomplish the task.

b. Comment: We received numerous comments suggesting that "Double keying" be deleted as a viable "Data authentication" mechanism, since this practice was generally associated with the use of punched cards.

Response: We agree that the process of "Double keying" is outdated. This final rule omits any reference to "Double keying."

4. *Person or Entity Authentication (§ 164.312(d)).* We proposed that an organization implement the requirement for "Entity authentication", the corroboration that an entity is who it claims to be. "Automatic logoff" and "Unique user identification" were specified as mandatory features, and were to be coupled with at least one of the following features: (1) A "biometric" identification system; (2) a "password" system; (3) a "personal identification number"; and (4) "telephone callback," or a "token" system that uses a physical device for user identification.

In this final rule, we provide a general requirement for person or entity authentication without the specifics of the proposed rule.

Comment: We received comments from a number of organizations requesting that the implementation features for entity authentication be either deleted in their entirety or at least be made optional. On the other hand, comments were received requesting that the use of digital signatures and soft tokens be added to the list of implementation features.

Response: We agree with the commenters that many different mechanisms may be used to authenticate entities, and this final rule now reflects this fact by not incorporating a list of implementation specifications, in order to allow covered entities to use whatever is reasonable and appropriate. "Digital signatures" and "soft tokens" may be used, as well as many other mechanisms, to implement this standard.

The proposed mandatory implementation feature, "Unique user identification," has been moved from this standard and is now a required implementation specification under "Access control" at § 164.312(a)(1).

"Automatic logoff" has also been moved from this standard to the "Access control" standard and is now an addressable implementation specification.

5. *Transmission Security (§ 164.312(e)(1)).* Under "Technical Security Mechanisms to Guard Against Unauthorized Access to Data that is Transmitted Over a Communications Network," we proposed that "Communications/network controls" be required to protect the security of health information when being transmitted electronically from one point to another over open networks, along with a combination of mandatory and optional implementation features. We proposed that some form of encryption must be employed on "open" networks such as the Internet or dial-up lines.

In this final rule, we adopt integrity controls and encryption, as addressable implementation specifications.

a. Comment: We received a number of comments asking for overall clarification as well as a definition of terms used in this section. A definition for the term "open networks" was the most requested action, but there was a general expression of dislike for the manner in which we approached this section, with some comments suggesting that the entire section be rewritten. A significant number of comments were received on the question of encryption requirements when dial-up lines were to be employed as a means of connectivity. The overwhelming majority strongly urged that encryption not be mandatory when using any transmission media other than the Internet, but rather be considered optional based on individual entity risk assessment/analysis. Many comments noted that there are very few known breaches of security over dial-up lines and that nonjudicious use of encryption can adversely affect processing times and become both financially and technically burdensome. Only one commenter suggested that "most" external traffic should be encrypted.

Response: In general, we agree with the commenters who asked for clarification and revision. This final rule has been significantly revised to reflect a much simpler and more direct requirement. The term "Communications/network controls" has been replaced with "Transmission security" to better reflect the requirement that, when electronic protected health information is transmitted from one point to another, it must be protected in a manner commensurate with the associated risk.

We agree with the commenters that switched, point-to-point connections, for example, dial-up lines, have a very small probability of interception.

Thus, we agree that encryption should not be a mandatory requirement for transmission over dial-up lines. We also agree with commenters who mentioned the financial and technical burdens associated with the employment of encryption tools. Particularly when considering situations faced by small and rural providers, it became clear that there is not yet available a simple and interoperable solution to encrypting e-mail communications with patients. As a result, we decided to make the use of encryption in the transmission process an addressable implementation specification. Covered entities are encouraged, however, to consider use of encryption technology for transmitting electronic protected health information, particularly over the internet.

As business practices and technology change, there may arise situations where electronic protected health information being transmitted from a covered entity would be at significant risk of being accessed by unauthorized entities. Where risk analysis showed such risk to be significant, we would expect covered entities to encrypt those transmissions, if appropriate, under the addressable implementation specification for encryption.

We do not use the term "open network" in this final rule because its meaning is too broad. We include as an addressable implementation specification the requirement that transmissions be encrypted when appropriate based on the entity's risk analysis.

b. Comment: We received comments requesting that the implementation features be deleted or made optional. Three commenters asked that the requirement for an alarm be deleted.

Response: This final rule has been revised to reflect deletion of the following implementation features: (1) The alarm capability; (2) audit trail; (3) entity authentication; and (4) event reporting. These features were associated with a proposed requirement for "Communications/network controls" and have been deleted since they are normally incorporated by telecommunications providers as part of network management and control functions that are included with the provision of network services. A health care entity would not expect to be responsible for these technical telecommunications features. "Access controls" has also been deleted from the implementation features since the consideration of the use of encryption will satisfy the intent of this feature. We retain as addressable implementation specifications two features: (1) "Integrity controls" and "encryption". "Message authentication" has been deleted as an implementation feature because the use of data authentication codes (called for in the "integrity controls" implementation specification) satisfies the intent of "Message authentication."

c. Comment: A number of comments were received asking that this final rule establish a specific (or at least a minimum) cryptographic algorithm strength. Others recommended that the rule not specify an encryption strength since technology is changing so rapidly. Several commenters requested guidelines and minimum encryption standards for the Internet. Another stated that, since an example was included (small or rural providers for example), the government should feel free to name a specific encryption package. One commenter stated that the requirement for encryption on the Internet should reference the "CMS Internet Security Policy."

Response: We remain committed to the principle of technology neutrality and agree with the comment that rapidly changing technology makes it impractical and inappropriate to name a specific technology. Consistent with this principle, specification of an algorithm strength or specific products would be inappropriate. Moreover, rapid advances in the success of "brute force" cryptanalysis techniques suggest that any minimum specification would soon be outmoded. We maintain that it is much more appropriate for this final rule to state a general requirement for encryption protection when necessary and depend on covered entities to specify technical details, such as algorithm types and strength. Because "CMS Internet Security Policy" is the policy of a single organization and applies only to information sent to CMS, and not between all covered entities, we have not referred to it here.

d.   Comment: The proposed definition of "Integrity controls" generated comments that asked that the word "validity" be changed to "Integrity." Commenters were concerned about the ability of an entity to ensure that information was "valid."

Response: We agree with the commenters about the meaning of the word "validity" in the context of the proposed definition of "Integrity controls." We have named "integrity controls" as an implementation specification in this final rule to require mechanisms to ensure that electronically transmitted information is not improperly modified without detection (see § 164.312(c)(1)).

e.   Comment: Three commenters asked for clarification and guidance regarding the unsolicited electronic receipt of health information in an unsecured manner, for example, when the information was submitted by a patient via e-mail over the Internet. Commenters asked for guidance as to what was their obligation to protect data received in this manner.

Response: The manner in which electronic protected health information is received by a covered entity does not affect the requirement that security protection must subsequently be afforded to that information by the covered entity once that information is in possession of the covered entity.

6.   *Proposed Requirements Not Adopted in This Final Rule*

a.   Authorization Control   We proposed, under "Technical Security Services to Guard Data Integrity, Confidentiality, and Availability," that a mechanism be required for obtaining consent for the use and disclosure of health information using either "Role-based access" or "User-based access" controls. In this final rule, we do not adopt this requirement.

Comment: We received a large number of comments regarding use of the word "consent." It was pointed out that this could be construed to mean patient consent to the use or disclosure of patient information, which would make this a privacy issue, rather than one of security. Other comments suggested deletion of the requirement in its entirety. We received a comment asking for clarification about the distinction between "Access control" and "Authorizations."

Response: These requirements were intended to address authorization of workforce members and others for the use and disclosure of health information, not patient consent.

Upon reviewing the differences between "Access control" and "Authorization control," we found it to be unnecessary to retain "Authorization control" as a separate requirement. Both the access control and the authorization control proposed requirements involved implementation of types of automated access controls, that is, role-based access and user-based access. It can be argued that the process of managing access involves allowing and restricting access to those individuals that have been authorized to access the data. The intent of the proposed authorization control implementation feature is now incorporated in the access authorization implementation specification under the information access management standard in § 164.308(a)(4). Under the information access management standard, a covered entity must implement, if appropriate and reasonable to its situation, policies and procedures first to authorize a person to access electronic protected health information and then to actually establish such access. These policies and procedures will enable entities to follow the Privacy Rule minimum necessary requirements, which provide when persons should have access to information.

H. Organizational Requirements (§ 164.314)

We proposed that each health care clearinghouse must comply with the security standards to ensure all health information and activities are protected from unauthorized access. If the clearinghouse is part of a larger organization, then unauthorized access by the larger organization must be prevented. We also proposed that parties processing data through a third party would be required to enter into a chain of trust partner agreement, a contract in which the parties agree to electronically exchange data and to protect the transmitted data in accordance with the security standards.

In this final rule, we have adopted the concepts of hybrid and affiliated entities, as previously defined in § 164.504, and now defined in § 164.103, and business associates as defined in § 160.103, to be consistent with the Privacy Rule. General organizational requirements related to affiliated covered entities and hybrid entities are now contained in a new § 164.105. The proposed chain of trust partner agreement has been replaced by the standards for business associate contracts or other arrangements and the standards for group health plans. Consistent with the statute and the policy of the Privacy Rule, this final rule does not require noncovered entities to comply with the security standards.

1. Health Care Clearinghouses. The proposed rule proposed that if a health care clearinghouse were part of a larger organization, it would be required to ensure that all health information pertaining to an individual is protected from unauthorized access by the larger organization; this statement closely tracked the statutory language in section 1173(d)(1)(B) of the Act. Since the point of the statutory language is to ensure that health care information in the possession of a health care clearinghouse is not inappropriately accessed by the larger organization of which it is a part, this final rule implements the statutory language through the information access management provision of § 164.308(a)(4)(ii)(A).

The final rule, at § 164.105, makes the health care component and affiliated entity standards of the Privacy Rule applicable to the security standards. Therefore, we have not changed those standards substantively. In pertaining to the Privacy Rule, we have simply moved them to a new location in part 164. Any differences between § 164.105 and § 164.504(a) through (d) reflects the addition of requirements specific to the security standards.

The health care component approach was developed in response to extensive comment received principally on the Privacy Rule. See 65 FR 82502 through 82503 and 82637 through 82640 for a discussion of the policy concerns underlying the health care component approach. Since the security standards are intended to support the protection of electronic information protected by the Privacy Rule, it makes sense to incorporate organizational requirements that parallel those required of covered entities by the Privacy Rule. This policy will also minimize the burden of complying with both rules.

a. *Comment:* Relative to the following preamble statement (63 FR 43258): "If the clearinghouse is part of a larger organization, then security must be imposed to prevent unauthorized access by the larger organization." One commenter asked what is considered to be "the larger organization." For example, if a clearinghouse function occurs in a department of a larger business entity, will the regulation cover all internal electronic communication, such as e-mail, within the larger business and all external electronic communication, such as e-mail with its owners?

Response: The "larger organization" is the overall business entity that a clearinghouse would be part of. Under the Security Rule, the larger organization must assure that the health care clearinghouse function has instituted measures to ensure only that electronic protected health information that it processes is not improperly accessed by unauthorized persons or other entities, including the larger organization. Internal electronic communication within the larger organization will not be covered by the rule if it does not involve the clearinghouse, assuming that it has designated health care components, of which the health care clearinghouse is one. External communication must be protected as sent by the clearinghouse, but need not be protected once received.

b. *Comment:* One commenter asked that the first sentence in § 142.306(b) of the proposed rule, "If a health care clearinghouse is part of a larger organization, it must assure all health information is protected from unauthorized access by the larger organization" be expanded to read, "If a health care clearinghouse or any other health care entity is part of a larger organization …"

Response: The Act specifically provides, at section 1173(d)(1)(B), that the Secretary must adopt standards to ensure that a health care clearinghouse, if part of a larger organization, has policies and security procedures to protect information from unauthorized access by the larger organization.

Health care providers and health plans are often part of larger organizations that are not themselves health care providers or health plans. The security measures implemented by health plans and covered health care providers should protect electronic protected health information in circumstances such as the one identified by the commenter. Therefore, we agree with the comment that the requirement should be expanded as suggested by the commenter. In this final rule, those components of a hybrid entity that are designated as health care components must comply with the security standards and protect against unauthorized access with respect to the other components of the larger entity in the same way as they must deal with separate entities.

2. *Business Associate Contracts and Other Arrangements.* We proposed that parties processing data through a third party would be required to enter into a chain of trust partner agreement, a contract in which the parties agree to electronically exchange data and to protect the transmitted data. This final rule narrows the scope of agreements required. It essentially tracks the provisions in § 164.502(e) and § 164.504(e) of the Privacy Rule, although appropriate modifications have been made in this rule to the required elements of the contract.

In this final rule, a contract between a covered entity and a business associate must provide that the business associate must—(1) implement safeguards that reasonably and appropriately protect the confidentiality, integrity, and availability of the electronic protected health information that it creates, receives, maintains, or transmits on behalf of the covered entity; (2) ensure that any agent, including a subcontractor, to whom it provides this information agrees to implement reasonable and appropriate safeguards; (3) report to the covered entity any security incident of which it becomes aware; (4) make its policies and procedures, and documentation required by this subpart relating to such safeguards, available to the Secretary for purposes of determining the covered entity's compliance with this subpart; and (5) authorize termination of the contract by the covered entity if the covered entity determines that the business associate has violated a material term of the contract.

When a covered entity and its business associate are both governmental entities, an "other arrangement" is sufficient. The covered entity is in compliance with this standard if it enters into a memorandum of understanding with the business associate that contains terms that accomplish the objectives of the above-described business associate contract. However, the covered entity may omit from this memorandum the termination authorization required by the business associate contract provisions if this authorization is inconsistent with the statutory obligations of the covered entity or its business associate. If other law (including regulations adopted by the covered entity or its business associate) contains requirements applicable to the business associate that accomplish the objectives of the above-described business associate contract, a contract or agreement is not required. If a covered entity enters into other arrangements with another governmental entity that is a business associate, such arrangements may omit provisions equivalent to the termination authorization required by the business associate contract, if inconsistent with the statutory obligation of the covered entity or its business associate.

If a business associate is required by law to perform a function or activity on behalf of a covered entity or to provide a service described in the definition of business associate in § 160.103 of this subchapter to a covered entity, the covered entity may permit the business associate to receive, create, maintain, or transmit electronic protected health information on its behalf to the extent necessary to comply with the legal mandate without meeting the requirements of the above-described business associate contract, *provided that* the covered entity attempts in good faith to obtain satisfactory assurances as required by the above described business associate contract and documents the attempt and the reasons that these assurances cannot be obtained.

We have added a standard for group health plans that parallels the provisions of the Privacy Rule. It became apparent during the course of the security and privacy rulemaking that our original chain of trust approach was both overly broad in scope and failed to address appropriately the circumstances of certain covered entities, particularly the ERISA group health plans. These latter considerations and the solutions arrived at in the Privacy Rule are described in detail in the Privacy Rule at 65 FR 82507 through 82509. Because the purpose of the security standards is in part to reinforce privacy protections, it makes sense to align the organizational policies of the two rules. This decision should also make compliance less burdensome for covered entities than would a decision to have different organizational requirements for the two sets of rules.

Thus, we have added at § 164.314(b) a standard for group health plan that tracks the standard at § 164.504(f) very closely. The purpose of these provisions is to ensure that, except when the electronic protected health information disclosed to a plan sponsor is summary health information or enrollment or disenrollment information as provided for by § 164.504(f), group health plan documents provide that the plan sponsor will reasonably and appropriately safeguard electronic protected health information created, received, maintained or transmitted to or by the plan sponsor on behalf of the group health plan. The plan documents of the group health plan must be amended to incorporate provisions to require the plan sponsor to implement reasonable and appropriate safeguards to protect the confidentiality, integrity, and availability of the electronic protected health information that it creates, receives, maintains, or transmits on behalf of the group health plan; ensure that the adequate separation required by § 164.504(f)(2)(iii) is supported by reasonable and appropriate security measures; ensure that any agents, including a subcontractor, to whom it provides this information agrees to implement reasonable and appropriate safeguards to protect the information; report to the group health plan any security incident of which it becomes aware; and make its policies and procedures and documentation relating to these safeguards available to the Secretary for purposes of determining the group health plan's compliance with this subpart.

a. Comment: Several commenters expressed confusion concerning the applicability of proposed § 142.104 to security.

Response: The proposed preamble included language generally applicable to most of the proposed standards under HIPAA. Proposed § 142.104 concerned general requirements for health plans relative to processing transactions. We proposed that plans could not refuse to conduct a transaction as a standard transaction, or delay or otherwise adversely affect a transaction on the grounds that it was a standard transaction; health information transmitted and received in connection with a transaction must be in the form of standard data elements; and plans conducting transactions through an agent must ensure that the agent met all the requirements that applied to the health plan. Except for the statement that a plan's agent ("business associate" in the final rule) must meet the requirements (which would include security) that apply to the health plan, this proposed section did not pertain to the security standards and was addressed in the Transaction Rule.

b. Comment: The majority of comments concerned proposed rule language stating "the same level of security will be maintained at all links in the chain * * *" Commenters believed the current language will have an adverse impact on one of the security standard's basic premises, which is scalability. It was requested that the language be changed to indicate that, while appropriate security must be maintained, all partners do not need to maintain the same level of security.

A number of commenters expressed some confusion concerning their responsibility for the security of information once it has passed from their control to their trading partner's control, and so on down the trading partner chain. Requests were made that we clarify that chain of trust partner agreements were really between two parties, and that, if a trading partner agreement has been entered into, any given partner would not be responsible, or liable, for the security of data once it is out of his or her control.

In line with this concern, several commenters were concerned that they would have some responsibility to ensure the level of security maintained by their trading partner.

Several commenters believe a chain of trust partner agreement should not be a security requirement. One commenter stated that because covered entities must already conform to the regulation requirements, a "chain of trust" agreement does not add to overall security. Compliance with the regulation should be sufficient.

Response: We believe the commenters are correct that the rule as proposed would—(1) not allow for scalability; and (2) would lead an entity to believe it is responsible, and liable, for making sure all entities down the line maintain the same level of security. The confusion here seems to come from the phrase "same level of security." Our intention was that each trading partner would maintain reasonable and appropriate safeguards to protect the information. We did not mean that partners would need to implement the same security technology or measures and procedures.

We have replaced the proposed "Chain of trust" standard with a standard for "Business associate contracts and other arrangements."

When another entity is acting as a business associate of a covered entity, we require the covered entity to require the other entity to protect the electronic protected health information that it creates, receives, maintains or transmits on the covered entity's behalf. The level of security afforded particular electronic protected health information should not decrease just because the covered entity has made the business decision to entrust a business associate with using or disclosing that information in connection with the performance of certain functions instead of doing those functions itself. Thus, the rule below requires covered entities to require their business associates to implement certain safeguards and take other measures to ensure that the information is safeguarded (see § 164.308(b)(1) and § 164.314(a)(1)).

The specific requirements of § 164.314(a)(1) are drawn from the analogous requirements at 45 CFR 164.504(e) of the Privacy Rule, although they have been adapted to reflect the objectives and context of the security standards. Compare, in particular, 45 CFR 164.504(e)(2)(ii) with § 164.314(a)(1). We have not imported all of the requirements of 45 CFR 164.504(e), however, as many have no clear analog in the security context (see, for example, 45 CFR 164.504(e)(2)(i) regarding permitted and required uses and disclosures made by a business associate). HHS had previously committed to reconciling its security and privacy policies regarding business associates (see 65 FR 82643). The close relationship of many of the organizational requirements in section 164.314 with the analogous requirements of the Privacy Rule should facilitate the implementation and coordination of security and privacy policies and procedures by covered entities.

In contrast, when another entity is not acting as a business associate for the covered entity, but rather is acting in the capacity of some other sort of trading partner, we do not require the covered entity to require the other entity to adopt particular security measures, as previously proposed. This policy is likewise consistent with the general approach of the Privacy Rule (see the discussion in the Privacy Rule at 65 FR 82476). The covered entity is free to negotiate security arrangements with its non-business associate trading partners, but this rule does not require it to do so.

A similar approach underlies § 164.314(b) below. These provisions are likewise drawn from, and intended to support, the analogous privacy protections provided for by 45 CFR 164.504(f) (see the discussion of § 164.504(f) of the Privacy Rule at 65 FR 82507 through 82509, and 82646 through 82648). As with the business associate contract provisions, however, they are imported and adapted only to the extent they make sense in the security context. Thus, for example, the requirement at § 164.504(f)(2)(ii)(C) prohibits the plan documents from permitting disclosure of protected health information to the plan sponsor for employment-related purposes. As this prohibition goes entirely to the permissibility of a particular type of disclosure, it has no analog in § 164.314(b).

c. Comment: Several commenters stated that if security features are determined by agreements established between "trading partners," as stated in the proposed regulations, there should be some guidelines or boundaries for those agreements so that extreme or unusual provisions are not permitted.

Response: This final rule sets a baseline, or minimum level, of security measures that must be taken by a covered entity and stipulates that a business associate must also implement reasonable and appropriate safeguards. This final rule does not, however, prohibit a covered entity from employing more stringent security measures or from requiring a business associate to employ more stringent security measures. A covered entity may determine that, in order to do business with it, a business associate must also employ equivalent measures. This would be a business decision and would not be governed by the provisions of this rule. Security mechanisms relative to the transmission of electronic protected health information between entities may need to be agreed upon by both parties in order to successfully complete the transmission. However, the determination of the specific transmission mechanisms and the specific security features to be implemented remains a business decision.

d. Comment: Several commenters asked whether existing contracts could be used to meet the requirement for a trading partner agreement, or does the rule require entry into a new contract specific to this purpose. Also, the commenters want to know about those whose working agreements do not involve written contractual agreement: Do they now need to set up formal agreements and incur the additional expense that would entail?

Response: This final rule requires written agreements between covered entities and business associates. New contracts do not have to be entered into specifically for this purpose, if existing written contracts adequately address the applicable requirements (or can be amended to do so).

e. Comment: Several commenters asked whether covered entities are responsible for the security of all individual health information sent to them, or only information sent by chain of trust partners. They also asked if they can refuse to process standard transactions sent to them in an unsecured fashion. In addition, they inquired if they can refuse to send secured information in standard transactions to entities not required by law to secure the information. One commenter asked if there is a formula for understanding in any particular set of relationships where the ultimate responsibility for compliance with the standards would lie.

Response: Pursuant to the Transactions Rule, if a health plan receives an unsecured standard transaction, it may not refuse to process that transaction simply because it was sent in an unsecured manner. The health plan is not responsible under this rule, for how the transaction was sent to it (unless the transmission was made by a business associate, in which case different considerations apply); however, once electronic protected health information is in the possession of a covered entity, the covered entity is responsible for the security of the electronic protected health information received. The covered entity must implement technical security mechanisms to guard against unauthorized access to electronic protected health information that is transmitted over an electronic communication network. In addition, the rule requires the transmitting covered entity to obtain written assurance from a business associate receiving the transmission that it will provide an adequate level of protection to the information. For the business associate provisions, see § 164.308(b) and § 164.314(a) of this final rule.

f. Comment: One commenter asked what security standards a vendor having access to a covered entity's health information during development, testing, and repair must meet and wanted to know whether the rule anticipates having a double layer of security compliance (one at the user level and one at the vendor level). If so, the commenter believes this will cause duplication of work.

Response: In the situation described, the vendor would be acting as a business associate. The covered entity must require the business associate to implement reasonable and appropriate security protections of electronic protected health information. This requirement, however, does not impose detailed requirements for how that level of protection must be achieved. The resulting flexibility should permit entities and their business associates to adapt their security safeguards in ways that make sense in their particular environments.

g. *Comment:* A number of commenters requested sample contract language or models of contracts. We also received one comment that suggested that we should not dictate the contents of contracted agreements.

Response: We will consider developing sample contract language as part of our guideline development.

I. Policies and Procedures and Documentation Requirements (§ 164.316)

We proposed requiring documented policies and procedures for the routine and nonroutine receipt, manipulation, storage, dissemination, transmission, and/or disposal of health information. We proposed that the documentation be reviewed and updated periodically.

We have emphasized throughout this final rule the scalability allowed by the security standards. This final rule requires covered entities to implement policies and procedures that are reasonably designed, taking into account the size and type of activities of the covered entity that relate to electronic protected health information, and requires that the policies and procedures must be documented in written form, which may be in electronic form. This final rule also provides that a covered entity may change its policies and procedures at any time, provided that it documents and implements the changes in accordance with the applicable requirements. Covered entities must also document designations, for example, of affiliation between covered entities (see § 164.105(b)), and other actions, as required by other provisions of the subpart.

1. *Comment:* One commenter wanted development of written policies regarding such things as confidentiality and privacy rights for access to medical records, and approval of research by a review board when appropriate.

Response: These issues are covered in the Privacy Rule (65 FR 82462) (see, in particular, § 164.512(i), § 164.524, and § 164.530(i)).

2. *Comment:* One commenter asked if standards will override agreements that require others to maintain hardcopy documentation (for example, signature on file) and no longer require submitters to maintain hardcopy documentation.

Response: The security standards will require a minimum level of documentation of security practices. Any agreements between trading partners for the exchange of electronic protected health information that impose additional documentation requirements will not be overridden by this final rule.

3. *Comment:* One commenter stated that there should be a requirement to document only applications deemed necessary by an applications and data criticality assessment.

Response: Electronic protected health information must be afforded security protection under this rule regardless of what application it resides in. The measures taken to protect that information must be documented.

4. Comment: One commenter asked how detailed the documentation must be. Another commenter asked what "kept current" meant.

Response: Documentation must be detailed enough to communicate the security measures taken and to facilitate periodic evaluations pursuant to § 164.308(a)(8). While the term "current" is not in the final rule, this concept has been adopted in the requirement that documentation must be updated as needed to reflect security measures currently in effect.

5. Comment: We received one comment concerning review and updating of implementing documentation suggesting that "periodically" be changed to "at least annually."

Response: We believe that the requirement should remain as written, in order to allow individual entities to establish review and update cycles as deemed necessary. The need for review and update will vary dependent upon a given entity's size, configuration, environment, operational changes, and the security measures implemented.

J. Compliance Dates for Initial Implementation (§ 164.318)

We proposed that how the security standard would be implemented by each covered entity would be dependent upon industry trading partner agreements for electronic transmissions. Covered entities would be able to adapt the security matrix to meet business needs. We suggested that requirements of the security standard may be implemented earlier than the compliance date. However, we would require implementation to be complete by the applicable compliance date, which is 24 months after adoption of the standard, and 36 months after adoption of the standard for small health plans, as provided by the Act. In the proposed rule, we suggested that an entity choosing to convert from paper to standard EDI transactions, before the effective date of the security standard, consider implementing the security standard at the same time.

In this final rule the dates by which entities must be in compliance with the standards are called "compliance dates," consistent with our practice in the Transactions, Privacy, and Employer Identifier Rules. Section 164.318 in this final rule is also organized consistent with the format of those rules. The substantive requirements, which are statutory, remain unchanged.

Many of the comments received concerning effective dates and compliance dates, including the compliance dates for modifications of standards, were addressed in the Transactions Rule. Those that were not addressed in that publication are presented below.

1. Comment: A number of commenters expressed support for the effective dates of the rules and stated that they should not be delayed. In contrast, one commenter stated that we should delay this rule to allow for an open consensus building debate to occur concerning security. One commenter asked that the rule be delayed until after implementation of the ICD-CM changes.

A number of comments were received expressing the opinion that the security regulation should not be published until either the Congress has enacted legislation governing standards with respect to the privacy of individually identifiable health information, or the Secretary of HHS has promulgated final regulations containing these standards. One commenter stated, "we find ourselves in the difficult position of reacting to proposed rules setting the standards for how information should be physically and electronically protected, without having reached agreement on the larger issues of consent for and disclosure of individual medical information."

Response: The effective date of the final rule is 60 days after this final rule is published in the **Federal Register**. The statute sets forth the compliance dates for the standards. Covered entities must comply with this final rule no later than 24 months (36 months for small plans) after the effective date.

The final Privacy Rule has already been published. We note that numerous comments concerning the timing of the adoption of privacy and security standards were also received in the privacy rulemaking and are discussed in the Privacy Rule at 65 FR 82752.

2. *Comment:* One commenter asked that proposed § 142.312 be rewritten to separate the effective dates for the Security Rule and the Transactions Rule.

Response: The proposed rule incorporated general language applicable to all the proposed Administrative Simplification standards. Language concerning standards other than Security is not included in § 164.318. Because this final rule is adopted after the Transactions Rule was adopted, the compliance dates for the security standards differ from those for the transactions standards. Comments concerning general effective dates were addressed in the Transactions Rule. Comments specific to the security standards are addressed here.

3. *Comment:* Several commenters suggested that we not allow early implementation of the Security Rules. A number of others asked that we allow, but not require, early implementation by willing trading partners. Another commenter suggested that early implementation by willing trading partners be allowed as long as the data content transmitted is equal to that required by statute. Another commenter requested that it be stipulated that entities cannot implement less than 1 year from the date of this final rule and then only after successful testing, and that a "start testing by" date be defined.

Response: Whether or not to implement before the compliance date is a business decision that each covered entity must make. Moreover, the vast majority of the standards address internal policies and procedures that can be implemented at any time without any impact on trading partners.

4. *Comment:* One commenter asked us to establish a research site or test laboratory for a trial implementation.

Response: The concept of a "trial implementation" that would have widespread relevance is inconsistent with our basic principles of flexibility, scalability, and technology-neutrality.

5. *Comment:* One commenter stated that the 2-year time frame for implementation of a contingency plan is too short for health plans that serve multiple regions of the country.

Response: The Congress mandated that entities must be in compliance 2 years from the initial standard's adoption date (3 years for small plans).

K. Appendix

The proposed rule contained three addenda. Addendum 1 set out in matrix form the proposed requirements and related implementation features of the proposed rule. Addendum 2 set out in list form a glossary of terms with citations to the sources of those terms. Addendum 3 identified and mapped areas of overlap in the proposed security standard and implementation features.

This final rule retains only the first proposed addendum, the matrix, as an appendix, that is modified to reflect the changes in the administrative, physical, and technical safeguard portions of the rule below. Numerous terms in the glossary now appear in the rule below, typically (but not always) as definitions.

1. Comment: Over two-thirds of the comments received on this topic asked that the matrix be incorporated into the final rule. One commenter asked that a simplified version be made part of the final rule. Six commenters wanted it kept in this final rule as an addendum. One commenter stated that it should be in an appendix to the rule, while others stated that it should not be included in this final rule.

Response: Since a significant majority of commenters requested retention of the matrix, it has been incorporated into this final rule as an appendix. The matrix displays, in tabular form, the administrative, physical, and technical safeguard standards and relating implementation specifications described in this final rule in § 164.308, § 164.310, and § 164.312. It should be noted that the requirements of § 164.105, § 164.314, and § 164.316 are not presented in the matrix.

2. Comment: A large majority of commenters stated that the glossary located in Addendum 2 of the proposed rule should be included as part of the final rule. Several commenters asked that it be incorporated into the definitions section of the final rule. One commenter stated that the glossary should not be part of this final rule.

Response: The terms defined in the glossary in Addendum 2 of the proposed rule are found throughout this final rule, either as part of the text of § 164.306 through § 164.312 or under § 164.304, as appropriate. We included only terms relevant to the particular standards and implementation specifications being adopted.

3. Comment: Several commenters requested that the mapped matrix located in Addendum 3 of the proposed rule be included in this final rule, either as part of the rule or as an addendum, while others stated that it should not be part of this final rule. Several commenters cited items to be added to the mapped matrix.

Response: The mapped matrix was merely a snapshot of current standards and guidelines that the implementation team was able to obtain for review during the development of the security and electronic signature requirements and was provided in the proposed rule as background material. Since this matrix has not been fully populated or kept up-to-date, it is not being published as part of this final rule. Where relevant, we do reference various standards and guidelines indicated in the matrix in this preamble.

L. Miscellaneous Issues

1. **Preemption.** The statute requires generally that the security standards supersede contrary provisions of State law including State law requiring medical or health plan records to be maintained or transmitted in written rather than electronic formats. The statute provides certain exceptions to the general rule; section 1178(a)(2) of the Act identifies conditions under which an exception applies. The proposed rule did not provide for a process for making exception determinations; rather, a process was proposed in the privacy rulemaking and was adopted with the Privacy Rule (see part 160, subpart B). This process applies to exception determinations for all of the Administrative Simplification rules, including this rule.

a. Comment: Several commenters stated that the proposed rule does not include substantive protections for the privacy rights of patients' electronic medical records, while the rule attempts to preempt State privacy laws with respect to these records. Comments stated that, by omitting a clarification of State privacy law applicability, the proposed rule creates confusion. They believe that the rule must contain express and specific exemptions of State laws with respect to medical privacy.

Response: The Privacy Rule establishes standards for the rights of patients in regard to the privacy of their medical records and for the allowable uses and disclosures of protected health information. The identified concerns were discussed in the Privacy Rule (see 65 FR 82587 through 82588). The security standards do not specifically address privacy but will safeguard electronic protected health information against unauthorized access or modification.

b. Comment: One commenter asked how these regulations relate to confidentiality laws, which vary from State to State.

Response: It is difficult to respond to this question in the abstract without the benefit of reference to a specific State statute. However, in general, these security standards will preempt contrary State laws. Per section 1178(a)(2) of the Act, this general rule would not hold if the Secretary determines that a contrary provision of State law is necessary for certain identified purposes to prevent fraud and abuse; to ensure appropriate State regulation of insurance and health plans; for State reporting on health care delivery costs; or if it addresses controlled substances. See 45 CFR part 160 subpart B. In such case, the contrary provision of State law would preempt a Federal provision of these security standards. State laws that are related but not contrary to this final rule, will not be affected.
Section 1178 of the Act also limits the preemptive effect of the Federal requirements on certain State laws other than where the Secretary makes certain determinations. Section 1178(b) of the Act provides that State laws for reporting of disease and other conditions and for public health surveillance, investigation, or intervention are not invalidated or limited by the Administrative Simplification rules. Section 1178(c) of the Act provides that the Federal requirements do not limit States' abilities to require that health plans report or provide access to certain information.

c. Comment: Several commenters stated that allowing State law to establish additional security restrictions conflicts with the purpose of the Federal rule and/or would make implementation very difficult. One commenter asked for clarification as to whether additional requirements tighter than the requirements outlined in the proposed rule may be imposed.

Response: The general rule is that the security standards in this final rule supersede contrary State law. Only where the Secretary has granted an exception under section 1178(a)(2)(A) of the Act, or in situations under section 1178(b) or (c) of the Act, will the general rule not hold true. Covered entities may be required to adhere to stricter State-imposed security measures that are not contrary to this final rule.

2. Enforcement. The proposed rule did not contain specific enforcement provisions. This final rule likewise does not contain specific enforcement provisions; it is expected that enforcement provisions applicable to all Administrative Simplification rules will be proposed in a future rulemaking.

a. Comment: One commenter voiced support for the proposed rule's approach. Another stated that the process is poorly defined. One commenter stated that fines should be eliminated, or the scope of activity subject to fines should be more narrowly defined.

While a number of commenters were of the opinion that HHS must retain enforcement responsibility, stating that it would be unconstitutional to give it to a private entity, several others stated that it may not be practical for HHS to retain the responsibility for determining violations and imposing penalties specified by the statute. A concern was voiced over HHS's ability to fairly and consistently apply the rules due to budget constraints. Several commenters support industry solutions to enforcement with some level of government involvement. One commenter recommended a single audit process using accrediting bodies already in place. Another stated that entities providing accreditation services should not be involved in enforcement as this would result in a conflict of interest.

Clarification was requested, including the use of examples, concerning what constitutes a violation, and how a penalty applies to a "person." Commenters asked if the term "person" referred to the people responsible for the system and how penalties would apply to corporations and other entities.

Response: It is expected that enforcement of HIPAA standards will be addressed in regulations to be issued at a later date.

b. Comment: Several commenters stated that enforcement of the security standards will be arbitrarily delegated to private businesses that compete with physicians and with each other.

Response: These comments are premature for the reasons stated above.

3. Comment Period. The comment period on the proposed rule was 60 days.

Comment: We received comments suggesting that significant changes to the standards could occur in the final rule as a result of changes made in response to comments. The commenter believes such changes could adversely affect payers and providers, and suggested that the rule should be republished as a proposed rule with a new comment period to allow additional comments concerning any changes. A "work-in-progress" approach was also suggested, to give all stakeholders time to read, analyze, and comment upon evolving versions of a particular proposed rule.

Response: We have not accepted these suggestions. The numerous comments received were thoughtful, analytical, detailed, and addressed every area of the proposed rule. This response to the proposed rule indicates that the public had ample time to read, analyze, and comment upon the proposed rule. If we were to treat the rule as a "work-in-progress" and issue evolving versions, allowing for comments to each version, we would never implement the statute and achieve administrative simplification as directed by the Congress.

M. Proposed Impact Analysis

The preamble to the Transactions Rule contains comments and responses on the impact of all the administrative simplification standards in general except privacy. Comments and responses specific to the relative impact of implementing this final rule are presented below.

a. Comment: Several commenters stated that the proposed security standards are complex, costly, administratively burdensome, and could result in decreased use of EDI. One commenter stated that this rule runs counter to the explicit intent of Administrative Simplification that requires, "any standard adopted under this part shall be consistent with the objective of reducing the administrative costs of providing and paying for health care."

Several commenters expressed concern that there was no cost benefit analysis provided for these proposed regulations, stating that, faced with increasingly limited resources, it is essential that a security standards cost/benefit analysis for all health care trading partners be provided. Another said an independent cost estimate by the General Accounting Office (GAO) should be performed on these rules and HHS cost estimates should be publicized for comparison purposes.

Still another commenter stated that HHS must provide accurate public sector implementation cost figures and provide funds to offset the cost burden.

One commenter asked for cost benefit evaluations to understand the relationship between competing technologies, levels of security and potential threats to be guarded against. These would demonstrate the costs and the benefits to be gained for both large and small organizations and would provide an understanding of how the levels of security vary by organization size and what the inducements and support available to facilitate adoption are. One commenter suggested that we establish a workgroup to more fully assess the costs and provide Federal funds to offset implementation costs.

One commenter noted a seeming disconnect between two statements in the preamble. Section A, Security standards, states, "no individual small entity is expected to experience direct costs that exceed benefits as a result of this rule." In contrast, section E, Factors in establishing the security standards reads, "We cannot estimate the per-entity cost of implementation because there is no information available regarding the extent to which providers', plans', and clearinghouses' current security practices are deficient."

Response: We are unable to estimate, of the nation's 2 million-plus health plans and 1 million-plus providers that conduct electronic transactions, the number of entities that would require new or modified security safeguards and procedures beyond what they currently have in place. Nor are we able to estimate the number of entities that neither conduct electronic transactions nor maintain individually identifiable electronic health information but may become covered entities at some future time. As we are unable to estimate the number of entities and what measures are or are not already in place, or what specific implementation will be chosen to meet the requirements of the regulation, we are also unable to estimate the cost to those entities.

However, the use of electronic technology to maintain or transmit health information results in many new and potentially large risks. These risks represent expected costs, both monetary and social. Leaving risk assessment up to individual entities will minimize the impact and ensure that security effort is proportional to security risk.

As discussed earlier, the security requirements are both scalable and technically flexible. We have made significant changes to this final rule, reducing the number of required implementation features and providing for greater flexibility in satisfaction of the requirements. In other words, we have focused more on what needs to be done and less on how it should be accomplished.

We have removed the statement regarding the extent of costs versus benefits for small entities.

b. Comment: One commenter stated that on page 43262 of the proposed rule, it indicate that complexity of conversion to the security standards would be affected by the choice to use a clearinghouse. The commenter stated that this choice would have little effect on implementation of security standards. Another commenter stated that the complexity (and cost) of the conversion to meet the security standards is affected by far more than just the "volume of claims health plans process electronically and the desire to transmit the claims or to use the services of a VAN or clearinghouse" as is stated on page 43262. Because the security standards apply to internal systems as well as to transactions between entities, a number of additional factors must be considered, for example, modification of existing security mechanisms, legacy systems, architecture, and culture.

Response: We agree. We have modified the Regulatory Impact Analysis section to take into account that there are other factors involved, such as the architecture and technology limitations of existing systems.

c. Comment: One commenter stated that States will need 90 percent funding of development and implementation, without the burden of an advanced planning documents requirement, from us for this costly process to succeed. Any new operational obligation should be 100 percent funded. Also human resource obligations will be significant. Some States believe they will have difficulty obtaining the budget funds for the State share of the costs. State Medicaid agencies, as purchasers, may also face paying the implementation costs of health care providers, clearinghouses, and health plans in the form of higher rates.

Response: The statute does not authorize any new or special funding for implementation of the regulations. Medicaid system changes, simply because they are "HIPAA related" do not automatically qualify for 90 percent Federal funding participation. As with any systems request, the usual rules will be applied to determine funding eligibility for State HIPAA initiatives. Nevertheless, HHS recognizes that there are significant issues regarding the funding and implementation of HIPAA by Medicaid State agencies, and intends to address them through normal channels of communication with States.

d. Comment: One commenter stated that the proposed rule does not establish how the security standards will contribute to reduced cost for providers. One commenter expected the unintended result of this regulation will be impediment of EDI growth and perhaps even a decline in EDI use by providers. Another stated that the proposed rule actively discourages physician EDI participation by suggesting a fallback to paper processing for those unable to meet the cost of highly complex security compliance.

Response: Ensuring the integrity of an electronic message, its delivery to the correct person, and its confidentiality must be an integral part of conducting electronic commerce. We believe that the consistent application of the measures provided in this rule will actually encourage use of EDI because it will provide increased confidence in the reliability and confidentiality of health information to all parties involved. Also, the implementation of these security requirements will reduce the potential overall cost of risk to a greater extent than additional security controls will increase costs. Put another way, the potential cost of not reasonably addressing security risks could substantially exceed the cost of compliance.

e. Comment: One commenter stated that the implementation impact of the technical safeguards is clearly understated for physicians who use digitally-based equipment that has been in place for some time. The commenter believes that the rule will likely have greatest impact on the installed base of digital systems, including imaging modalities and other medical devices that store or transmit patient information because software for legacy systems will likely require retrofitting or replacement to come into compliance. The commenter believes that this is a negative impact and would outweigh any benefits derived from the potential risk of security breaches. The commenter recommended compliance for digital imaging devices be extended by an additional 3 years to allow time to upgrade systems and defray the associated costs.

Response: Compliance dates for the initial implementation of the initial standards are statutorily prescribed; therefore, we are unable to allow additional time outside of the statutory timeframes for compliance.

f. Comment: A commenter stated that, as a new regulatory mandate, HIPAA costs must be factored into any base year calculations for the proposed prospective payment system. Without an adjustment, this will be another regulatory mandate that comes at the cost of patient care.

Response: Costs included in the prospective payment system are legislatively mandated. The Congress did not direct the inclusion of HIPAA costs into the system, so they are not included. However, the Department believes that the HIPAA standards will provide savings to the provider community over the next 10 years.

g. Comment: One commenter suggested that we include requirements for how a compliant business could dually operate—(1) in a HIPAA compliant manner; and (2) in their former noncompliant manner in order to accommodate doing business with other organizations that are not yet compliant.

Response: The statute imposes a 2-year implementation period between the adoption of the initial standards and the date by which covered entities (except small health plans) must be in compliance. An entity may come into compliance at any point in time during the 2 years. Therefore, the rule does not require a covered entity to comply before the established compliance date. Those entities that come into compliance before the 2-year deadline should decide how best to deal with entities that are not yet compliant. Further, we note that, generally speaking, compliance by a covered entity with these security rules will not hinge on compliance by other entities.

h. Comment: One commenter stated that privacy legislation could impose significant changes to written policies and procedures on authorization, access to health information, and how sensitive information is disclosed to others. The commenter believes these changes could mean the imposition of security requirements different from those contained in the proposed rule, and money spent complying with the security provisions could be ill spent if significant new requirements result from the privacy legislation.

Response: The privacy standards at subpart E of 42 CFR part 164 are now in effect, and this final rule is compatible with them. If, in the future, the Congress passes a law whose provisions differ from these standards, the standards would have to be modified.

i. Comment: One commenter stated that the private sector should develop educational tools or models in order to assist physicians, other providers, and health plans to comply with the security regulations.

Response: We agree. The health care industry is striving to do this. HHS is also considering provider outreach and education activities.

IV. PROVISIONS OF THE FINAL REGULATION

We have made the following changes to the provisions of the August 12, 1998 proposed rule. Specifically, we have—

- Changed the CFR part from 142 to 164.

- Removed information throughout the document pertaining to electronic signature standards. Electronic signature standards will be published in a separate final rule.

- Replaced the word "requirement," when referring to a standard, with "standard." Replaced "Implementation feature" with "Implementation specification."

- Made minor modifications to the text throughout the document for purposes of clarity.

- Modified numerous implementation features so that they are now addressable rather than mandatory.

- Removed the word "formal" when referring to documentation.

- Revised the phrase "health information pertaining to an individual" to "electronic protected health information."

- Added the following definitions to § 160.103: "Disclosure," "Electronic protected health information," "Electronic media," "Organized health care arrangement," and "Use."

- Removed proposed § 142.101 as this information is conveyed in § 160.101 and § 160.102 of the Privacy Rule (65 FR 82798). Removed proposed § 142.102 as it is redundant.

- Removed the following definitions from proposed § 142.103 since they are pertinent to other administrative simplification regulations and are defined elsewhere: code set, health care clearinghouse, health care provider, health information, health plan, medical care, small health plan, standard, and transaction.

- Moved the following definitions from § 164.501 to § 164.103 (proposed § 142.103): "Plan sponsor" and "Protected health information." Added definitions of "Covered functions" and "Required by law."

- Removed proposed § 142.104, "General requirements for health plans," and proposed § 142.105, "Compliance using a health care clearinghouse," since these sections are not pertinent to the security standards.

- Removed proposed § 142.106, "Effective dates of a modification to a standard or implementation specification," since this information is covered in the "Standards for Electronic Transactions" final rule (65 FR 50312),

- Moved proposed § 142.302 to § 164.302. Changed the section heading from "Applicability and scope" to "Applicability." Modified language to state that covered entities must comply with the security standards.

- Moved proposed § 142.304 to § 164.304. Modified language to remove definitions of words and concepts not used in this final rule: "Access control," "Contingency plan," "Participant," "Role-based access control," "Token," and "User-based access."

- Moved proposed § 142.304 to § 164.304. Modified language to add definitions requested by commenters; previously published in Addendum 2 but not in the draft regulation itself; or necessitated by the change of scope to electronic protected health information and alignment with the Privacy Rule to include: "Administrative safeguards," "Availability," "Confidentiality," "Data," "Data authentication Code," "Integrity," "Electronic protected health information," "Facility," "Information System," "Security or security measures," "Security incident," "Technical safeguards," "User," and "Workstation."

- Moved definitions related to privacy from § 164.504 to new § 164.103: "Common control," "Common ownership," "Health care component," "Hybrid entity."

- Moved proposed § 142.306, "Rules for the security Standard," to § 164.306. Modified language to more clearly state the general requirements of the final rule relative to the standards and implementation specifications contained therein. Retitled the section as "Security standards: General Rules."

- Moved proposed § 142.308 to § 164.308. Where this section was proposed to contain all of the security standards in paragraphs (a) through (d), it now encompasses the Administrative safeguards.

- Moved and reorganized proposed § 142.308 (a) through (d) requirements to § 164.308, § 164.310, and § 164.312.

- Moved proposed § 142.308(a)(1), "Certification," to § 164.308(a)(8). Modified language to indicate both technical and nontechnical evaluation is involved and renamed "Evaluation".

- Moved proposed § 142.308(a)(2), "Chain of trust," to § 164.308(b)(1), renamed to "Business associate contracts and other arrangements," and revised language to redefine who must enter into a contract under this rule for the protection of electronic protected health information.

- Moved proposed § 142.308(a)(3), "Contingency plan," to § 164.308(a)(7)(i). Modified language to state that two implementation specifications, "Applications and data criticality analysis" and "Testing and revision procedures," are addressable.

- Removed "Formal mechanism for processing records" (proposed § 142.308(a)(4)) since this requirement was determined to be in part intrusive into business functions and in part redundant.

- Moved proposed § 142.308(a)(5), "Information access control," to § 164.308(a)(4)(i) and renamed as "Information access management." Removed the word "formal" from description. Modified language to state that two implementation specifications ("Access Authorization" and Access Establishment and Modification") are addressable.

- Moved proposed § 142.308(a)(6), "Internal audit," to § 164.308(a)(1)(ii)(D) as an implementation specification under the "Security management process" standard since this was determined to be a more logical placement of this item. Retitled, for clarity, "Information system activity review."

- Moved proposed § 142.308(a)(7), "Personnel security," to § 164.308(a)(3)(i) and retitled "Workforce security." Modified language to state that implementation specifications are addressable.

- Combined proposed § 142.308(a)(7)(i), and § 142.308(a)(7)(iii) ("Assuring supervision of maintenance personnel by an authorized, knowledgeable person" and "Assuring that operations and maintenance personnel have proper access authorization,") under § 164.308(a)(3)(ii)(A) and renamed to "Authorization and/or supervision." Modified description for clarity.

- Moved proposed § 142.308(a)(7)(iv), "Personnel clearance procedure," to § 164.308(a)(3)(ii)(B), renamed to "Workforce clearance procedure," and modified description for clarity.

- Removed proposed § 142.308(a)(7)(v), "Personnel security policies and procedures," as this feature was determined to require redundant effort.

- Removed proposed § 142.308(a)(7)(vi), "Security awareness training." Information concerning this subject has been incorporated under § 164.308(a)(5)(i), "Security awareness and training."

- Removed proposed § 142.308(a)(8), "Security configuration management," and all implementation features, except "Documentation" (hardware and/or software installation, Inventory, Security testing, and Virus checking), since this requirement was determined to be redundant. "Documentation" has been made a discrete standard at § 164.316.

- Moved proposed § 142.308(a)(9), "Security incident procedures," to § 164.308(a)(6)(i) and reworded for clarity. Combined "Report procedures" and "Response procedures" features into a single required implementation specification, named "Response and Reporting" at § 164.308(a)(6)(ii).

- Moved proposed § 142.308(a)(10), "Security management process," to § 164.308(a)(1).

- Moved proposed § 142.308(a)(10)(i), "Risk analysis," to § 164.308(a)(1)(ii)(A).

- Moved proposed § 142.308(a)(10)(ii), "Risk management," to § 164.308(a)(1)(ii)(B).

- Moved proposed § 142.308(a)(10)(iii), "Sanction policy," to § 164.308(a)(1)(ii)(C).

- Removed proposed § 142.308(a)(10)(iv), "Security policy," since this requirement was determined to be redundant.

- Moved proposed § 142.308(a)(11), "Termination," to § 164.308(a)(3)(ii)(C) as an addressable implementation specification under the "Workforce security" standard, and renamed as "Termination procedures". Removed "Termination" implementation features (changing locks, removal from access lists, removal of user accounts, turning in of keys, tokens, or cards) since these were determined to be too specific.

- Moved proposed § 142.308(a)(12), "Training," to § 164.308(a)(5)(i) and renamed as "Security awareness and training." Language modified to incorporate all training information under this one standard. Revised and made addressable all implementation specifications under this standard.

- Moved proposed § 142.308(b), "Physical safeguards to guard data integrity, confidentiality and availability," to § 164.310 and renamed as "Physical safeguards." Removed specific reference to locks and keys.

- Moved proposed § 142.308(b)(1), "Assigned security responsibility requirement," to § 164.308(a)(2) since this has been determined to be an administrative procedure. Modified language to clarify that responsibility could be assigned to more than one individual.

- Moved proposed § 142.308(b)(2), "Media controls," to § 164.310(d)(1) and renamed as "Device and media controls." Removed the word "formal." Added "Media re-use" as a required implementation specification at § 164.310(d)(2)(ii).

- Removed proposed § 142.308(b)(2)(i), "Access control," implementation feature as it was determined to be redundant.

- Moved proposed § 142.308(b)(2)(ii), "Accountability" implementation feature to § 164.310(d)(2)(iii), and made it an addressable implementation specification.

- Combined proposed § 142.308(b)(2)(iii), "Data backup," implementation feature with proposed § 142.308(b)(2)(iv), "Data storage" implementation feature, renamed as "Data backup and storage", moved to § 164.310(d)(2)(iv), and made it an addressable implementation specification.

- Moved proposed § 142.308(b)(2)(v), "Data disposal," implementation feature to § 164.310(d)(2)(i) and made it a required implementation specification.
- Moved proposed § 142.308(b)(3),"Physical access controls," to § 164.310(a)(1) and renamed as "Facility access controls." Removed word "formal."
- Moved proposed § 142.308(b)(3)(i), "Disaster recovery," implementation feature to § 164.310(a)(2)(i). It is now part of the "Contingency operations" implementation specification.
- Moved proposed § 142.308(b)(3)(ii), "Emergency mode operations," implementation feature to § 164.310(a)(2)(i). It is now part of the "Contingency operations" implementation specification.
- Removed proposed § 142.308(b)(3)(iii), "Equipment control (into and out of site)," as this information is now covered under § 164.310(d)(1), "Device and media controls."
- Moved proposed § 142.308(b)(3)(iv), "A facility security plan," to § 164.310(a)(2)(ii).
- Moved proposed § 142.308(b)(3)(v), "Procedure for verifying access authorizations," to § 164.310(a)(2)(iii) and renamed as "Access control and validation procedures." Removed the word "formal" from text.
- Moved proposed § 142.308(b)(3)(vi), "Maintenance records," to § 164.310(a)(2)(iv).
- Moved proposed § 142.308(b)(3)(vii), "Need to know procedures for personnel access," to sect; 164.310(a)(2)(iii) and renamed as "Access control and validation procedures."
- Moved proposed § 142.308(b)(3)(viii), "Procedures to sign in visitors and provide escort, if appropriate," to § 164.310(a)(2)(iii) and renamed as "Access control and validation procedures."
- Moved proposed § 142.308(b)(3)(ix), "Testing and revision," to § 164.310(a)(2)(iii) and renamed as "Access control and validation procedures."
- Moved proposed § 142.308(b)(4), "Policy and guidelines on workstation use," to § 164.310(b) and renamed as "Workstation use."
- Moved proposed § 142.308(b)(5), "Secure work station location," to § 164.310(c) and renamed as "Workstation security."
- Removed proposed § 142.308(b)(6), "Security awareness training," as a separate requirement. This requirement has been incorporated under § 164.308(a)(5)(i), "Security awareness and training."
- Combined and moved proposed § 142.308(c) and § 142.308(d), "Technical security services to guard data integrity, confidentiality and availability" and "Technical security mechanisms," to § 164.312 and renamed as "Technical safeguards."
- Removed proposed § 142.308(c)(1) since it is no longer pertinent.
- Moved proposed § 142.308(c)(1)(i), "Access control," to § 164.312(a)(1).
- Moved proposed § 142.308(c)(1)(i)(A), "Procedure for emergency access," to § 164.312(a)(2)(ii), and renamed as "Emergency access procedures."
- Removed proposed § 142.308(c)(1)(i)(B).
- Removed proposed § 142.308(c)(1)(i)(B)(1), "Context-based access," § 142.308(c)(1)(i)(B)(2), "Role-based access," and § 142.308(c)(1)(i)(B)(3), "User-based access," since these features were deemed too specific and were perceived as the only options permissible.
- Moved proposed § 142.308(c)(1)(i)(C), "Optional use of encryption," to § 164.312(a)(2)(iv) and retitled "Encryption and decryption."
- Moved proposed § 142.308(c)(1)(ii), "Audit controls," to § 164.312(b).

- Removed proposed § 142.308(c)(1)(iii), "Authorization control," and all implementation features (Role-based access, User-based access) since this function has been incorporated into § 164.308(a)(4), "Information access management."
- Moved proposed § 142.308(c)(1)(iv), "Data authentication," to § 164.312(c)(1), and retitled as "Integrity." Reworded part of description and placed in § 164.312(c)(2), "Mechanism to authenticate data," a new, addressable implementation specification. Removed reference to double keying.
- Moved proposed § 142.308(c)(1)(v), "Entity authentication," to § 164.312(d) and retitled as "Person or entity authentication."
- Moved proposed § 142.308(c)(1)(v)(A), "Automatic logoff," to § 164.312(a)(2)(iii).
- Moved proposed § 142.308(c)(1)(v)(B), "Unique user identification," to § 164.312(a)(2)(i).
- Removed proposed § 142.308(c)(1)(v)(C) since text is no longer pertinent.
- Removed proposed § 142.308(c)(1)(v)(C)(2), "Password," as too specific.
- Removed proposed § 142.308(c)(1)(v)(C)(3), "PIN," as too specific.
- Removed proposed § 142.308(c)(1)(v)(C)(4), "Telephone callback," as too specific.
- Removed proposed § 142.308(c)(1)(v)(C)(5), "Token," as too specific.
- Removed proposed § 142.308(c)(2), as no longer relevant.
- Moved proposed § 142.308(d)(1), "Communications or network controls," to § 164.312(e)(1) and renamed as "Transmission security."
- Removed proposed § 142.308(d)(1)(i), since it is no longer pertinent.
- Moved proposed § 142.308(d)(1)(i)(A), "Integrity controls," to § 164.312(e)(2)(i) and reworded for clarity.
- Removed proposed § 142.308(d)(1)(i)(B), "Message authentication," since this subject is now covered under § 164.312(e)(2)(i), "Integrity controls."
- Removed proposed § 142.308(d)(1)(ii) text since it is no longer pertinent.
- Removed proposed § 142.308(d)(1)(ii)(A), "Access controls."
- Moved proposed § 142.308(d)(1)(ii)(B), "Encryption," to § 164.312(e)(2)(ii) and reworded to enhance flexibility and scalability.
- Removed proposed § 142.308(d)(2) text regarding: "Network controls," and all implementation features ("Alarm," "Audio trail," "Entity authentication," "Event reporting").
- Removed proposed § 142.310, "Electronic signature," and all subheadings. This section will be issued as a separate future regulation.
- Moved proposed § 142.310 "Electronic signature Standard," to § 164.310. Where this section was proposed to contain the electronic signature standard, it now encompasses the "Physical safeguards."
- Moved proposed § 142.312, "Effective date of the implementation of the security and electronic signature standards," to § 164.318 and retitled as "Compliance dates for the initial implementation of the security standards." Reworded and retitled subsections.
- Added § 164.105, "Organizational requirements," with two standards, "Health care component and "Affiliated covered entities" with related implementation specifications.
- Added § 164.310(d)(2)(ii), "Media re-use procedures," implementation specification.
- Added § 164.312, "Technical safeguards," encompassing the combined technical services and technical mechanisms standards (proposed § 142.308(c) and (d)).

- Added § 164.314, "Organizational requirements."
- Added § 164.314(a)(1), "Business associate contracts or other arrangements" standard and related implementation specifications.
- Added § 164.314(b)(1), "Requirements for group health plans" standard and related implementation specifications.
- Added § 164.316, "Policies and procedures and documentation requirements."
- Added § 164.316(a), "Policies and procedures" standard.
- Added § 164.316(b)(1), "Documentation" standard and related implementation specifications.
- Added § 164.318, "Compliance dates for the initial implementation of the security standards."
- Renamed Addendum 1 as Appendix A.
- Removed Addendum 2. Definitions of terms used in this final rule are now incorporated into § 164.103 and § 164.304, or within the rule itself.
- Removed Addendum 3.

V. COLLECTION OF INFORMATION REQUIREMENTS

Under the Paperwork Reduction Act of 1995 (PRA), we are required to provide 30-day notice in the **Federal Register** and solicit public comment before a collection of information requirement is submitted to the Office of Management and Budget (OMB) for review and approval. In order to fairly evaluate whether an information collection should be approved by OMB, section 3506(c)(2)(A) of the Paperwork Reduction Act of 1995 (PRA) requires that we solicit comment on the following issues:

- The need for the information collection and its usefulness in carrying out the proper functions of our agency.
- The accuracy of our estimate of the information collection burden.
- The quality, utility, and clarity of the information to be collected.
- Recommendations to minimize the information collection burden on the affected public, including automated collection techniques.

As discussed below, we are soliciting comment on the recordkeeping requirements, as referenced in § 164.306, § 164.308, § 164.310, § 164.314, and § 164.316 of this document.

Section 164.306   Security Standards: General Rules

Under paragraph (d), a covered entity must, if implementing the implementation specification is not reasonable and appropriate, document why it would not be reasonable and appropriate to implement the implementation specification.

We estimate that 75,000 entities will be affected by this requirement and that they will have to create documentation 3 times for this requirement. We estimate each instance of documentation will take. 25 hours, for a one-time total burden of 56,250 hours.

Section 164.308   Administrative Safeguards

Under this section, a covered entity must document known security incidents and their outcomes.

We estimate that there will be 50 known incidents annually and that it will take 8 hours to document this requirement, for an annual burden of 400 hours.

This section further requires that each entity have a contingency plan, with specified components.

We estimate that there will be 60,000 entities affected by this requirement and that it will take each entity 8 hours to comply, for a total one-time burden of 480,000 hours.

This section also requires that the written contract or other arrangement with a business associate document the satisfactory assurances that the business associate will appropriately safeguard the information through a written contract or other arrangement with the business associate that meets the applicable requirements of § 164.314(a).

We believe that the burden associated with this requirement is not subject to the PRA. It is good business practice for entities to document their arrangements via written contracts and as such is usual and customary among the entities subject to them. A burden associated with a requirement conducted in the normal course of business is exempt from the PRA as defined in 5 CFR 1320.3(b)(2).

Section 164.310   Physical Safeguards

This section requires that a covered entity implement policies and procedures to document repairs and modifications to the physical components of a facility that are related to security (for example, hardware, walls, doors, and locks).

We believe that 15,500 entities will have to repair or modify physical components, most of which will need to be done in the first year of implementation. In the following years, we estimate that 500 entities will need to make repairs or modifications. We estimate that it will take 10 minutes to document each repair or modification for a burden of 2,583 hours the first year and 83 hours annually subsequently.

This section requires that a covered entity create a retrievable, exact copy of electronic protected health information, where needed, before movement of equipment.

We believe that the burden associated with this requirement is not subject to the PRA. It is good business practice for entities to backup their data files, and as such is usual and customary among the entities subject to them. A burden associated with a requirement conducted in the normal course of business is exempt from the PRA as defined in 5 CFR 1320.3(b)(2).

Section 164.314   Organizational Requirements

This section requires that a covered entity report to the Secretary problems with a business associate's pattern of an activity or practice of the business associate that constitute a material breach or violation of the business associate's obligation under the contract or other arrangement if it is not feasible to terminate the contract or arrangement.

We believe that 10 entities will need to comply with this reporting requirement and that it will take them 60 minutes to comply with this requirement for an annual burden of 10 hours.

This section also requires that a covered entity may, if a business associate is required by law to perform a function or activity on behalf of a covered entity or to provide a service described in the definition of business associate as specified in § 160.103 of this subchapter to a covered entity, permit the business associate to create, receive, maintain, or transmit electronic protected health information on its behalf to the extent necessary to comply with the legal mandate without meeting the requirements of paragraph (a)(2)(i) of this section, provided that the covered entity attempts in good faith to obtain satisfactory assurances as required by paragraph (a)(2)(ii)(A) of this section, and documents the attempt and the reasons that these assurances cannot be obtained.

We believe that this situation will affect 20 entities and that it will take 60 minutes to document attempts to obtain assurances and the reasons they cannot be obtained for an annual burden of 20 hours.

This section further requires that business associate contracts or other arrangements and group health plans must require the business entity and plan sponsor, respectively, to report to the covered entity any security incident of which it becomes aware.

We believe that the burden associated with this requirement is not subject to the PRA. It is good business practice for entities to document their agreements via written contracts, and as such is usual and customary among the entities subject to them. A burden associated with a requirement conducted in the normal course of business is exempt from the PRA as defined in 5 CFR 1320.3(b)(2).

Section 164.316   Policies and Procedures and Documentation Requirements

Paragraph (b)(1), *Standard: Documentation*, of this section requires a covered entity to—

(i) Maintain the policies and procedures implemented to comply with this subpart in written (which may be electronic) form; and

(ii) If an action, activity, assessment, or designation is required by this subpart to be documented, maintain a written (which may be electronic) record of the action, activity, assessment, or designation.

We estimate that it will take the 4,000,000 entities covered by this final rule 16 hours to document their policies and procedures, for a total one-time burden of 64,000,000 hours.

The total annual burden of the information collection requirements contained in this final rule is 64,539,264 hours. These information collection requirements will be submitted to OMB for review under the PRA and will not become effective until approved by OMB.

If you comment on these information collection and recordkeeping requirements, please mail copies directly to the following:

> Centers for Medicare and Medicaid Services, Office of Strategic Operations and Regulatory Affairs, Regulations Development and Issuances Group, Attn: Reports Clearance Officer, 7500 Security Boulevard, Baltimore, MD 21244–1850, Attn: Julie Brown, CMS–0049–F; and

> Office of Information and Regulatory Affairs, Office of Management and Budget, Room 10235, New Executive Office Building, Washington, DC 20503, Attn: Brenda Aguilar, CMS Desk Officer.

IV.   REGULATORY IMPACT ANALYSIS

A.   Overall Impact

We have examined the impacts of this rule as required by Executive Order 12866 (September 1993, Regulatory Planning and Review), the Regulatory Flexibility Act (RFA) (September 16, 1980, Pub. L. 96–354), section 1102(b) of the Social Security Act, the Unfunded Mandates Reform Act of 1995 (Pub. L. 104–4), and Executive Order 13132.

Executive Order 12866 (as amended by Executive Order 13258, which merely reassigns responsibility of duties) directs agencies to assess all costs and benefits of available regulatory alternatives and, if regulation is necessary, to select regulatory approaches that maximize net benefits (including potential economic, environmental, public health and safety effects, distributive impacts, and equity). A regulatory impact analysis (RIA) must be prepared for major rules with economically significant effects ($100 million or more in any 1 year). Although we cannot determine the specific economic impact of the standards in this final rule (and individually each standard may not have a significant impact), the overall impact analysis makes clear that, collectively, all the standards will have a significant impact of over $100 million on the economy. Because this rule affects over 2 million entities, a requirement as low as $50 per entity would render this rule economically significant. This rule requires each of these entities to engage in, for example, at least some risk assessment activity; thus, this rule is almost certainly economically significant even though we do not have an estimate of the marginal impact of the additional security standards. However, the standards adopted in this rule are considerably more flexible than those anticipated in the overall impact analysis. Therefore, their implementation costs should be lower than those assumed in the impact analysis.

The RFA requires agencies to analyze options for regulatory relief of small businesses. For purposes of the RFA, small entities include small businesses, nonprofit organizations, and government agencies. Most hospitals and most other providers and suppliers are small entities, either by nonprofit status or by having revenues of $6 million to $29 million in any 1 year. While each standard may not have a significant impact on a substantial number of small entities, the combined effects of all the standards are likely to have a significant effect on a substantial number of small entities. Although we have certified this rule as having a significant impact, we have previously discussed the impact of small entities in the RFA published as part of the August 17, 2000 final regulation for the Standards for Electronic Transactions (65 FR 50312), on pages 50359 through 50360. That analysis included the impact of the set of HIPAA standards regulations (transactions and code sets, identifiers, and security). Although we discussed the impact on small entities in the previous analysis, we would like to discuss how this final rule has been structured to minimize the impact on small entities, compared to the proposed rule.

The proposed rule mandated 69 implementation features for all entities. A large number of commenters indicated that mandating such a large number would be burdensome for all entities. As a result, we have restructured this final rule to permit greater flexibility. While all standards must be met, we are now only requiring 13 implementation specifications. The remainder of the implementation specifications is "addressable." For addressable specifications, an entity decides whether each specification is a reasonable and appropriate security measure to apply within its particular security framework. This decision is based on a variety of factors, for example, the entity's risk analysis, what measures are already in place, the particular interest to small entities, and the cost of implementation.

Based on the decision, an entity can—(1) implement the specification if reasonable and appropriate; (2) implement an alternative security measure to accomplish the purposes of the standard; or (3) not implement anything if the specification is not reasonable and appropriate and the standard can still be met.

This approach will provide flexibility for all entities, and especially small entities that would be most concerned about the cost and complexity of the security standards. Small entities can look at the addressable implementation specifications and tailor their compliance based on their risks and capabilities of addressing those risks.

The required risk analysis is also a tool to allow flexibility for entities in meeting the requirements of this final rule. The risk analysis requirement is designed to allow entities to look at their own operations and determine the security risks involved. The degree of response is determined by the risks identified. We assume that smaller entities, who deal with smaller amounts of information would have smaller physical facilities, smaller work forces, and therefore, would assume less risk. The smaller amount of risk involved means that the response to that risk can be developed on a smaller scale than that for larger organizations.

Individuals and States are not included in the definition of a small entity. However, the security standards will affect small entities, such as providers and health plans, and vendors in much the same way as they affect any larger entities. Small providers who conduct electronic transactions and small health plans must meet the provisions of this regulation and implement the security standards. A more detailed analysis of the impact on small entities is part of the impact analysis published on August 17, 2000 (65 FR 50312), which provided the impact for all of the HIPAA standards, except privacy. As we discussed above, the scalability factor of the standards means that the requirements placed upon small providers and plans would be consistent with the complexity of their operations. Therefore, small providers and plans with appropriate security processes in place would need to do relatively little in order to comply with the standards. Moreover, small plans will have an additional year to come into compliance.

In addition, section 1102(b) of the Act requires us to prepare a regulatory impact analysis if a rule may have a significant impact on the operations of a substantial number of small rural hospitals. This analysis must conform to the provisions of section 604 of the RFA. For purposes of section 1102(b) of the Act, we define a small rural hospital as a hospital that is located outside of a Metropolitan Statistical Area and has fewer than 100 beds. While this rule may have a significant impact on small rural hospitals, the impact should be minimized by the scalability factors of the standards, as discussed above in the impact on all small entities. In addition, we have previously discussed the impact of small entities in the RIA published as part of the August 17, 2000 final regulation for the Standards for Electronic Transactions.

Section 202 of the Unfunded Mandates Reform Act (UMRA) of 1995 also requires that agencies assess anticipated costs and benefits before issuing any rule that may result in expenditure in any 1 year by State, local, or tribal governments, in the aggregate, or by the private sector, of $110 million. We estimate that implementation of all the standards will require the expenditure of more than $110 million by the private sector. Therefore, the rule establishes a Federal private sector mandate and is a significant regulatory action within the meaning of section 202 of UMRA (2 U.S.C. 1532). We have included the statements to address the anticipated effects of these rules under section 202.

These standards also apply to State and local governments in their roles as health plans or health care providers. Because these entities, in their roles as health plans or providers, must implement the requirements in these rules, the rules impose unfunded mandates on them. Further discussion of this issue can be found in the previously published impact analysis for all standards (65 FR 50360 through 50361).

The anticipated benefits and costs of the security standards, and other issues raised in section 202 of the UMRA, are addressed in the analysis below, and in the combined impact analysis. In addition, as required under section 205 of the UMRA (2 U.S.C. 1535), having considered a reasonable number of alternatives as outlined in the preamble to this rule, HHS has concluded that this final rule is the most cost-effective alternative for implementation of HHS's statutory objective of administrative simplification.

Executive Order 13132 establishes certain requirements that an agency must meet when it promulgates a proposed rule (and subsequent final rule) that imposes substantial direct requirement costs on State and local governments, preempts State law, or otherwise has Federalism implications. The proposed rule was published before the enactment of Executive

Order 13132 of August 4, 1999, Federalism (published in the **Federal Register** on August 10, 1999 (64 FR 43255)), which required meaningful and timely input by State and local officials in the development of rules that have Federalism implications). However, we received and considered comments on the proposed rule from State agencies and from entities who conduct transactions with State agencies. Several of the comments referred to the costs that will result from implementation of the HIPAA standards. As we stated in the impact analysis, we are unable to estimate the cost of implementing security features as implementation needs will vary dependent upon a risk assessment and upon what is already in place. However, the previously referenced impact analysis in the August 17, 2000 final rule (65 FR 50312) showed that Administrative Simplification costs will be offset by future savings.

In complying with the requirements of part C of title XI, the Secretary established interdepartmental implementation teams who consulted with appropriate State and Federal agencies and private organizations. These external groups consisted of the National Committee on Vital and Health Statistics (NCVHS) Subcommittee on Standards and Security, the Workgroup for Electronic Data Interchange (WEDI), the National Uniform Claim Committee (NUCC), the National Uniform Billing Committee (NUBC), and the American Dental Association (ADA). The teams also received comments on the proposed regulation from a variety of organizations, including State Medicaid agencies and other Federal agencies.

B. Anticipated Effects

The analysis in the August 2000, Transaction Rule included the expected costs and benefits of the administrative simplification regulations related to electronic systems for 10 years. Although only the electronic transaction standards were promulgated in the transaction rule, HHS expected affected parties to make systems compliance investments collectively because the regulations are so integrated. Moreover, the data available to us were also based on the collective requirements of this regulation. It is not feasible to identify the incremental technological and computer costs for each regulation. Although HHS is issuing rules under HIPAA sequentially, affected entities and vendors are bundling services, that is, they have been anticipating the various needs and are designing relatively comprehensive systems as they develop hardware and software. For example, a vendor developing a system for electronic billing would also anticipate and include security features, even in the absence of any regulation. Moreover, a draft of the security rule was first published in 1998. Even though the final is different (and less burdensome), vendors had a reasonable indication of the direction policy would go. Thus, in preparing the electronic transaction rule, we recognized and included costs that might theoretically be associated with security or other HIPPA rules. Hence, some of the "costs" of security have already been accounted for in the Standards for Electronic Transactions cost estimate (45 CFR parts 160 and 162), which was published in the **Federal Register** on August 17, 2000 (65 FR 50312).

This analysis showed that the combined impact of the Administrative Simplification standards is expected to save the industry $29.9 billion over 10 years. We are including in each subsequent rule an impact analysis that is specific to the standard or standards in that rule, but the impact analysis will assess only the incremental cost of implementing a given standard over another. Thus, the following discussion contains the impact analysis for the marginal costs of the security standards in this final rule.

The following describes the specific impacts that relate to the security standards. The security of electronic protected health information is, and has been for some time, a basic business requirement that health care entities ignore at their peril. Instances of "hacking" and other security violations may be widely publicized, and can seriously damage an institution's community standing. Appropriate security protections are crucial for encouraging the growth and use of electronic data interchange. The synergistic effect of the employment of the security standards will enhance all aspects of HIPAA's Administrative Simplification requirements. In addition, it is important to recognize that security is not a one-time project, but rather an ongoing, dynamic process.

C. Changes From the 1998 Impact Analysis

The overall impact analysis for Administrative Simplification was first published on May 7, 1998 (63 FR 25320) in the proposed rule for the National Provider Identifier standard (45 CFR part 142), the first of the proposed Administrative Simplification rules. That impact analysis was based on the industry situation at that time, used statistics which were current at that time, and assumed that all of the HIPAA standards would be implemented at roughly the same time, which would permit software changes to be made less expensively. While the original impact analysis represented our best information at that time, we realize that the state of the industry, and of security technology, has changed since 1998. We discuss several of those changes and how they affect the impact of this regulation.

1. Changes in Technology. The state of technology for health care security has changed since 1998. New technologies to protect information have been developed over the past several years. As a result, HHS has consulted with the Gartner Group, a leading technology assessment organization, regarding what impact these changes in the industry might have on the expected impact of this regulation. The Gartner analysis indicated that the cost of meeting the requirements of a reasonable interpretation of the security rule in 2002 is probably less than 10 percent higher in 2002 than it was in 1998. This increase is mainly driven by more active threats and increased personnel costs offsetting decreases in technology costs over the past 4 years. However, spending by companies who have anticipated the security rule or who have independently made business decisions to implement security policies and procedures as good business practice(s) has already occurred, and probably will cancel out the increased costs of implementation. Therefore, Gartner expects the cost of complying with the HIPAA security standards to be about the same now as it was in 1998.

2. Synchronizing Standards. The timelines for the implementation of the initial HIPAA standards (transactions, identifiers, and security) are no longer closely synchronized. However, we do not believe that this lack of synchronization will have a significant impact on the cost of implementing security. The analysis provided by the Gartner group indicated that implementing security standards is being viewed by entities as a separate task from implementing the transaction standards, and that this is not having a significant impact on costs. As with other HIPAA standards, most current entities will have a 2-year implementation period before compliance with the standards is required. Covered entities will develop their own implementation schedules, and may phase in various security measures over that time period.

3. Relationship to Privacy Standards. The publication of the final Privacy Rules (45 CFR parts 160 and 164) on December 28, 2000 in the **Federal Register** (65 FR 82462) and on August 14, 2002 (67 FR 53182) has affected the impact of this regulation significantly. Covered entities must implement the privacy standards by April 14, 2003 (April 14, 2004 for small health plans). The implementation of privacy standards reduces the cost of implementing the security standards in two significant areas.

First, we have made substantial efforts to ensure that the many requirements in the security standards parallel those for privacy, and can easily be satisfied using the solutions for privacy. Administrative requirements like the need for written policies, responsible officers, and business associate agreements that are already required by the Privacy Rule can also serve to meet the security standards without significant additional cost. The analysis of data flows and data uses that covered entities are doing so as to comply with the Privacy Rule should also serve as the starting point for parallel analysis required by this final rule.

Second, it is likely that covered entities will meet a number of the requirements in the security standards through the implementation of the privacy requirements. For example, in order to comply with the Privacy Rule requirements to make reasonable efforts to limit the access of members of the work force to specified categories of protected health information, covered entities may implement some of the administrative, physical, and technical safeguards that the entity's risk analysis and assessment would require under the Security Rule. E-mail authentication procedures put into place for privacy protection may also meet the security standards, thereby eliminating the need for additional investments to meet these standards. As a result, covered entities that have moved forward in implementing the privacy standards are also implementing security measures at the same time. Since the proposed security standards proposed rule represents the most authoritative guidance now available on the nature of these standards, some entities have been using them to develop their security measures. Those entities should face minimal incremental costs in implementing the final version of these standards.

We are unable to quantify these overlaps, but we believe they may reduce the cost of implementing these security standards. The analysis provided to the HHS by the Gartner Group also stated that compliance with the Privacy Rule will have a moderate effect on the cost of compliance with the Security Rule, reducing it slightly.

4. *Sensitivity to Security Concerns as a Result of September 11, 2001.* In our discussions with the Gartner Group, they indicated that they saw little evidence of increased security awareness in health care organizations as a result of the events of September 11, 2001. However, a survey conducted by Phoenix Health Systems in the winter of 2002 showed that 65 percent of the respondents to the survey (hospitals, payers, vendors, and clearinghouses) have moderately to greatly increased their attention on overall security. If these organizations have already made investments in security that meet some of the requirements of this rule, it will reduce their added costs of compliance. However, HHS can make no clear statement of the impact of this attention.

D. Guiding Principles for Standard Selection

The implementation teams charged with designating standards under the statute have defined, with significant input from the health care industry, a set of common criteria for evaluating potential standards. These criteria are based on direct specifications in the HIPAA, the purpose of the law, and principles that support the regulatory philosophy set forth in the E.O. 12866 of September 30, 1993, and the Paperwork Reduction Act of 1995. In order to be designated as such, a standard should do the following:

- Improve the efficiency and effectiveness of the health care system by leading to cost reductions for or improvements in benefits from electronic health care transactions. This principle supports the regulatory goals of cost-effectiveness and avoidance of burden.

- Meet the needs of the health data standards user community, particularly health care providers, health plans, and health care clearinghouses. This principle supports the regulatory goal of cost-effectiveness.

- Be consistent and uniform with the other HIPAA standards (that is, their data element definitions and codes, and their privacy and security requirements) and, secondarily, with other private and public sector health data standards. This principle supports the regulatory goals of consistency and avoidance of incompatibility, and it establishes a performance objective for the standard.

- Have low additional development and implementation costs relative to the benefits of using the standard. This principle supports the regulatory goals of cost-effectiveness and avoidance of burden.

- Be supported by an ANSI-accredited standards developing organization or other private or public organization that would ensure continuity and efficient updating of the standard over time. This principle supports the regulatory goal of predictability.

- Have timely development, testing, implementation, and updating procedures to achieve administrative simplification benefits faster. This principle establishes a performance objective for the standard.

- Be technologically independent of the computer platforms and transmission protocols used in health transactions, except when they are explicitly part of the standard. This principle establishes a performance objective for the standard and supports the regulatory goal of flexibility.

- Be precise and unambiguous but as simple as possible. This principle supports the regulatory goals of predictability and simplicity.

- Keep data collection and paperwork burdens on users as low as is feasible. This principle supports the regulatory goals of cost-effectiveness and avoidance of duplication and burden.

- Incorporate flexibility to adapt more easily to changes in the health care infrastructure (for example, new services, organizations, and provider types) and information technology. This principle supports the regulatory goals of flexibility and encouragement of innovation.

We assessed a wide variety of security standards and guidelines against the principles listed above, with the overall goal of achieving the maximum benefit for the least cost. As we stated in the proposed rule, we found that no single standard for security exists that encompasses all the requirements that were listed in the law. However, we believe that the standards we are adopting in this final rule collectively accomplish these goals.

E.   Affected Entities

1.   Health Care Providers.   Covered health care providers may incur implementation costs for establishing or updating their security systems. The majority of costs to implement the security standard (purchase and installation of appropriate computer hardware and software, and physical safeguards) would generally be incurred in the initial implementation period for the specific requirements of the security standard. Health care providers that do not conduct electronic transactions for which standards have been adopted are not affected by these regulations.

2.   Health Plans.   All health plans, as the term is defined in regulation at 45 CFR 160.103, must comply with these security standards. In addition, health plans that engage in electronic health care transactions may have to modify their systems to meet the security standards. Health plans that maintain electronic health information may also have to modify their systems to meet the security standards. This conversion would have a one-time cost impact on Federal, State, and private plans alike.

We recognize that this conversion process has the potential to cause business disruption of some health plans. However, health plans would be able to schedule their implementation of the security standards and other standards in a way that best fits their needs, as long as they meet the deadlines specified in the HIPAA law and regulations. Moreover, small plans (many of which are employer-sponsored) will have an additional year in which to achieve compliance. Small health plans are defined at 45 CFR 160.103 as health plans with annual receipts of $5 million or less.

3. Clearinghouses. All health care clearinghouses must meet the requirements of this regulation. Health care clearinghouses would face effects similar to those experienced by health care providers and health plans. However, because clearinghouses represent one way in which providers and plans can achieve compliance, the clearinghouses' costs of complying with these standards would probably be passed along to those entities, to be shared over the entire customer base.

4. System Vendors. Systems vendors that provide computer software applications to health care providers and other billers of health care services would likely be affected. These vendors would have to develop software solutions that would allow health plans, providers, and other users of electronic transactions to protect these transactions and the information in their databases from unauthorized access to their systems. Their costs would also probably be passed along to their customer bases.

F. Factors in Establishing the Security Standard

1. General Effect. In assessing the impact of these standards, it is first necessary to focus on the general nature of the standards, their scalability, and the fact that they are not dependent upon specific technologies. These factors will make it possible for covered entities to implement them with the least possible impact on resources. Because there is no national security standard in widespread use throughout the industry, adopting any of the candidate standards would require most health care providers, health plans, and health care clearinghouses to at least conduct an assessment of how their current security measures conform to the new standards. However, we assume that most, if not all, covered entities already have at least some rudimentary security measures in place. Covered entities that identify gaps in their current measures would need to establish or revise their security precautions.

It is also important to note that the standards specify what goals are to be achieved, but give the covered entity some flexibility to determine how to meet those goals. This is different from the transaction standards, where all covered entities must use the exact same implementation guide. With respect to security, covered entities will be able to blend security processes now in place with new processes. This should significantly reduce compliance costs.

Based on our analysis and comments received, the security standards adopted in this rule do not impose a greater burden on the industry than the options we did not select, and they present significant advantages in terms of universality and flexibility.

We understand that some large health plans, health care providers, and health care clearinghouses that currently exchange health information among trading partners may already have security systems and procedures in place to protect the information from unauthorized access. These entities may not incur significant costs to meet the security standards. Large entities that have sophisticated security systems in place may only need minor revisions or updates to their systems to meet the security standards, or indeed, may not need to make any changes in their systems.

While small providers are not likely to have implemented sophisticated security measures, they are also not as likely to need them as larger covered entities. The scalability principle allows providers to adopt measures that are appropriate to their own circumstances.

2. Complexity of Conversion.   The complexity of the conversion to the security standards could be significantly affected by the volume of transactions that covered entities transmit and process electronically and the desire to transmit directly or to use the services of a Value Added Network (VAN) or a clearinghouse. If a VAN or clearinghouse is used, some of the conversion activities would be carried out by that organization, rather than by the covered entity. This would simplify conversion for the covered entity, but makes the covered entity dependent on the success of its business associate. The architecture, and specific technology limitations of existing systems could also affect the complexity of the conversion (for example, certain practice management software that does not contain password protection will require a greater conversion effort than software that has a password protection option already built into it).

3. Cost of Conversion.   Virtually all providers, health plans, and clearinghouses that transmit or store data electronically have already implemented some security measures and will need to assess existing security, identify areas of risk, and implement additional measures in order to come into compliance with the standards adopted in this rule. We cannot estimate the per-entity cost of implementation because there is no information available regarding the extent to which providers', plans', and clearinghouses' current security practices are deficient. Moreover, some security solutions are almost cost-free to implement (for example, reminding employees not to post passwords on their monitors), while others are not.

Affected entities will have many choices regarding how they will implement security. Some may choose to assess security using in-house staff, while others will use consultants. Practice management software vendors may also provide security consultation services to their customers. Entities may also choose to implement security measures that require hardware and/or software purchases at the time they do routine equipment upgrades.

The security standards we adopt in this rule were developed with considerable input from the health care industry, including providers, health plans, clearinghouses, vendors, and standards organizations. Industry members strongly advocated the flexible approach we adopt in this rule, which permits each affected entity to develop cost-effective security measures appropriate to their particular needs. We believe that this approach will yield the lowest implementation cost to industry while ensuring that electronic protected health information is safeguarded.

All of the nation's health plans (over 2 million) and providers (over 600,000) will need to conduct some level of gap analysis to assess current procedures against the standards. However, we cannot estimate the number of covered entities that would have to implement additional security systems and procedures to meet the adopted standards. Also, we are not able to estimate the number of providers that do not conduct electronic transactions today but may choose to do so at some future time (these would be entities that send and receive paper transactions and maintain paper records and thus would not be affected). We believe that the security standards represent the minimum necessary for adequate protection of health information in an electronic format and as such should be implemented by all covered entities. As discussed earlier in this preamble, the security requirements are both scalable and technically flexible; and while the law requires each health plan that is not a small plan to comply with the security and electronic signature requirements no later than 24 months after the effective date of the final rule, small plans will be allowed an additional 12 months to comply.

Since we are unable to estimate the number of entities that may need to make changes to meet the security standards, we are also unable to estimate the cost for those entities. However, we believe that the cost of establishing security systems and procedures is a portion of the costs associated with converting to the administrative simplification standards that are required under HIPAA, which are estimated in the previously referenced impact analysis.

This discussion on conversion costs relates only to health plans, health care providers, and health care clearinghouses that are required to implement the security standards. The cost of implementing security systems and procedures for entities that do not transmit, receive, or maintain health information electronically is not a cost imposed by the rule, and thus, is not included in our estimates.

G. Alternatives Considered

In developing this final rule, the Department considered some alternatives. One alternative was to not issue a final rule. However, this would not meet the Department's obligations under the HIPAA statute. It would also leave the health industry without a set of standards for protecting the security of health information. The vast majority of commenters supported our efforts in developing a set of standards. Thus, we concluded that not publishing a final rule was not in the best interests of the industry and not in the best interests of persons whose medical information will be protected by these measures.

A second alternative was to publish the final rule basically unchanged from the proposed rule. Although most commenters supported the approach of the proposed rule, there were significant objections to the number of required specifications, concerns about the scope of certain requirements, duplication and ambiguity of some requirements, and the overall complexity of the approach. Based on those comments, it was clear that revisions had to be made. In addition, the proposed rule was developed before the Privacy Rule requirements were developed. Thus, it did not allow for any alignment of requirements between the Privacy and Security standards.

As a result, the Department determined that an approach that modified the proposed rule and aligned the requirements with the Privacy standards was the preferred alternative.

V. FEDERALISM

Executive Order 13132 of August 4, 1999, Federalism, published in the **Federal Register** on August 10, 1999 (64 FR 43255), requires us to ensure meaningful and timely input by State and local officials in the development of rules that have Federalism implications. Although the proposed rule for security standards was published before the enactment of this Executive Order, the Department consulted with State and local officials as part of an outreach program in the process of developing the proposed regulation. The Department received comments on the proposed rule from State agencies and from entities that conduct transactions with State agencies. Many of these comments were concerned with the burden that the proposed security standards would place on their organizations. In response to those comments, we have modified the security standards to make them more flexible and less burdensome.

In complying with the requirements of part C of Title XI, the Secretary established an interdepartmental team who consulted with appropriate State and Federal agencies and private organizations. These external groups included the NCVHS Workgroup on Standards and Security, the Workgroup for Electronic Data Interchange, the National Uniform Claim Committee, and the National Uniform Billing Committee. Most of these groups have State officials as members. We also received comments on the proposed regulation from these organizations.

In accordance with the provisions of Executive Order 12866, this rule has been reviewed by the Office of Management and Budget.

LIST OF SUBJECTS

45 CFR Part 160

Electronic transactions, Employer benefit plan, Health, Health care, Health facilities, Health insurance, Health records, Medicaid, Medical research, Medicare, Privacy, Reporting and record keeping requirements.

45 CFR Part 162

Administrative practice and procedure, Health facilities, Health insurance, Hospitals, Medicaid, Medicare, report and recordkeeping requirement.

45 CFR Part 164

Administrative practice and procedure, Health facilities, Health insurance, Hospitals, Medicaid, Medicare, Electronic Information System, Security, Report and recordkeeping requirement.

For the reasons set forth in the preamble, the Department of Health and Human Services amends title 45, subtitle A, subchapter C, parts 160, 162, and 164 as set forth below:

PART 160 — GENERAL ADMINISTRATIVE REQUIREMENTS

1. The authority citation for part 160 continues to read as follows:

Authority: Sec. 1171 through 1179 of the Social Security Act, (42 U.S.C. 1320d–1329d–8) as added by sec. 262 of Pub. L. 104–191, 110 Stat. 2021–2031 and sec. 264 of Pub. L. 104–191 (42 U.S.C. 1320d–2(note)).

2. In § 160.103, the definitions of "disclosure", "electronic media", "electronic protected health information," "individual," "organized health care arrangement", "protected health information," and "use" are added in alphabetical order to read as follows:

§ 160.103 DEFINITIONS

*   *   *   *   *

Disclosure means the release, transfer, provision of, access to, or

divulging in any other manner of information outside the entity holding the information.

*   *   *   *   *

Electronic media means:

(1) Electronic storage media including memory devices in computers (hard drives) and any removable/transportable digital memory medium, such as magnetic tape or disk, optical disk, or digital memory card; or

(2) Transmission media used to exchange information already in electronic storage media. Transmission media include, for example, the internet (wide-open), extranet (using internet technology to link a business with information accessible only to collaborating parties), leased lines, dial-up lines, private networks, and the physical movement of removable/ transportable electronic storage media. Certain transmissions, including of paper, via facsimile, and of voice, via telephone, are not considered to be transmissions via electronic media, because the information being exchanged did not exist in electronic form before the transmission.

Electronic protected health information means information that comes within paragraphs (1)(i) or (1)(ii) of the definition of *protected health information* as specified in this section.

* * * * *

Individual means the person who is the subject of protected health information.

* * * * *

Organized health care arrangement means:

(1) A clinically integrated care setting in which individuals typically receive health care from more than one health care provider;

(2) An organized system of health care in which more than one covered entity participates and in which the participating covered entities:

(i) Hold themselves out to the public as participating in a joint arrangement; and

(ii) Participate in joint activities that include at least one of the following:

(A) Utilization review, in which health care decisions by participating covered entities are reviewed by other participating covered entities or by a third party on their behalf;

(B) Quality assessment and improvement activities, in which treatment provided by participating covered entities is assessed by other participating covered entities or by a third party on their behalf; or

(C) Payment activities, if the financial risk for delivering health care is shared, in part or in whole, by participating covered entities through the joint arrangement and if protected health information created or received by a covered entity is reviewed by other participating covered entities or by a third party on their behalf for the purpose of administering the sharing of financial risk.

(3) A group health plan and a health insurance issuer or HMO with respect to such group health plan, but only with respect to protected health information created or received by such health insurance issuer or HMO that relates to individuals who are or who have been participants or beneficiaries in such group health plan;

(4) A group health plan and one or more other group health plans each of which are maintained by the same plan sponsor; or

(5) The group health plans described in paragraph (4) of this definition and health insurance issuers or HMOs with respect to such group health plans, but only with respect to protected health information created or received by such health insurance issuers or HMOs that relates to individuals who are or have been participants or beneficiaries in any of such group health plans.

Protected health information means individually identifiable health information:

(1) Except as provided in paragraph (2) of this definition, that is:

    (i) Transmitted by electronic media;

    (ii) Maintained in electronic media; or

    (iii) Transmitted or maintained in any other form or medium.

(2) *Protected health information* excludes individually identifiable health information in:

    (i) Education records covered by the Family Educational Rights and Privacy Act, as amended, 20 U.S.C. 1232g;

    (ii) Records described at 20 U.S.C. 1232g(a)(4)(B)(iv); and

    (iii) Employment records held by a covered entity in its role as employer.

    *   *   *   *   *

Use means, with respect to individually identifiable health information, the sharing, employment, application, utilization, examination, or analysis of such information within an entity that maintains such information.

    *   *   *   *   *

PART 162—ADMINISTRATIVE REQUIREMENTS

1. The authority citation for part 162 is revised to read as follows:

    Authority: Secs. 1171 through 1179 of the Social Security Act (42 U.S.C. 1320d–1320d–8), as added by sec. 262 of Pub. L. 104–191, 110 Stat. 2021–2031, and sec. 264 of Pub. L. 104–191, 110 Stat. 2033–2034 (42 U.S.C. 1320d–2 (note)).

§ 162.103 [AMENDED]

2. In § 162.103, the definition of "electronic media" is removed.

PART 164—SECURITY AND PRIVACY

1. The authority citation for part 164 is revised to read as follows:

    Authority: Secs. 1171 through 1179 of the Social Security Act (42 U.S.C. 1320d–1320d–8), as added by sec. 262 of Pub. L. 104–191, 110 Stat. 2021–2031, and 42 U.S.C. 1320d–2 and 1320d–4, sec. 264 of Pub. L. 104–191, 110 Stat. 2033–2034 (42 U.S.C. 1320d–2 (note)).

    2. A new § 164.103 is added to read as follows:

§ 164.103 DEFINITIONS

As used in this part, the following terms have the following meanings:

Common control exists if an entity has the power, directly or indirectly, significantly to influence or direct the actions or policies of another entity.

Common ownership exists if an entity or entities possess an ownership or equity interest of 5 percent or more in another entity.

Covered functions means those functions of a covered entity the performance of which makes the entity a health plan, health care provider, or health care clearinghouse.

Health care component means a component or combination of components of a hybrid entity designated by the hybrid entity in accordance with § 164.105(a)(2)(iii)(C).

Hybrid entity means a single legal entity:

(1) That is a covered entity;

(2) Whose business activities include both covered and non-covered functions; and

(3) That designates health care components in accordance with paragraph § 164.105(a)(2) (iii)(C).

Plan sponsor is defined as defined at section 3(16)(B) of ERISA, 29 U.S.C. 1002(16)(B).

Required by law means a mandate contained in law that compels an entity to make a use or disclosure of protected health information and that is enforceable in a court of law. *Required by law* includes, but is not limited to, court orders and court-ordered warrants; subpoenas or summons issued by a court, grand jury, a governmental or tribal inspector general, or an administrative body authorized to require the production of information; a civil or an authorized investigative demand; Medicare conditions of participation with respect to health care providers participating in the program; and statutes or regulations that require the production of information, including statutes or regulations that require such information if payment is sought under a government program providing public benefits.

3. Section 164.104 is revised to read as follows:

§ 164.104   APPLICABILITY

(a) Except as otherwise provided, the standards, requirements, and implementation specifications adopted under this part apply to the following entities:

(1) A health plan.

(2) A health care clearinghouse.

(3) A health care provider who transmits any health information in electronic form in connection with a transaction covered by this subchapter.

(b) When a health care clearinghouse creates or receives protected health information as a business associate of another covered entity, or other than as a business associate of a covered entity, the clearinghouse must comply with § 164.105 relating to organizational requirements for covered entities, including the designation of health care components of a covered entity.

4. A new § 164.105 is added to read as follows:

§ 164.105   ORGANIZATIONAL REQUIREMENTS

(a) **(1)** *Standard: Health care component.* If a covered entity is a hybrid entity, the requirements of subparts C and E of this part, other than the requirements of this section, § 164.314, and § 164.504, apply only to the health care component(s) of the entity, as specified in this section.

(2) *Implementation specifications:*

(i) *Application of other provisions.* In applying a provision of subparts C and E of this part, other than the requirements of this section, § 164.314, and § 164.504, to a hybrid entity:

(A) A reference in such provision to a "covered entity" refers to a health care component of the covered entity;

(B) A reference in such provision to a "health plan," "covered health care provider," or "health care clearinghouse," refers to a health care component of the covered entity if such health care component performs the functions of a health plan, health care provider, or health care clearinghouse, as applicable;

(C) A reference in such provision to "protected health information" refers to protected health information that is created or received by or on behalf of the health care component of the covered entity; and

(D) A reference in such provision to "electronic protected health information" refers to electronic protected health information that is created, received, maintained, or transmitted by or on behalf of the health care component of the covered entity.

(ii) *Safeguard requirements.* The covered entity that is a hybrid entity must ensure that a health care component of the entity complies with the applicable requirements of this section and subparts C and E of this part. In particular, and without limiting this requirement, such covered entity must ensure that:

(A) Its health care component does not disclose protected health information to another component of the covered entity in circumstances in which subpart E of this part would prohibit such disclosure if the health care component and the other component were separate and distinct legal entities;

(B) Its health care component protects electronic protected health information with respect to another component of the covered entity to the same extent that it would be required under subpart C of this part to protect such information if the health care component and the other component were separate and distinct legal entities;

(C) A component that is described by paragraph (a)(2)(iii)(C)(2) of this section does not use or disclose protected health information that it creates or receives from or on behalf of the health care component in a way prohibited by subpart E of this part;

(D) A component that is described by paragraph (a)(2)(iii)(C)(2) of this section that creates, receives, maintains, or transmits electronic protected health information on behalf of the health care component is in compliance with subpart C of this part; and

(E) If a person performs duties for both the health care component in the capacity of a member of the workforce of such component and for another component of the entity in the same capacity with respect to that component, such workforce member must not use or disclose protected health information created or received in the course of or incident to the member's work for the health care component in a way prohibited by subpart E of this part.

(iii) *Responsibilities of the covered entity.* A covered entity that is a hybrid entity has the following responsibilities:

 (A) For purposes of subpart C of part 160 of this subchapter, pertaining to compliance and enforcement, the covered entity has the responsibility of complying with subpart E of this part.

 (B) The covered entity is responsible for complying with § 164.316(a) and § 164.530(i), pertaining to the implementation of policies and procedures to ensure compliance with applicable requirements of this section and subparts C and E of this part, including the safeguard requirements in paragraph (a)(2)(ii) of this section.

 (C) The covered entity is responsible for designating the components that are part of one or more health care components of the covered entity and documenting the designation in accordance with paragraph (c) of this section, provided that, if the covered entity designates a health care component or components, it must include any component that would meet the definition of covered entity if it were a separate legal entity. Health care component(s) also may include a component only to the extent that it performs:

  (1) Covered functions; or

  (2) Activities that would make such component a business associate of a component that performs covered functions if the two components were separate legal entities.

(b) **(1)** *Standard: Affiliated covered entities.* Legally separate covered entities that are affiliated may designate themselves as a single covered entity for purposes of subparts C and E of this part.

 (1) *Implementation specifications:*

  (i) *Requirements for designation of an affiliated covered entity.*

   (A) Legally separate covered entities may designate themselves (including any health care component of such covered entity) as a single affiliated covered entity, for purposes of subparts C and E of this part, if all of the covered entities designated are under common ownership or control.

   (B) The designation of an affiliated covered entity must be documented and the documentation maintained as required by paragraph (c) of this section.

  (ii) *Safeguard requirements.* An affiliated covered entity must ensure that:

   (A) The affiliated covered entity's creation, receipt, maintenance, or transmission of electronic protected health information complies with the applicable requirements of subpart C of this part;

   (B) The affiliated covered entity's use and disclosure of protected health information comply with the applicable requirements of subpart E of this part; and

   (C) If the affiliated covered entity combines the functions of a health plan, health care provider, or health care clearinghouse, the affiliated covered entity complies with § 164.308(a)(4)(ii)(A) and § 164.504(g), as applicable.

(c) (1) *Standard: Documentation.* A covered entity must maintain a written or electronic record of a designation as required by paragraphs (a) or (b) of this section.

(2) *Implementation specification: Retention period.* A covered entity must retain the documentation as required by paragraph (c)(1) of this section for 6 years from the date of its creation or the date when it last was in effect, whichever is later.

5. A new subpart C is added to part 164 to read as follows:

SUBPART C—SECURITY STANDARDS FOR THE PROTECTION OF ELECTRONIC PROTECTED HEALTH INFORMATION

Sec.

APPENDIX A TO SUBPART C OF PART 164— SECURITY STANDARDS: MATRIX

Authority: 42 U.S.C. 1320d–2 and 1320d–4.

§ 164.302   APPLICABILITY

A covered entity must comply with the applicable standards, implementation specifications, and requirements of this subpart with respect to electronic protected health information.

§ 164.304   DEFINITIONS

As used in this subpart, the following terms have the following meanings:

Access means the ability or the means necessary to read, write, modify, or communicate data/information or otherwise use any system resource. (This definition applies to "access" as used in this subpart, not as used in subpart E of this part.)

Administrative safeguards are administrative actions, and policies and procedures, to manage the selection, development, implementation, and maintenance of security measures to protect electronic protected health information and to manage the conduct of the covered entity's workforce in relation to the protection of that information.

Authentication means the corroboration that a person is the one claimed.

Availability means the property that data or information is accessible and useable upon demand by an authorized person.

Confidentiality means the property that data or information is not made available or disclosed to unauthorized persons or processes.

Encryption means the use of an algorithmic process to transform data into a form in which there is a low probability of assigning meaning without use of a confidential process or key.

Facility means the physical premises and the interior and exterior of a building(s).

Information system means an interconnected set of information resources under the same direct management control that shares common functionality. A system normally includes hardware, software, information, data, applications, communications, and people.

Integrity means the property that data or information have not been altered or destroyed in an unauthorized manner.

Malicious software means software, for example, a virus, designed to damage or disrupt a system.

Password means confidential authentication information composed of a string of characters.

Physical safeguards are physical measures, policies, and procedures to protect a covered entity's electronic information systems and related buildings and equipment, from natural and environmental hazards, and unauthorized intrusion.

Security or Security measures encompass all of the administrative, physical, and technical safeguards in an information system.

Security incident means the attempted or successful unauthorized access, use, disclosure, modification, or destruction of information or interference with system operations in an information system.

Technical safeguards means the technology and the policy and procedures for its use that protect electronic protected health information and control access to it.

User means a person or entity with authorized access.

Workstation means an electronic computing device, for example, a laptop or desktop computer, or any other device that performs similar functions, and electronic media stored in its immediate environment.

§ 164.306   SECURITY STANDARDS: GENERAL RULES

(a) *General requirements.* Covered entities must do the following:

(1) Ensure the confidentiality, integrity, and availability of all electronic protected health information the covered entity creates, receives, maintains, or transmits.

(2) Protect against any reasonably anticipated threats or hazards to the security or integrity of such information.

(3) Protect against any reasonably anticipated uses or disclosures of such information that are not permitted or required under subpart E of this part.

(4) Ensure compliance with this subpart by its workforce.

(b) *Flexibility of approach.*

(1) Covered entities may use any security measures that allow the covered entity to reasonably and appropriately implement the standards and implementation specifications as specified in this subpart.

(2) In deciding which security measures to use, a covered entity must take into account the following factors:

(i) The size, complexity, and capabilities of the covered entity.

(ii) The covered entity's technical infrastructure, hardware, and software security capabilities.

(iii) The costs of security measures.

(iv) The probability and criticality of potential risks to electronic protected health information.

(c) *Standards.* A covered entity must comply with the standards as provided in this section and in § 164.308, § 164.310, § 164.312, § 164.314, and § 164.316 with respect to all electronic protected health information.

(d) *Implementation specifications.*

In this subpart:

(1) Implementation specifications are required or addressable. If an implementation specification is required, the word "Required" appears in parentheses after the title of the implementation specification. If an implementation specification is addressable, the word "Addressable" appears in parentheses after the title of the implementation specification.

(2) When a standard adopted in § 164.308, § 164.310, § 164.312, § 164.314, or § 164.316 includes required implementation specifications, a covered entity must implement the implementation specifications.

(1) When a standard adopted in § 164.308, § 164.310, § 164.312, § 164.314, or § 164.316 includes addressable implementation specifications, a covered entity must—

(i) Assess whether each implementation specification is a reasonable and appropriate safeguard in its environment, when analyzed with reference to the likely contribution to protecting the entity's electronic protected health information; and

(ii) As applicable to the entity—

(A) Implement the implementation specification if reasonable and appropriate; or

(B) If implementing the implementation specification is not reasonable and appropriate—

(1) Document why it would not be reasonable and appropriate to implement the implementation specification; and

(2) Implement an equivalent alternative measure if reasonable and appropriate.

(e) *Maintenance.* Security measures implemented to comply with standards and implementation specifications adopted under § 164.105 and this subpart must be reviewed and modified as needed to continue provision of reasonable and appropriate protection of electronic protected health information as described at § 164.316.

§ 164.308   ADMINISTRATIVE SAFEGUARDS

(a) A covered entity must, in accordance with § 164.306:

(1) (i) *Standard: Security management process.* Implement policies and procedures to prevent, detect, contain, and correct security violations.

(ii) *Implementation specifications:*

(A) *Risk analysis* (Required). Conduct an accurate and thorough assessment of the potential risks and vulnerabilities to the confidentiality, integrity, and availability of electronic protected health information held by the covered entity.

(B) *Risk management* (Required). Implement security measures sufficient to reduce risks and vulnerabilities to a reasonable and appropriate level to comply with § 164.306(a).

(C) *Sanction policy* (Required). Apply appropriate sanctions against workforce members who fail to comply with the security policies and procedures of the covered entity.

(D) *Information system activity review* (Required). Implement procedures to regularly review records of information system activity, such as audit logs, access reports, and security incident tracking reports.

(2) *Standard: Assigned security responsibility.* Identify the security official who is responsible for the development and implementation of the policies and procedures required by this subpart for the entity.

(3) (i) *Standard: Workforce security.* Implement policies and procedures to ensure that all members of its workforce have appropriate access to electronic protected health information, as provided under paragraph (a)(4) of this section, and to prevent those workforce members who do not have access under paragraph (a)(4) of this section from obtaining access to electronic protected health information.

(ii) *Implementation specifications:*

(A) *Authorization and/or supervision* (Addressable). Implement procedures for the authorization and/or supervision of workforce members who work with electronic protected health information or in locations where it might be accessed.

(B) *Workforce clearance procedure* (Addressable). Implement procedures to determine that the access of a workforce member to electronic protected health information is appropriate.

(C) *Termination procedures* (Addressable). Implement procedures for terminating access to electronic protected health information when the employment of a workforce member ends or as required by determinations made as specified in paragraph (a)(3)(ii)(B) of this section.

(4) (i) *Standard: Information access management.* Implement policies and procedures for authorizing access to electronic protected health information that are consistent with the applicable requirements of subpart E of this part.

(ii) *Implementation specifications*:

(A) *Isolating health care clearinghouse functions* (Required). If a health care clearinghouse is part of a larger organization, the clearinghouse must implement policies and procedures that protect the electronic protected health information of the clearinghouse from unauthorized access by the larger organization.

(B) *Access authorization* (Addressable). Implement policies and procedures for granting access to electronic protected health information, for example, through access to a workstation, transaction, program, process, or other mechanism.

(C) *Access establishment and modification* (Addressable). Implement policies and procedures that, based upon the entity's access authorization policies, establish, document, review, and modify a user's right of access to a workstation, transaction, program, or process.

(5) (i) *Standard: Security awareness and training.* Implement a security awareness and training program for all members of its workforce (including management).

(ii) *Implementation specifications.* Implement:

(A) *Security reminders* (Addressable). Periodic security updates.

(B) *Protection from malicious software* (Addressable). Procedures for guarding against, detecting, and reporting malicious software.

(C) *Log-in monitoring* (Addressable). Procedures for monitoring log-in attempts and reporting discrepancies.

(D) *Password management* (Addressable). Procedures for creating, changing, and safeguarding passwords.

(6) (i) *Standard: Security incident procedures.* Implement policies and procedures to address security incidents.

(ii) *Implementation specification: Response and Reporting* (Required). Identify and respond to suspected or known security incidents; mitigate, to the extent practicable, harmful effects of security incidents that are known to the covered entity; and document security incidents and their outcomes.

(7) (i) *Standard: Contingency plan.* Establish (and implement as needed) policies and procedures for responding to an emergency or other occurrence (for example, fire, vandalism, system failure, and natural disaster) that damages systems that contain electronic protected health information.

(ii) *Implementation specifications:*

(A) *Data backup plan* (Required). Establish and implement procedures to create and maintain retrievable exact copies of electronic protected health information.

(B) *Disaster recovery plan* (Required). Establish (and implement as needed) procedures to restore any loss of data.

(C) *Emergency mode operation plan* (Required). Establish (and implement as needed) procedures to enable continuation of critical business processes for protection of the security of electronic protected health information while operating in emergency mode.

(D) *Testing and revision procedures* (Addressable). Implement procedures for periodic testing and revision of contingency plans.

(E) *Applications and data criticality analysis* (Addressable). Assess the relative criticality of specific applications and data in support of other contingency plan components.

(8) *Standard: Evaluation.* Perform a periodic technical and nontechnical evaluation, based initially upon the standards implemented under this rule and subsequently, in response to environmental or operational changes affecting the security of electronic protected health information, that establishes the extent to which an entity's security policies and procedures meet the requirements of this subpart.

(b) (1) *Standard: Business associate contracts and other arrangements.* A covered entity, in accordance with § 164.306, may permit a business associate to create, receive, maintain, or transmit electronic protected health information on the covered entity's behalf only if the covered entity obtains satisfactory assurances, in accordance with § 164.314(a) that the business associate will appropriately safeguard the information.

(2) This standard does not apply with respect to—

(i) The transmission by a covered entity of electronic protected health information to a health care provider concerning the treatment of an individual.

(ii) The transmission of electronic protected health information by a group health plan or an HMO or health insurance issuer on behalf of a group health plan to a plan sponsor, to the extent that the requirements of § 164.314(b) and § 164.504(f) apply and are met; or

(iii) The transmission of electronic protected health information from or to other agencies providing the services at § 164.502(e)(1)(ii)(C), when the covered entity is a health plan that is a government program providing public benefits, if the requirements of § 164.502(e)(1)(ii)(C) are met.

(3) A covered entity that violates the satisfactory assurances it provided as a business associate of another covered entity will be in noncompliance with the standards, implementation specifications, and requirements of this paragraph and § 164.314(a).

(4) *Implementation specifications: Written contract or other arrangement* (Required). Document the satisfactory assurances required by paragraph (b)(1) of this section through a written contract or other arrangement with the business associate that meets the applicable requirements of § 164.314(a).

§ 164.310   PHYSICAL SAFEGUARDS

A covered entity must, in accordance with § 164.306:

(a) (1) *Standard: Facility access controls.* Implement policies and procedures to limit physical access to its electronic information systems and the facility or facilities in which they are housed, while ensuring that properly authorized access is allowed.

(2) *Implementation specifications:*

(i) *Contingency operations* (Addressable). Establish (and implement as needed) procedures that allow facility access in support of restoration of lost data under the disaster recovery plan and emergency mode operations plan in the event of an emergency.

(ii) *Facility security plan* (Addressable). Implement policies and procedures to safeguard the facility and the equipment therein from unauthorized physical access, tampering, and theft.

(iii) *Access control and validation procedures* (Addressable). Implement procedures to control and validate a person's access to facilities based on their role or function, including visitor control, and control of access to software programs for testing and revision.

(iv) *Maintenance records* (Addressable). Implement policies and procedures to document repairs and modifications to the physical components of a facility which are related to security (for example, hardware, walls, doors, and locks).

(b) *Standard: Workstation use.* Implement policies and procedures that specify the proper functions to be performed, the manner in which those functions are to be performed, and the physical attributes of the surroundings of a specific workstation or class of workstation that can access electronic protected health information.

(c) *Standard: Workstation security.* Implement physical safeguards for all workstations that access electronic protected health information, to restrict access to authorized users.

(d) (1) *Standard: Device and media controls.* Implement policies and procedures that govern the receipt and removal of hardware and electronic media that contain electronic protected health information into and out of a facility, and the movement of these items within the facility.

(2) *Implementation specifications:*

(i) *Disposal* (Required). Implement policies and procedures to address the final disposition of electronic protected health information, and/or the hardware or electronic media on which it is stored.

(ii) *Media re-use* (Required). Implement procedures for removal of electronic protected health information from electronic media before the media are made available for re-use.

(iii) *Accountability* (Addressable). Maintain a record of the movements of hardware and electronic media and any person responsible therefore.

(iv) *Data backup and storage* (Addressable). Create a retrievable, exact copy of electronic protected health information, when needed, before movement of equipment.

§ 164.312   TECHNICAL SAFEGUARDS

A covered entity must, in accordance with § 164.306:

(a) (1) *Standard: Access control.* Implement technical policies and procedures for electronic information systems that maintain electronic protected health information to allow access only to those persons or software programs that have been granted access rights as specified in § 164.308(a)(4).

(2) *Implementation specifications:*

(i) *Unique user identification* (Required). Assign a unique name and/ or number for identifying and tracking user identity.

(ii) *Emergency access procedure* (Required). Establish (and implement as needed) procedures for obtaining necessary electronic protected health information during an emergency.

(iii) *Automatic logoff* (Addressable). Implement electronic procedures that terminate an electronic session after a predetermined time of inactivity.

(iv) *Encryption and decryption* (Addressable). Implement a mechanism to encrypt and decrypt electronic protected health information.

(b) *Standard: Audit controls.* Implement hardware, software, and/or procedural mechanisms that record and examine activity in information systems that contain or use electronic protected health information.

(c) (1) *Standard: Integrity.* Implement policies and procedures to protect electronic protected health information from improper alteration or destruction.

(2) *Implementation specification: Mechanism to authenticate electronic protected health information* (Addressable). Implement electronic mechanisms to corroborate that electronic protected health information has not been altered or destroyed in an unauthorized manner.

(d) *Standard: Person or entity authentication.* Implement procedures to verify that a person or entity seeking access to electronic protected health information is the one claimed.

(e) (1) *Standard: Transmission security.* Implement technical security measures to guard against unauthorized access to electronic protected health information that is being transmitted over an electronic communications network.

(2) *Implementation specifications:*

(i) *Integrity controls* (Addressable). Implement security measures to ensure that electronically transmitted electronic protected health information is not improperly modified without detection until disposed of.

(ii) *Encryption* (Addressable). Implement a mechanism to encrypt electronic protected health information whenever deemed appropriate.

§ 164.314   ORGANIZATIONAL REQUIREMENTS

(a) (1) *Standard: Business associate contracts or other arrangements.*

(i) The contract or other arrangement between the covered entity and its business associate required by § 164.308(b) must meet the requirements of paragraph (a)(2)(i) or (a)(2)(ii) of this section, as applicable.

(ii) A covered entity is not in compliance with the standards in § 164.502(e) and paragraph (a) of this section if the covered entity knew of a pattern of an activity or practice of the business associate that constituted a material breach or violation of the business associate's obligation under the contract or other arrangement, unless the covered entity took reasonable steps to cure the breach or end the violation, as applicable, and, if such steps were unsuccessful—

(A) Terminated the contract or arrangement, if feasible; or

(B) If termination is not feasible, reported the problem to the Secretary.

(2) *Implementation specifications* (Required).

(i) *Business associate contracts.* The contract between a covered entity and a business associate must provide that the business associate will—

(A) Implement administrative, physical, and technical safeguards that reasonably and appropriately protect the confidentiality, integrity, and availability of the electronic protected health information that it creates, receives, maintains, or transmits on behalf of the covered entity as required by this subpart;

(B) Ensure that any agent, including a subcontractor, to whom it provides such information agrees to implement reasonable and appropriate safeguards to protect it;

(C) Report to the covered entity any security incident of which it becomes aware;

(D) Authorize termination of the contract by the covered entity, if the covered entity determines that the business associate has violated a material term of the contract.

(ii) *Other arrangements.*

(A) When a covered entity and its business associate are both governmental entities, the covered entity is in compliance with paragraph (a)(1) of this section, if—

(1) It enters into a memorandum of understanding with the business associate that contains terms that accomplish the objectives of paragraph (a)(2)(i) of this section; or

(2) Other law (including regulations adopted by the covered entity or its business associate) contains requirements applicable to the business associate that accomplish the objectives of paragraph (a)(2)(i) of this section.

(B) If a business associate is required by law to perform a function or activity on behalf of a covered entity or to provide a service described in the definition of business associate as specified in § 160.103 of this subchapter to a covered entity, the covered entity may permit the business associate to create, receive, maintain, or transmit electronic protected health information on its behalf to the extent necessary to comply with the legal mandate without meeting the requirements of paragraph (a)(2)(i) of this section, provided that the covered entity attempts in good faith to obtain satisfactory assurances as required by paragraph (a)(2)(ii)(A) of this section, and documents the attempt and the reasons that these assurances cannot be obtained.

(C) The covered entity may omit from its other arrangements authorization of the termination of the contract by the covered entity, as required by paragraph (a)(2)(i)(D) of this section if such authorization is inconsistent with the statutory obligations of the covered entity or its business associate.

(b) **(1)** *Standard: Requirements for group health plans.* Except when the only electronic protected health information disclosed to a plan sponsor is disclosed pursuant to § 164.504(f)(1)(ii) or (iii), or as authorized under § 164.508, a group health plan must ensure that its plan documents provide that the plan sponsor will reasonably and appropriately safeguard electronic protected health information created, received, maintained, or transmitted to or by the plan sponsor on behalf of the group health plan.

(2) *Implementation specifications* (Required). The plan documents of the group health plan must be amended to incorporate provisions to require the plan sponsor to—

(i) Implement administrative, physical, and technical safeguards that reasonably and appropriately protect the confidentiality, integrity, and availability of the electronic protected health information that it creates, receives, maintains, or transmits on behalf of the group health plan;

(ii) Ensure that the adequate separation required by § 164.504(f)(2)(iii) is supported by reasonable and appropriate security measures;

(iii) Ensure that any agent, including a subcontractor, to whom it provides this information agrees to implement reasonable and appropriate security measures to protect the information; and

(iv) Report to the group health plan any security incident of which it becomes aware.

§ 164.316   POLICIES AND PROCEDURES AND DOCUMENTATION REQUIREMENTS

A covered entity must, in accordance with § 164.306:

(a) *Standard: Policies and procedures.* Implement reasonable and appropriate policies and procedures to comply with the standards, implementation specifications, or other requirements of this subpart, taking into account those factors specified in § 164.306(b) (2)(i), (ii), (iii), and (iv). This standard is not to be construed to permit or excuse an action that violates any other standard, implementation specification, or other requirements of this subpart. A covered entity may change its policies and procedures at any time, provided that the changes are documented and are implemented in accordance with this subpart.

(b) (1) *Standard: Documentation.*

(i) Maintain the policies and procedures implemented to comply with this subpart in written (which may be electronic) form; and

(ii) If an action, activity or assessment is required by this subpart to be documented, maintain a written (which may be electronic) record of the action, activity, or assessment.

(2) *Implementation specifications:*

(i) *Time limit* (Required). Retain the documentation required by paragraph (b)(1) of this section for 6 years from the date of its creation or the date when it last was in effect, whichever is later.

(ii) *Availability* (Required). Make documentation available to those persons responsible for implementing the procedures to which the documentation pertains.

(iii) *Updates* (Required). Review documentation periodically, and update as needed, in response to environmental or operational changes affecting the security of the electronic protected health information.

§ 164.318   COMPLIANCE DATES FOR THE INITIAL IMPLEMENTATION OF THE SECURITY STANDARDS

(a) *Health plan.*

> **(1)** A health plan that is not a small health plan must comply with the applicable requirements of this subpart no later than April 20, 2005.
>
> **(2)** A small health plan must comply with the applicable requirements of this subpart no later than April 20, 2006.

(b) *Health care clearinghouse.* A health care clearinghouse must comply with the applicable requirements of this subpart no later than April 20, 2005.

(c) *Health care provider.* A covered health care provider must comply with the applicable requirements of this subpart no later than April 20, 2005.

Appendix A to Subpart C of Part 164 — Security Standards: Matrix

Standards	Sections	Implementation Specifications (R) = Required, (A) = Addressable
Administrative Safeguards		
Security Management Process	164.308(a)(1)	Risk Analysis (R)
		Risk Management (R)
		Sanction Policy (R)
		Information System Activity Review (R)
Assigned Security Responsibility	164.308(a)(2)	(R)
Workforce Security	164.308(a)(3)	Authorization and/or Supervision (A)
		Workforce Clearance Procedure
		Termination Procedures (A)
Information Access Management	164.308(a)(4)	Isolating Health care Clearinghouse Function (R)
		Access Authorization (A)
		Access Establishment and Modification (A)
Security Awareness and Training	164.308(a)(5)	Security Reminders (A)
		Protection from Malicious Software (A)
		Log-in Monitoring (A)
		Password Management (A)
Security Incident Procedures	164.308(a)(6)	Response and Reporting (R)
Contingency Plan	164.308(a)(7)	Data Backup Plan (R)
		Disaster Recovery Plan (R)
		Emergency Mode Operation Plan (R)
		Testing and Revision Procedure (A)
		Applications and Data Criticality Analysis (A)
Evaluation	164.308(a)(8)	(R)
Business Associate Contracts and Other Arrangement.	164.308(b)(1)	Written Contract or Other Arrangement (R)

Appendix A to Subpart C of Part 164—Security Standards: Matrix (*Continued*)

Standards	Sections	Implementation Specifications (R) = Required, (A) = Addressable
Physical Safeguards		
Facility Access Controls	164.310(a)(1)	Contingency Operations (A)
		Facility Security Plan (A)
		Access Control and Validation Procedures (A)
		Maintenance Records (A)
Workstation Use	164.310(b)	(R)
Workstation Security	164.310(c)	(R)
Device and Media Controls	164.310(d)(1)	Disposal (R)
		Media Re-use (R)
		Accountability (A)
		Data Backup and Storage (A)
Technical Safeguards (see § 164.312)		
Access Control	164.312(a)(1)	Unique User Identification (R)
		Emergency Access Procedure (R)
		Automatic Logoff (A)
		Encryption and Decryption (A)
Audit Controls	164.312(b)	(R)
Integrity	164.312(c)(1)	Mechanism to Authenticate Electronic Protected Health Information (A)
Person or Entity Authentication	164.312(d)	(R)
Transmission Security	164.312(e)(1)	Integrity Controls (A)
		Encryption (A)

§164.500   [AMENDED]

6. § In 164.500(b)(1)(iv), remove the words "including the designation of health care components of a covered entity".

§ 165.501   [AMENDED]

7. In §164.501, the definitions of the following terms are removed: *Covered functions, Disclosure, Individual, Organized health care arrangement, Plan sponsor Protected health information, Required by law,* and *Use.*

§ 164.504   [AMENDED]

8. In §164.504, the following changes are made:

a. The definitions of the following terms are removed: *Common control, Common ownership, Health care component,* and *Hybrid entity.*

b. Paragraphs (b) through (d) are removed and reserved.

Authority: Sections 1173 and 1175 of the Social Security Act (42 U.S.C. 1329d–2 and 1320–4).

Dated: January 13, 2003.

Tommy G. Thompson,

Secretary.

[FR Doc. 03–3877 Filed 2–13–03; 8:45 am]

BILLING CODE 4120–01–P

GUIDANCE FOR INDUSTRY
Part 11, Electronic Records; Electronic Signatures—Scope and Application

Division of Drug Information, HFD-240
Center for Drug Evaluation and Research (CDER)
(Tel) 301-827-4573
http://www.fda.gov/cder/guidance/index.htm
or
Office of Communication, Training and
Manufacturers Assistance, HFM-40
Center for Biologics Evaluation and Research (CBER)
http://www.fda.gov/cber/guidelines.htm
Phone: the Voice Information System at 800-835-4709 or 301-827-1800
or
Communications Staff (HFV-12),
Center for Veterinary Medicine (CVM)
(Tel) 301-594-1755
http://www.fda.gov/cvm/guidance/guidance.html
or
Division of Small Manufacturers Assistance (HFZ-220)
http://www.fda.gov/cdrh/ggpmain.html
Manufacturers Assistance Phone Number: 800.638.2041 or 301.443.6597
Internt'l Staff Phone: 301.827.3993
or
Center for Food Safety and Applied Nutrition (CFSAN)
http://www.cfsan.fda.gov/~dms/guidance.html.

U.S. Department of Health and Human Services
Food and Drug Administration
Center for Drug Evaluation and Research (CDER)
Center for Biologics Evaluation and Research (CBER)
Center for Devices and Radiological Health (CDRH)
Center for Food Safety and Applied Nutrition (CFSAN)
Center for Veterinary Medicine (CVM)
Office of Regulatory Affairs (ORA)

August 2003
Pharmaceutical CGMPs

TABLE OF CONTENTS

Guidance for Industry[1]
Part 11, Electronic Records; Electronic Signatures—Scope and Application

This guidance represents the Food and Drug Administration's (FDA's) current thinking on this topic. It does not create or confer any rights for or on any person and does not operate to bind FDA or the public. You can use an alternative approach if the approach satisfies the requirements of the applicable statutes and regulations. If you want to discuss an alternative approach, contact the FDA staff responsible for implementing this guidance. If you cannot identify the appropriate FDA staff, call the appropriate number listed on the title page of this guidance.

I. INTRODUCTION

This guidance is intended to describe the Food and Drug Administration's (FDA's) current thinking regarding the scope and application of part 11 of Title 21 of the Code of Federal Regulations; Electronic Records; Electronic Signatures (21 CFR Part 11).[2]

This document provides guidance to persons who, in fulfillment of a requirement in a statute or another part of FDA's regulations to maintain records or submit information to FDA,[3] have chosen to maintain the records or submit designated information electronically and, as a result, have become subject to part 11. Part 11 applies to records in electronic form that are created, modified, maintained, archived, retrieved, or transmitted under any records requirements set forth in Agency regulations. Part 11 also applies to electronic records submitted to the Agency under the Federal Food, Drug, and Cosmetic Act (the Act) and the Public Health Service Act (the PHS Act), even if such records are not specifically identified in Agency regulations (§ 11.1). The underlying requirements set forth in the Act, PHS Act, and FDA regulations (other than part 11) are referred to in this guidance document as *predicate rules*.

[1]This guidance has been prepared by the Office of Compliance in the Center for Drug Evaluation and Research (CDER) in consultation with the other Agency centers and the Office of Regulatory Affairs at the Food and Drug Administration.

[2]62 FR 13430.

[3]These requirements include, for example, certain provisions of the Current Good Manufacturing Practice regulations (21 CFR Part 211), the Quality System regulation (21 CFR Part 820), and the Good Laboratory Practice for Nonclinical Laboratory Studies regulations (21 CFR Part 58).

As an outgrowth of its current good manufacturing practice (CGMP) initiative for human and animal drugs and biologics,[4] FDA is re-examining part 11 as it applies to all FDA regulated products. We anticipate initiating rulemaking to change part 11 as a result of that re-examination. This guidance explains that we will narrowly interpret the scope of part 11. While the re-examination of part 11 is under way, we intend to exercise enforcement discretion with respect to certain part 11 requirements. That is, we do not intend to take enforcement action to enforce compliance with the validation, audit trail, record retention, and record copying requirements of part 11 as explained in this guidance. However, records must still be maintained or submitted in accordance with the underlying predicate rules, and the Agency can take regulatory action for noncompliance with such predicate rules.

In addition, we intend to exercise enforcement discretion and do not intend to take (or recommend) action to enforce any part 11 requirements with regard to systems that were operational before August 20, 1997, the effective date of part 11 (commonly known as legacy systems) under the circumstances described in section III.C.3 of this guidance.

Note that part 11 remains in effect and that this exercise of enforcement discretion applies only as identified in this guidance.

FDA's guidance documents, including this guidance, do not establish legally enforceable responsibilities. Instead, guidances describe the Agency's current thinking on a topic and should be viewed only as recommendations, unless specific regulatory or statutory requirements are cited. The use of the word *should* in Agency guidances means that something is suggested or recommended, but not required.

II. BACKGROUND

In March of 1997, FDA issued final part 11 regulations that provide criteria for acceptance by FDA, under certain circumstances, of electronic records, electronic signatures, and handwritten signatures executed to electronic records as equivalent to paper records and handwritten signatures executed on paper. These regulations, which apply to all FDA program areas, were intended to permit the widest possible use of electronic technology, compatible with FDA's responsibility to protect the public health.

After part 11 became effective in August 1997, significant discussions ensued among industry, contractors, and the Agency concerning the interpretation and implementation of the regulations. FDA has (1) spoken about part 11 at many conferences and met numerous times with an industry coalition and other interested parties in an effort to hear more about potential part 11 issues; (2) published a compliance policy guide, CPG 7153.17: Enforcement Policy: 21 CFR Part 11; Electronic Records; Electronic Signatures; and (3) published numerous draft guidance documents including the following:

- 21 CFR Part 11; Electronic Records; Electronic Signatures, Validation
- 21 CFR Part 11; Electronic Records; Electronic Signatures, Glossary of Terms
- 21 CFR Part 11; Electronic Records; Electronic Signatures, Time Stamps
- 21 CFR Part 11; Electronic Records; Electronic Signatures, Maintenance of Electronic Records
- 21 CFR Part 11; Electronic Records; Electronic Signatures, Electronic Copies of Electronic Records

[4]See *Pharmaceutical CGMPs for the 21st Century: A Risk-Based Approach; A Science and Risk-Based Approach to Product Quality Regulation Incorporating an Integrated Quality Systems Approach* at www.fda.gov/oc/guidance/gmp.html.

Throughout all of these communications, concerns have been raised that some interpretations of the part 11 requirements would (1) unnecessarily restrict the use of electronic technology in a manner that is inconsistent with FDA's stated intent in issuing the rule, (2) significantly increase the costs of compliance to an extent that was not contemplated at the time the rule was drafted, and (3) discourage innovation and technological advances without providing a significant public health benefit. These concerns have been raised particularly in the areas of part 11 requirements for validation, audit trails, record retention, record copying, and legacy systems.

As a result of these concerns, we decided to review the part 11 documents and related issues, particularly in light of the Agency's CGMP initiative. In the *Federal Register* of February 4, 2003 (68 FR 5645), we announced the withdrawal of the draft guidance for industry, *21 CFR Part 11; Electronic Records; Electronic Signatures, Electronic Copies of Electronic Records.* We had decided we wanted to minimize industry time spent reviewing and commenting on the draft guidance when that draft guidance may no longer represent our approach under the CGMP initiative. Then, in the *Federal Register* of February 25, 2003 (68 FR 8775), we announced the withdrawal of the part 11 draft guidance documents on validation, glossary of terms, time stamps,[5] maintenance of electronic records, and CPG 7153.17. We received valuable public comments on these draft guidances, and we plan to use that information to help with future decision-making with respect to part 11. We do not intend to re-issue these draft guidance documents or the CPG.

We are now re-examining part 11, and we anticipate initiating rulemaking to revise provisions of that regulation. To avoid unnecessary resource expenditures to comply with part 11 requirements, we are issuing this guidance to describe how we intend to exercise enforcement discretion with regard to certain part 11 requirements during the re-examination of part 11. As mentioned previously, part 11 remains in effect during this re-examination period.

III. DISCUSSION

A. Overall Approach to Part 11 Requirements

As described in more detail below, the approach outlined in this guidance is based on three main elements:

- Part 11 will be interpreted narrowly; we are now clarifying that fewer records will be considered subject to part 11.

- For those records that remain subject to part 11, we intend to exercise enforcement discretion with regard to part 11 requirements for validation, audit trails, record retention, and record copying in the manner described in this guidance and with regard to all part 11 requirements for systems that were operational before the effective date of part 11 (also known as legacy systems).

- We will enforce all predicate rule requirements, including predicate rule record and recordkeeping requirements.

---

[5]Although we withdrew the draft guidance on time stamps, our current thinking has not changed in that when using time stamps for systems that span different time zones, we do not expect you to record the signer's local time. When using time stamps, they should be implemented with a clear understanding of the time zone reference used. In such instances, system documentation should explain time zone references as well as zone acronyms or other naming conventions.

It is important to note that FDA's exercise of enforcement discretion as described in this guidance is limited to specified part 11 requirements (setting aside legacy systems, as to which the extent of enforcement discretion, under certain circumstances, will be more broad). We intend to enforce all other provisions of part 11 including, but not limited to, certain controls for closed systems in § 11.10. For example, we intend to enforce provisions related to the following controls and requirements:

- limiting system access to authorized individuals
- use of operational system checks
- use of authority checks
- use of device checks
- determination that persons who develop, maintain, or use electronic systems have the education, training, and experience to perform their assigned tasks
- establishment of and adherence to written policies that hold individuals accountable for actions initiated under their electronic signatures
- appropriate controls over systems documentation
- controls for open systems corresponding to controls for closed systems bulleted above (§ 11.30)
- requirements related to electronic signatures (e.g., §§ 11.50, 11.70, 11.100, 11.200, and 11.300)

We expect continued compliance with these provisions, and we will continue to enforce them. Furthermore, persons must comply with applicable predicate rules, and records that are required to be maintained or submitted must remain secure and reliable in accordance with the predicate rules.

B.   Details of Approach—Scope of Part 11

1.   Narrow Interpretation of Scope.   We understand that there is some confusion about the scope of part 11. Some have understood the scope of part 11 to be very broad. We believe that some of those broad interpretations could lead to unnecessary controls and costs and could discourage innovation and technological advances without providing added benefit to the public health. As a result, we want to clarify that the Agency intends to interpret the scope of part 11 narrowly.

Under the narrow interpretation of the scope of part 11, with respect to records required to be maintained under predicate rules or submitted to FDA, when persons choose to use records in electronic format in place of paper format, part 11 would apply. On the other hand, when persons use computers to generate paper printouts of electronic records, and those paper records meet all the requirements of the applicable predicate rules and persons rely on the paper records to perform their regulated activities, FDA would generally not consider persons to be "using electronic records in lieu of paper records" under §§ 11.2(a) and 11.2(b). In these instances, the use of computer systems in the generation of paper records would not trigger part 11.

2.   Definition of Part 11 Records.   Under this narrow interpretation, FDA considers part 11 to be applicable to the following records or signatures in electronic format (part 11 records or signatures):

- Records that are required to be maintained under predicate rule requirements and that are maintained in electronic format *in place of paper format*. On the other hand, records (and any associated signatures) that are not required to be retained under predicate rules, but that are nonetheless maintained in electronic format, are not part 11 records.

We recommend that you determine, based on the predicate rules, whether specific records are part 11 records. We recommend that you document such decisions.

- Records that are required to be maintained under predicate rules, that are maintained in electronic format *in addition to paper format*, and that *are relied on to perform regulated activities.*

In some cases, actual business practices may dictate whether you are *using* electronic records instead of paper records under § 11.2(a). For example, if a record is required to be maintained under a predicate rule and you use a computer to generate a paper printout of the electronic records, but you nonetheless rely on the electronic record to perform regulated activities, the Agency may consider you to be *using* the electronic record instead of the paper record. That is, the Agency may take your business practices into account in determining whether part 11 applies.

Accordingly, we recommend that, for each record required to be maintained under predicate rules, you determine in advance whether you plan to rely on the electronic record or paper record to perform regulated activities. We recommend that you document this decision (e.g., in a Standard Operating Procedure (SOP), or specification document).

- Records submitted to FDA, under predicate rules (even if such records are not specifically identified in Agency regulations) in electronic format (assuming the records have been identified in docket number 92S-0251 as the types of submissions the Agency accepts in electronic format). However, a record that is not itself submitted, but is used in generating a submission, is not a part 11 record unless it is otherwise required to be maintained under a predicate rule and it is maintained in electronic format.

- Electronic signatures that are intended to be the equivalent of handwritten signatures, initials, and other general signings required by predicate rules. Part 11 signatures include electronic signatures that are used, for example, to document the fact that certain events or actions occurred in accordance with the predicate rule (e.g., *approved, reviewed*, and *verified*).

C. Approach to Specific Part 11 Requirements

1. ***Validation.*** The Agency intends to exercise enforcement discretion regarding specific part 11 requirements for validation of computerized systems (§ 11.10(a) and corresponding requirements in § 11.30). Although persons must still comply with all applicable predicate rule requirements for validation (e.g., 21 CFR 820.70(i)), this guidance should not be read to impose any additional requirements for validation.

We suggest that your decision to validate computerized systems, and the extent of the validation, take into account the impact the systems have on your ability to meet predicate rule requirements. You should also consider the impact those systems might have on the accuracy, reliability, integrity, availability, and authenticity of required records and signatures. Even if there is no predicate rule requirement to validate a system, in some instances it may still be important to validate the system.

We recommend that you base your approach on a justified and documented risk assessment and a determination of the potential of the system to affect product quality and safety, and record integrity. For instance, validation would not be important for a word processor used only to generate SOPs.

For further guidance on validation of computerized systems, see FDA's guidance for industry and FDA staff *General Principles of Software Validation* and also industry guidance such as the *GAMP 4 Guide* (See References).

2. *Audit Trail.*   The Agency intends to exercise enforcement discretion regarding specific part 11 requirements related to computer-generated, time-stamped audit trails (§ 11.10 (e), (k)(2) and any corresponding requirement in §11.30). Persons must still comply with all applicable predicate rule requirements related to documentation of, for example, date (e.g., § 58.130(e)), time, or sequencing of events, as well as any requirements for ensuring that changes to records do not obscure previous entries.

Even if there are no predicate rule requirements to document, for example, date, time, or sequence of events in a particular instance, it may nonetheless be important to have audit trails or other physical, logical, or procedural security measures in place to ensure the trustworthiness and reliability of the records.[6] We recommend that you base your decision on whether to apply audit trails, or other appropriate measures, on the need to comply with predicate rule requirements, a justified and documented risk assessment, and a determination of the potential effect on product quality and safety and record integrity. We suggest that you apply appropriate controls based on such an assessment. Audit trails can be particularly appropriate when users are expected to create, modify, or delete regulated records during normal operation.

3. *Legacy Systems.*[7]   The Agency intends to exercise enforcement discretion with respect to all part 11 requirements for systems that otherwise were operational prior to August 20, 1997, the effective date of part 11, under the circumstances specified below.

This means that the Agency does not intend to take enforcement action to enforce compliance with any part 11 requirements if all the following criteria are met for a specific system:

- The system was operational before the effective date.
- The system met all applicable predicate rule requirements before the effective date.
- The system currently meets all applicable predicate rule requirements.
- You have documented evidence and justification that the system is fit for its intended use (including having an acceptable level of record security and integrity, if applicable).

If a system has been changed since August 20, 1997, and if the changes would prevent the system from meeting predicate rule requirements, Part 11 controls should be applied to Part 11 records and signatures pursuant to the enforcement policy expressed in this guidance.

4. *Copies of Records.*   The Agency intends to exercise enforcement discretion with regard to specific part 11 requirements for generating copies of records (§ 11.10(b) and any corresponding requirement in §11.30). You should provide an investigator with reasonable and useful access to records during an inspection. All records held by you are subject to inspection in accordance with predicate rules (e.g., §§ 211.180(c),(d), and 108.35(c)(3)(ii)).

---

[6]Various guidance documents on information security are available (see References).

[7]In this guidance document, we use the term *legacy system* to describe systems already in operation before the effective date of part 11.

We recommend that you supply copies of electronic records by:

- Producing copies of records held in common portable formats when records are maintained in these formats

- Using established automated conversion or export methods, where available, to make copies in a more common format (examples of such formats include, but are not limited to, PDF, XML, or SGML)

In each case, we recommend that the copying process used produces copies that preserve the content and meaning of the record. If you have the ability to search, sort, or trend part 11 records, copies given to the Agency should provide the same capability if it is reasonable and technically feasible. You should allow inspection, review, and copying of records in a human readable form at your site using your hardware and following your established procedures and techniques for accessing records.

5. Record Retention. The Agency intends to exercise enforcement discretion with regard to the part 11 requirements for the protection of records to enable their accurate and ready retrieval throughout the records retention period (§ 11.10(c) and any corresponding requirement in §11.30). Persons must still comply with all applicable predicate rule requirements for record retention and availability (e.g., §§ 211.180(c),(d), 108.25(g), and 108.35(h)).

We suggest that your decision on how to maintain records be based on predicate rule requirements and that you base your decision on a justified and documented risk assessment and a determination of the value of the records over time.

FDA does not intend to object if you decide to archive required records in electronic format to nonelectronic media such as microfilm, microfiche, and paper, or to a standard electronic file format (examples of such formats include, but are not limited to, PDF, XML, or SGML). Persons must still comply with all predicate rule requirements, and the records themselves and any copies of the required records should preserve their content and meaning. As long as predicate rule requirements are fully satisfied and the content and meaning of the records are preserved and archived, you can delete the electronic version of the records. In addition, paper and electronic record and signature components can co-exist (i.e., a hybrid[8] situation) as long as predicate rule requirements are met and the content and meaning of those records are preserved.

REFERENCES

Food and Drug Administration References

Glossary of Computerized System and Software Development Terminology (Division of Field Investigations, Office of Regional Operations, Office of Regulatory Affairs, FDA 1995) (http://www.fda.gov/ora/inspect_ref/igs/gloss.html)

General Principles of Software Validation; Final Guidance for Industry and FDA Staff (FDA, Center for Devices and Radiological Health, Center for Biologics Evaluation and Research, 2002) (http://www.fda.gov/cdrh/comp/guidance/938.html)

Guidance for Industry, FDA Reviewers, and Compliance on Off-The-Shelf Software Use in Medical Devices (FDA, Center for Devices and Radiological Health, 1999) (http://www.fda.gov/cdrh/ode/guidance/585.html)

[8] Examples of hybrid situations include combinations of paper records (or other nonelectronic media) and electronic records, paper records and electronic signatures, or handwritten signatures executed to electronic records.

Pharmaceutical CGMPs for the 21st Century: A Risk-Based Approach; A Science and Risk-Based Approach to Product Quality Regulation Incorporating an Integrated Quality Systems Approach (FDA 2002) (http://www.fda.gov/oc/guidance/gmp.html)

Industry References

The Good Automated Manufacturing Practice (GAMP) Guide for Validation of Automated Systems, GAMP 4 (ISPE/GAMP Forum, 2001) (http://www.ispe.org/gamp/).

ISO/IEC 17799:2000 (BS 7799:2000) Information technology—Code of practice for information security management (ISO/IEC, 2000).

ISO 14971:2002 Medical Devices—Application of risk management to medical devices (ISO, 2001).

GUIDANCE FOR INDUSTRY
Computerized Systems Used in Clinical Investigations

Additional copies are available from:

Office of Training and Communication
Division of Drug Information
Center for Drug Evaluation and Research (CDER)
(Tel) 301-827-4573
http://www.fda.gov/cder/guidance/index.htm

or

Office of Communication, Training and
Manufacturers Assistance
Center for Biologics Evaluation and Research
http://www.fda.gov/cber/guidelines.htm
(Tel) 800-835-4709 or 301-827-1800

or

Office of Communication, Education, and Radiation Programs
Division of Small Manufacturers, International, and Consumer Assistance
Center for Devices and Radiological Health
http://www.fda.gov/cdrh/ggpmain.html
Email: dsmica@fda.hhs.gov
Fax: 240.276.3151
(Tel) Manufacturers and International Assistance: 800.638.2041 or 240.276.3150

or

Office of Food Additive Safety
Center for Food, Safety and Applied Nutrition
(Tel) 301-436-1200
http://www.cfsan.fda.gov/guidance.html

or

Communications Staff, HFV-12
Center for Veterinary Medicine
(Tel) 240-276-9300
http://www.fda.gov/cvm/guidance/published

or

Good Clinical Practice Programs
Office of the Commissioner

U.S. Department of Health and Human Services
Food and Drug Administration
Office of the Commissioner (OC)
May 2007

TABLE OF CONTENTS

Guidance for Industry[1]
Computerized Systems Used
in Clinical Investigations

This guidance represents the Food and Drug Administration's (FDA's) current thinking on this topic. It does not create or confer any rights for or on any person and does not operate to bind FDA or the public. You can use an alternative approach if the approach satisfies the requirements of the applicable statutes and regulations. If you want to discuss an alternative approach, contact the FDA staff responsible for implementing this guidance. If you cannot identify the appropriate FDA staff, call the appropriate number listed on the title page of this guidance.

I. INTRODUCTION

This document provides to sponsors, contract research organizations (CROs), data management centers, clinical investigators, and institutional review boards (IRBs), recommendations regarding the use of computerized systems in clinical investigations. The computerized system applies to records in electronic form that are used to create, modify, maintain, archive, retrieve, or transmit clinical data required to be maintained, or submitted to the FDA. Because the source data[2] are necessary for the reconstruction and evaluation of the study to determine the safety of food and color additives and safety and effectiveness of new human and animal drugs,[3] and medical devices, this guidance is intended to assist in ensuring confidence in the reliability, quality, and integrity of electronic source data and source documentation (i.e., electronic records).

This guidance supersedes the guidance of the same name dated April 1999; and supplements the guidance for industry on *Part 11, Electronic Records; Electronic Signatures—Scope and Application* and the Agency's international harmonization efforts[4] when applying these guidances to source data generated at clinical study sites.

---

[1] This guidance has been prepared by the Office of Critical Path Programs, the Good Clinical Practice Program, and the Office of Regulatory Affairs in cooperation with Bioresearch Monitoring Program Managers for each Center within the Food and Drug Administration.

[2] Under 21 CFR 312.62(b), reference is made to records that are part of case histories as "supporting data"; the ICH *E6 Good Clinical Practice* consolidated guidance uses the term "source documents." For the purpose of this guidance, these terms describe the same information and have been used interchangeably.

[3] Human drugs include biological drugs.

[4] In August 2003, FDA issued the guidance for industry entitled *Part 11, Electronic Records; Electronic Signatures-Scope and Application* clarifying that the Agency intends to interpret the scope of part 11 narrowly and to exercise enforcement discretion with regard to part 11 requirements for validation, audit trails, record retention, and record copying. In 1996, the International Conference on Harmonisation of Technical Requirements for Registration of Pharmaceuticals for Human Use (ICH) issued *E6 Good Clinical Practice: Consolidated Guidance.*

FDA's guidance documents, including this guidance, do not establish legally enforceable responsibilities. Instead, guidances describe the Agency's current thinking on a topic and should be viewed only as recommendations, unless specific regulatory or statutory requirements are cited. The use of the word *should* in Agency guidances means that something is suggested or recommended, but not required.

II. BACKGROUND

There is an increasing use of computerized systems in clinical trials to generate and maintain source data and source documentation on each clinical trial subject. Such electronic source data and source documentation must meet the same fundamental elements of data quality (e.g., attributable, legible, contemporaneous, original,[5] and accurate) that are expected of paper records and must comply with all applicable statutory and regulatory requirements. FDA's acceptance of data from clinical trials for decision-making purposes depends on FDA's ability to verify the quality and integrity of the data during FDA on-site inspections and audits. (21 CFR 312, 511.1(b), and 812).

In March 1997, FDA issued 21 CFR part 11, which provides criteria for acceptance by FDA, under certain circumstances, of electronic records, electronic signatures, and handwritten signatures executed to electronic records as equivalent to paper records and handwritten signatures executed on paper. After the effective date of 21 CFR part 11, significant concerns regarding the interpretation and implementation of part 11 were raised by both FDA and industry. As a result, we decided to reexamine 21 CFR part 11 with the possibility of proposing additional rulemaking, and exercising enforcement discretion regarding enforcement of certain part 11 requirements in the interim.

This guidance finalizes the draft guidance for industry entitled *Computerized Systems Used in Clinical Trials*, dated September 2004 and supplements the guidance for industry entitled *Part 11, Electronic Records; Electronic Signatures—Scope and Application* (Scope and Application Guidance), dated August 2003. The Scope and Application Guidance clarified that the Agency intends to interpret the scope of part 11 narrowly and to exercise enforcement discretion with regard to part 11 requirements for validation, audit trails, record retention, and record copying. However, other Part 11 provisions remain in effect.

The approach outlined in the Scope and Application Guidance, which applies to electronic records generated as part of a clinical trial, should be followed until such time as Part 11 is amended.

III. SCOPE

The principles outlined in this guidance should be used for computerized systems that contain any data that are relied on by an applicant in support of a marketing application, including computerized laboratory information management systems that capture analytical results of tests conducted during a clinical trial. For example, the recommendations in this guidance would apply to computerized systems that create source documents (electronic records) that satisfy the requirements in 21 CFR 312.62(b) and 812.140(b), such as case histories. This

---

[5]FDA is allowing original documents to be replaced by copies provided the copies are identical and have been verified as such (See, e.g., FDA Compliance Policy Guide # 7150.13). See Definitions section for a definition of original data.

guidance also applies to recorded source data transmitted from automated instruments directly to a computerized system (e.g., data from a chemistry autoanalyser or a Holter monitor to a laboratory information system). This guidance also applies when source documentation is created in hardcopy and later entered into a computerized system, recorded by direct entry into a computerized system, or automatically recorded by a computerized system (e.g., an ECG reading). The guidance does not apply to computerized medical devices that generate such data and that are otherwise regulated by FDA.

IV. RECOMMENDATIONS

This guidance provides the following recommendations regarding the use of computerized systems in clinical investigations.

A. Study Protocols

Each specific study protocol should identify each step at which a computerized system will be used to create, modify, maintain, archive, retrieve, or transmit source data. This information can be included in the protocol at the time the investigational new drug application (IND), Investigational Device Exemption (IDE), or Notice of Claimed Investigational Exemption for a New Animal Drug containing the protocols is submitted or at any time after the initial submission.

The computerized systems should be designed: (1) to satisfy the processes assigned to these systems for use in the specific study protocol (e.g., record data in metric units, blind the study), and (2) to prevent errors in data creation, modification, maintenance, archiving, retrieval, or transmission (e.g., inadvertently unblinding a study).

B. Standard Operating Procedures

There should be specific procedures and controls in place when using computerized systems to create, modify, maintain, or transmit electronic records, including when collecting source data at clinical trial sites. A list of recommended standard operating procedures (SOPs) is provided in Appendix A. Such SOPs should be maintained either on-site or be remotely accessible through electronic files as part of the specific study records, and the SOPs should be made available for use by personnel and for inspection by FDA.

C. Source Documentation and Retention

When original observations are entered directly into a computerized system, the electronic record is the source document. Under 21 CFR 312.62, 511.1(b)(7)(ii) and 812.140, the clinical investigator must retain records required to be maintained under part 312, § 511.1(b), and part 812, for a period of time specified in these regulations. This requirement applies to the retention of the original source document, or a copy of the source document.

When source data are transmitted from one system to another (e.g., from a personal data assistant to a sponsor's server), or entered directly into a remote computerized system (e.g., data are entered into a remote server via a computer terminal that is located at the clinical site), or an electrocardiogram at the clinical site is transmitted to the sponsor's computerized system, a copy of the data should be maintained at another location, typically at the clinical site but possibly at some other designated site. Copies should be made contemporaneously with data entry and should be preserved in an appropriate format, such as XML, PDF or paper formats.

D. Internal Security Safeguards

1. Limited Access. Access must be limited to authorized individuals (21 CFR 11.10(d)). This requirement can be accomplished by the following recommendations. We recommend that each user of the system have an individual account. The user should log into that account at the beginning of a data entry session, input information (including changes) on the electronic record, and log out at the completion of data entry session. The system should be designed to limit the number of log-in attempts and to record unauthorized access log-in attempts.

Individuals should work only under their own password or other access key and not share these with others. The system should not allow an individual to log onto the system to provide another person access to the system. We also recommend that passwords or other access keys be changed at established intervals commensurate with a documented risk assessment.

When someone leaves a workstation, the person should log off the system. Alternatively, an automatic log off may be appropriate for long idle periods. For short periods of inactivity, we recommend that a type of automatic protection be installed against unauthorized data entry (e.g., an automatic screen saver can prevent data entry until a password is entered).

2. Audit Trails. It is important to keep track of all changes made to information in the electronic records that document activities related to the conduct of the trial (audit trails). The use of audit trails or other security measures helps to ensure that only authorized additions, deletions, or alterations of information in the electronic record have occurred and allows a means to reconstruct significant details about study conduct and source data collection necessary to verify the quality and integrity of data. Computer-generated, time-stamped audit trails or other security measures can also capture information related to the creation, modification, or deletion of electronic records and may be useful to ensure compliance with the appropriate regulation.

The need for audit trails should be determined based on a justified and documented risk assessment that takes into consideration circumstances surrounding system use, the likelihood that information might be compromised, and any system vulnerabilities. Should it be decided that audit trails or other appropriate security measures are needed to ensure electronic record integrity, personnel who create, modify, or delete electronic records should not be able to modify the documents or security measures used to track electronic record changes. Computer-generated, time-stamped electronic audits trails are the preferred method for tracking changes to electronic source documentation.

Audit trails or other security methods used to capture electronic record activities should describe when, by whom, and the reason changes were made to the electronic record. Original information should not be obscured though the use of audit trails or other security measures used to capture electronic record activities.

3. Date/Time Stamps. Controls should be established to ensure that the system's date and time are correct. The ability to change the date or time should be limited to authorized personnel, and such personnel should be notified if a system date or time discrepancy is detected. Any changes to date or time should always be documented. We do not expect documentation of time changes that systems make automatically to adjust to daylight savings time conventions.

We recommend that dates and times include the year, month, day, hour, and minute and encourage synchronization of systems to the date and time provided by international standard-setting agencies (e.g., U.S. National Institute of Standards and Technology provides information about universal time, coordinated (UTC)).

Computerized systems are likely to be used in multi-center clinical trials and may be located in different time zones. For systems that span different time zones, it is better to implement time stamps with a clear understanding of the time zone reference used. We recommend that system documentation explain time zone references as well as zone acronyms or other naming conventions.

E. External Security Safeguards

In addition to internal safeguards built into a computerized system, external safeguards should be put in place to ensure that access to the computerized system and to the data is restricted to authorized personnel. Staff should be kept thoroughly aware of system security measures and the importance of limiting access to authorized personnel.

Procedures and controls should be put in place to prevent the altering, browsing, querying, or reporting of data via external software applications that do not enter through the protective system software.

You should maintain a cumulative record that indicates, for any point in time, the names of authorized personnel, their titles, and a description of their access privileges. That record should be kept in the study documentation, accessible for use by appropriate study personnel and for inspection by FDA investigators.

We also recommend that controls be implemented to prevent, detect, and mitigate effects of computer viruses, worms, or other potentially harmful software code on study data and software.

F. Other System Features

1. Direct Entry of Data. We recommend that you incorporate prompts, flags, or other help features into your computerized system to encourage consistent use of clinical terminology and to alert the user to data that are out of acceptable range. You should not use programming features that automatically enter data into a field when the field is bypassed (default entries). However, you can use programming features that permit repopulation of information specific to the subject. To avoid falsification of data, you should perform a careful analysis in deciding whether and when to use software programming instructions that permit data fields to be automatically populated.

2. Retrieving Data. The computerized system should be designed in such a way that retrieved data regarding each individual subject in a study is attributable to that subject. Reconstruction of the source documentation is essential to FDA's review of the clinical study submitted to the Agency. Therefore, the information provided to FDA should fully describe and explain how source data were obtained and managed, and how electronic records were used to capture data.

It is not necessary to reprocess data from a study that can be fully reconstructed from available documentation. Therefore, the actual application software, operating systems, and software development tools involved in the processing of data or records need not be retained.

*3. **Dependability System Documentation.*** For each study, documentation should identify what software and hardware will be used to create, modify, maintain, archive, retrieve, or transmit clinical data. Although it need not be submitted to FDA, this documentation should be retained as part of the study records and be available for inspection by FDA (either on-site or remotely accessible).

*4. **System Controls.*** When electronic formats are the only ones used to create and preserve electronic records, sufficient backup and recovery procedures should be designed to protect against data loss. Records should regularly be backed up in a procedure that would prevent a catastrophic loss and ensure the quality and integrity of the data. Records should be stored at a secure location specified in the SOP. Storage should typically be offsite or in a building separate from the original records.

We recommend that you maintain backup and recovery logs to facilitate an assessment of the nature and scope of data loss resulting from a system failure.

*5. **Change Controls.*** The integrity of the data and the integrity of the protocols should be maintained when making changes to the computerized system, such as software upgrades, including security and performance patches, equipment, or component replacement, or new instrumentation. The effects of any changes to the system should be evaluated and some should be validated depending on risk. Changes that exceed previously established operational limits or design specifications should be validated. Finally, all changes to the system should be documented.

G.   Training of Personnel

Those who use computerized systems must determine that individuals (e.g., employees, contractors) who develop, maintain, or use computerized systems have the education, training and experience necessary to perform their assigned tasks (21 CFR 11.10(i)).

Training should be provided to individuals in the specific operations with regard to computerized systems that they are to perform. Training should be conducted by qualified individuals on a continuing basis, as needed, to ensure familiarity with the computerized system and with any changes to the system during the course of the study.

We recommend that computer education, training, and experience be documented.

DEFINITIONS

The following is a list of definitions for terms used in, and for the purposes of, this guidance document.

Audit Trail: For the purpose of this guidance, an *audit trail* is a process that captures details such as additions, deletions, or alterations of information in an electronic record without obliterating the original record. An audit trail facilitates the reconstruction of the course of such details relating to the electronic record.

Certified Copy: A *certified copy* is a copy of original information that has been verified, as indicated by a dated signature, as an exact copy having all of the same attributes and information as the original.

Computerized System: A *computerized system* includes computer hardware, software, and associated documents (e.g., user manual) that create, modify, maintain, archive, retrieve, or transmit in digital form information related to the conduct of a clinical trial.

Direct Entry: *Direct entry* is recording data where an electronic record is the original means of capturing the data. Examples are the keying by an individual of original observations into a system, or automatic recording by the system of the output of a balance that measures subject's body weight.

Electronic Record: An *electronic record* is any combination of text, graphics, data, audio, pictorial, or other information representation in digital form that is created, modified, maintained, archived, retrieved, or distributed by a computer system.

Original data: For the purpose of this guidance, *original data* are those values that represent the first recording of study data. FDA is allowing original documents and the original data recorded on those documents to be replaced by copies provided the copies are identical and have been verified as such (see FDA Compliance Policy Guide # 7150.13).

Source Documents: Original documents and records including, but not limited to, hospital records, clinical and office charts, laboratory notes, memoranda, subjects' diaries or evaluation checklists, pharmacy dispensing records, recorded data from automated instruments, copies or transcriptions certified after verification as being accurate and complete, microfiches, photographic negatives, microfilm or magnetic media, x-rays, subject files, and records kept at the pharmacy, at the laboratories, and at medico-technical departments involved in a clinical trial.

Transmit: *Transmit* is to transfer data within or among clinical study sites, contract research organizations, data management centers, sponsors, or to FDA.

REFERENCES

FDA, *21 CFR Part 11*, "Electronic Records; Electronic Signatures; Final Rule." *Federal Register*, Vol. 62, No. 54, 13429, March 20, 1997.

FDA, *Compliance Program Guidance Manual*, "Compliance Program 7348.810—Bioresearch Monitoring—Sponsors, Contract Research Organizations and Monitors," February 21, 2001.

FDA, *Compliance Program Guidance Manual*, "Compliance Program 7348.811—Bioresearch Monitoring—Clinical Investigators," September 30, 2000.

FDA, *Good Clinical Practice VICH GL9*.

FDA, *Guideline for the Monitoring of Clinical Investigations*.

FDA, *Information Sheets for Institutional Review Boards and Clinical Investigators*. http://www.fda.gov/ic/ohrt/irbs/default.htm.

FDA, *E6 Good Clinical Practice: Consolidated Guidance*. http://www.fda.gov/cder/guidance/959fnl.pdf.

FDA, *Part 11, Electronic Records; Electronic Signatures—Scope and Application*, 2003.

FDA, *General Principles of Software Validation; Guidance for Industry and FDA Staff*.

APPENDIX A

Standard Operating Procedures

Standard operating procedures (SOPs) and documentation pertinent to the use of a computerized system should be made available for use by appropriate study personnel at the clinical site or remotely and for inspection by FDA. The SOPs should include, but are not limited to, the following processes.

- System setup/installation (including the description and specific use of software, hardware, and physical environment and the relationship)
- System operating manual
- Validation and functionality testing
- Data collection and handling (including data archiving, audit trails, and risk assessment)
- System maintenance (including system decommissioning)
- System security measures

- Change control
- Data backup, recovery, and contingency plans
- Alternative recording methods (in the case of system unavailability)
- Computer user training
- Roles and responsibilities of sponsors, clinical sites and other parties with respect to the use of computerized systems in the clinical trials

Index

Managing the Documentation Maze, By Janet Gough and David Nettleton
Copyright © 2010 John Wiley & Sons, Inc.